Dynamics of
Rotating Machines

Dynamics of Rotating Machines

Edited by
George Rivera

C WILLFORD PRESS

www.willfordpress.com

Published by Willford Press,
118-35 Queens Blvd., Suite 400,
Forest Hills, NY 11375, USA

ISBN: 978-1-68285-775-5

Cataloging-in-Publication Data

Dynamics of rotating machines / edited by George Rivera.
    p. cm.
Includes bibliographical references and index.
ISBN 978-1-68285-775-5
1. Rotors--Dynamics. 2. Rotational motion. 3. Machinery--Vibration. I. Rivera, George.
TJ1058 .D96 2020
621.82--dc23

For information on all Willford Press publications
visit our website at www.willfordpress.com

WILLFORD PRESS

Contents

Preface

This book aims to highlight the current researches and provides a platform to further the scope of innovations in this area. This book is a product of the combined efforts of many researchers and scientists, after going through thorough studies and analysis from different parts of the world. The objective of this book is to provide the readers with the latest information of the field.

Rotating machines are the machines which are made up of two main parts- the rotor and the stator. The non-moving segment of the machine is known as the stator and the rotating segment is known as the rotor. Rotating machines find a wide variety of uses in a number of domains such as domestic appliances, transportation vehicles and industrial manufacturing plants. AC and DC machines are prominent examples of rotating machines. In AC machines, the rotor is used as the field, and the stator is used as the armature, while the reverse is applicable for DC machines. This book provides significant information to help develop a good understanding of rotating machines and their dynamics. It is a valuable compilation of topics, ranging from the basic to the most complex advancements in this field. This book will serve as a valuable source of reference for graduate and post graduate students.

I would like to express my sincere thanks to the authors for their dedicated efforts in the completion of this book. I acknowledge the efforts of the publisher for providing constant support. Lastly, I would like to thank my family for their support in all academic endeavors.

Editor

Performance and Internal Flow of a Dental Air Turbine Handpiece

Yasuyuki Nishi ⓘ,[1] Hikaru Fushimi,[2] Kazuo Shimomura,[3] and Takeshi Hasegawa[4]

[1]*Department of Mechanical Engineering, Ibaraki University, 4-12-1 Nakanarusawa-cho, Hitachi-shi, Ibaraki 316-8511, Japan*
[2]*Graduate School of Science and Engineering, Ibaraki University, 4-12-1 Nakanarusawa-cho, Hitachi-shi, Ibaraki 316-8511, Japan*
[3]*Yoshida Seiko, Ltd., 660 Yakimaki, Namegata-shi, Ibaraki 311-3506, Japan*
[4]*The Yoshida Dental MFG Co., Ltd., 1-3-6 Kotobashi, Sumida-ku, Tokyo 130-8516, Japan*

Correspondence should be addressed to Yasuyuki Nishi; yasuyuki.nishi.fe@vc.ibaraki.ac.jp

Academic Editor: Ryoichi Samuel Amano

An air turbine handpiece is a dental abrasive device that rotates at high speed and uses compressed air as the driving force. It is characterized by its small size, light weight, and painless abrading due to its high-speed rotation, but its torque is small and noise level is high. Thus, to improve the performance of the air turbine handpiece, we conducted a performance test of an actual handpiece and a numerical analysis that modeled the whole handpiece; we also analyzed the internal flow of the handpiece. Results show that experimental and calculated values were consistent for a constant speed load method with the descending speed of 1 mm/min for torque and turbine output. When the tip of the blade was at the center of the nozzle, the torque was at its highest. This is likely because the jet from the nozzle entered the tip of the blade from a close distance that would not reduce the speed and exited along the blade.

1. Introduction

An air turbine handpiece is highly valued as a dental abrasive device that rotates at a high speed, and it is an essential device for dental treatment [1]. Since its development, many studies have been conducted to measure and evaluate rotation performance [2–6] and measure and evaluate noise [7, 8]. An air turbine handpiece uses compressed air as the driving force and is characterized by its small size, light weight, and painless abrading at high rotation speed. However, compared to an electric handpiece that uses a motor as the driving force, its torque is small and noise level is high.

As such, improved performance and reduced noise for an air turbine handpiece are desired. Considering that an air turbine handpiece has the equivalent rotation performance as turbine performance, it can be considered as a type of turbomachinery, and the fluid mechanics approach would be effective. In other words, it would be effective to first elucidate the internal flow of an air turbine handpiece and the relation between the performance and the noise characteristics and then control the flow. However, the main component of

an air turbine handpiece, the rotor, is small and rotates at high speeds (250,000–400,000 min^{-1}); thus, measuring the flow is not easy. In addition, the torque generated by the rotor is extremely small; thus, the measurement of torque is difficult. With the progress of computational fluid dynamics analysis, numerical analysis of internal flow to the head (turbine) of handpiece began recently [9–13], and numerical analysis that models the entire handpiece [14] has also been conducted. However, in these studies, details of internal flow are insufficiently elucidated, and there is no verification of numerical analysis results via experiment. As such, there are few reports on the study of internal flow in an air turbine handpiece.

In this study, as the first step toward improving the performance of air turbine handpiece, we conducted a performance test of an actual handpiece while conducting numerical analysis that models the whole handpiece, and we compared the turbine performances. Furthermore, we examined the torque fluctuation with the rotation of the rotor and examined its relation with the internal flow.

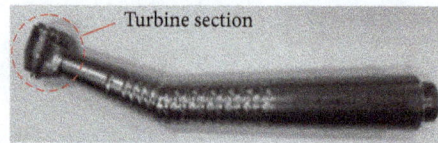

FIGURE 1: Air turbine handpiece.

TABLE 1: Specifications of turbine section.

Nozzle	d: nozzle inner diameter [mm]	1.3
	Number of nozzles	1
Rotor	D_1: rotor outer diameter [mm]	8.8
	D_2: rotor inner diameter [mm]	3.6
	b: blade width [mm]	3.3
	β: blade angle [°]	90
	Number of blades	8
Casing	D_3: casing inner diameter [mm]	9.2
	B: casing width [mm]	4.5

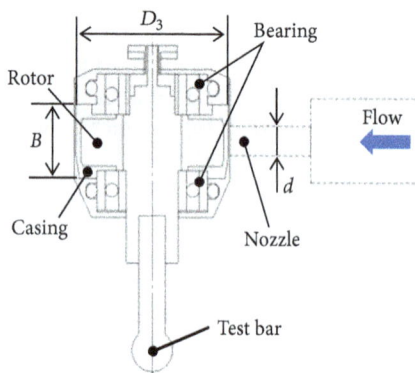

FIGURE 3: Rotor.

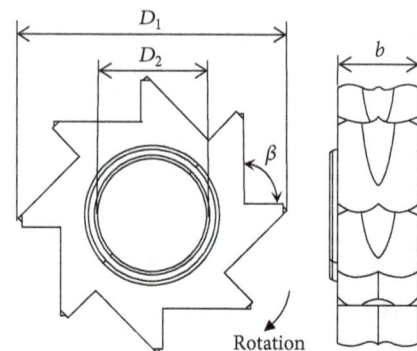

FIGURE 2: Schematic of turbine section.

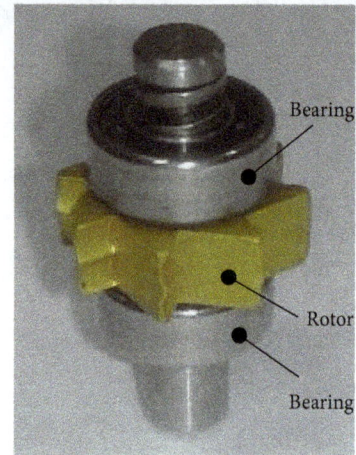

FIGURE 4: Schematic of rotor.

2. Experimental Apparatus and Methods

2.1. Test Air Turbine Handpiece. Figure 1 shows the test air turbine handpiece and Table 1 shows its specifications. In addition, Figure 2 shows the schematic of the turbine section. The turbine section, which forms the head section of the handpiece, comprises a nozzle, rotor, and casing and is a single-stage impulse turbine. Figure 3 shows the rotor, while Figure 4 shows its schematic. This air turbine handpiece jets out the compressed air that enters from the entrance through the nozzle, and the air is driven as the jet works on the rotor. The inner diameter of the nozzle is $d = 1.3$ mm, and there is one nozzle. The outer diameter of the rotor is $D_1 = 8.8$ mm, while the inner diameter is $D_2 = 3.6$ mm and the blade width is $b = 3.3$ mm, of which there are eight blades.

2.2. Experimental Apparatus and Methods. The rotation performance of the air turbine handpiece is assumed to be equivalent to turbine performance, and there are two methods to measure this value. One is when a test bar is loaded onto a work material at a constant speed, and the

other is when the test bar is loaded onto a work material at a constant load [2]. Experimental results from the constant speed load method with the descending speed of 1 mm/min are consistent with the constant load method [2]. Thus, in this study, we used the constant speed load method, which has a relatively easy mechanism and makes experiments simpler. The experimental apparatus is shown in Figure 5, and its schematic is shown in Figure 6. An air turbine handpiece with a test bar (cemented carbide with radius $r = 1$ mm) attached to the rotational axis of the rotor is fixed in the movable part and is lowered at a constant speed. Compressed air of 0.22 MPa is injected through the inlet of the handpiece, and when the rotor is turned on, the test bar is rotated. At this time, the pressure was adjusted according to the regulator (SMC Corporation; AW2000) installed at the handpiece inlet, and measured using a pressure gauge (Asahi Gauge Manufacturing Co., Ltd.; 101-F510, accuracy ±1.6% of full scale). The

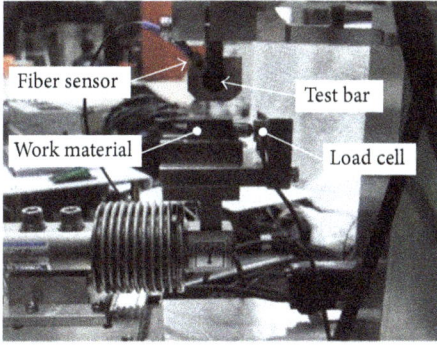

FIGURE 5: Experimental apparatus.

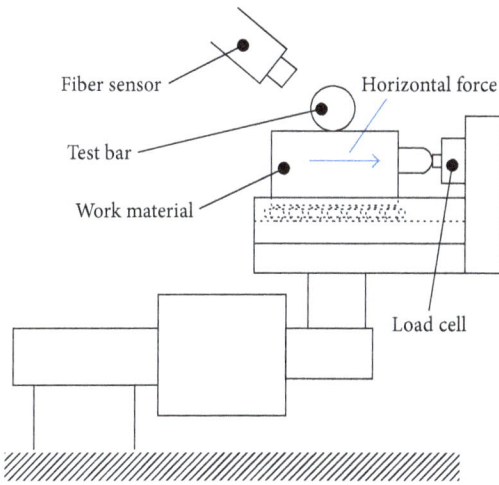

FIGURE 6: Schematic of experimental apparatus.

test bar is lowered at a descending speed of 1 mm/min, and the freely rotating test bar is pressed against the work material (cemented carbide) on the support stage. The horizontal force at this time F_H and rotational speed n were continuously measured with the load cell (Kyowa Electronic Instruments Co., Ltd.; LMB-A-100N, nonlinearity ±0.5% rated output) and fiber sensor (KEYENCE CORPORATION; FU-20/FS-N11N). The torque T and turbine output L were then obtained with the following equations using the horizontal force F_H and rotational speed n that were measured:

$$T = \frac{F_H r}{1000},$$
$$L = \frac{2\pi n T}{60}. \tag{1}$$

3. Numerical Analysis Method and Conditions

In this study, we performed a 3D compressible unsteady flow analysis by modeling the whole air turbine handpiece. For the analysis, we used general-purpose thermal fluid analysis software, ANSYS CFX14.5 (ANSYS, Inc.), and the working fluid was air. The governing equations were the mass conservation equation, momentum conservation equation,

energy conservation equation, and ideal gas law [15], which were solved with the finite volume method. Furthermore, we used the Shear Stress Transport (SST) model [16] as the turbulence model.

Figure 7 shows the computational model for the whole air turbine handpiece, and Figure 8 shows the turbine section region. The calculation regions are inlet flow region, turbine section region, and outlet flow region. The numbers of computational elements were approximately 391,000, 1,127,000, and 783,000 for the inlet flow region, the turbine section region, and the outlet flow region, respectively, corresponding to a total number of elements of approximately 2,301,000. Boundary conditions were a total pressure of 0.22 MPa and static temperature of 25°C for the inlet boundary and static pressure of 0 Pa for the outlet boundary. In addition, we provided arbitrary rotation speed to the rotor region, and the wall surface had adiabatic and nonslip conditions. The boundary between the rotational region and static region was attached using the Transient Rotor-Stator [17]. Time step size was determined the rotor rotates once with 120 steps at any speed. Calculations were continued until fluctuations in the torque stabilized. In terms of grid dependency, the number of computational elements was increased approximately 1.52 times, and it was analyzed with a rotation rate of $n = 250,000 \text{ min}^{-1}$. The result was that this decreased by approximately 2.2% in relation to the torque in this paper, and it was confirmed that the effect of the number of computational elements was comparatively small.

For the coordinate system, we defined x- and y-axes as in Figure 5. For the blade phase angle θ^*, the position at which the front surface of the reference Blade A is parallel to the y-axis (the position in Figure 5) is set as $\theta^* = 0°$, and the counterclockwise direction was set as positive. Furthermore, we defined the blade ahead of Blade A as Blade B. The cross-section at 3.7 mm from the rotation center in the perpendicular direction to the front surface of Blades A and B was defined as cross-section a–a, and the cross-section at 4.1 mm was defined as cross-section b–b.

4. Results and Discussion

4.1. Turbine Performance. Figure 9 shows the relation between torque T and rotational speed n, while Figure 10 shows the relation between turbine output L and rotational speed n. Figures 9 and 10 show that calculated T and L are slightly higher than the experimental values, but both T and L are consistent. Considering that calculated T and L ignore the loss of mechanical friction on the bearing, these analytical results are valid. Experimental and calculated T decreased with increase in rotational speed, and calculated L showed the maximum value at $n = 250,000 \text{ min}^{-1}$ ($L = 12.1 \text{ W}$).

Next, to search the locations where loss inside of the handpiece is high, we created a measurement surface in seven locations, as shown in Figure 11, and obtained entropy E for each measurement surface. Figure 12 shows the entropy E on each measurement surface for the handpiece when the maximum output rotational speed was $n = 250,000 \text{ min}^{-1}$. Figure 12 shows that E demonstrated the highest increase in locations that are equivalent to the rotor region inside of F4-F5

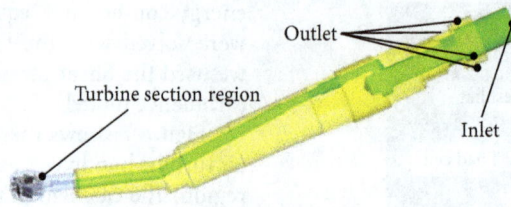

FIGURE 7: Air turbine handpiece model.

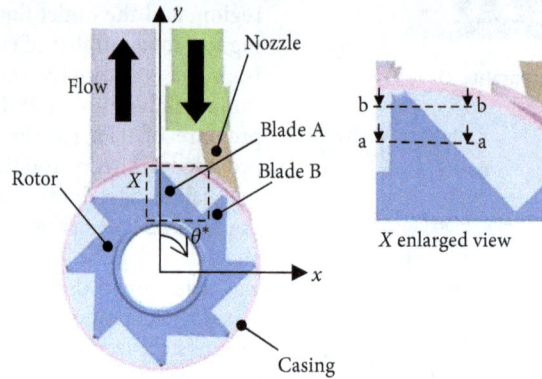

FIGURE 8: Turbine section region.

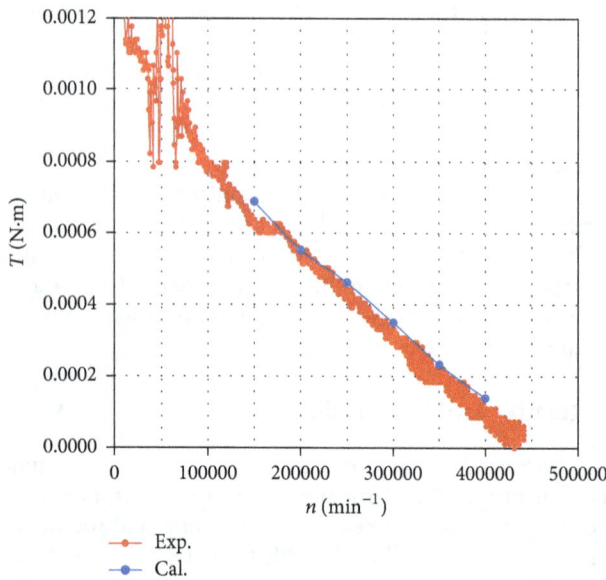

FIGURE 9: Correlation between rotational speed and torque.

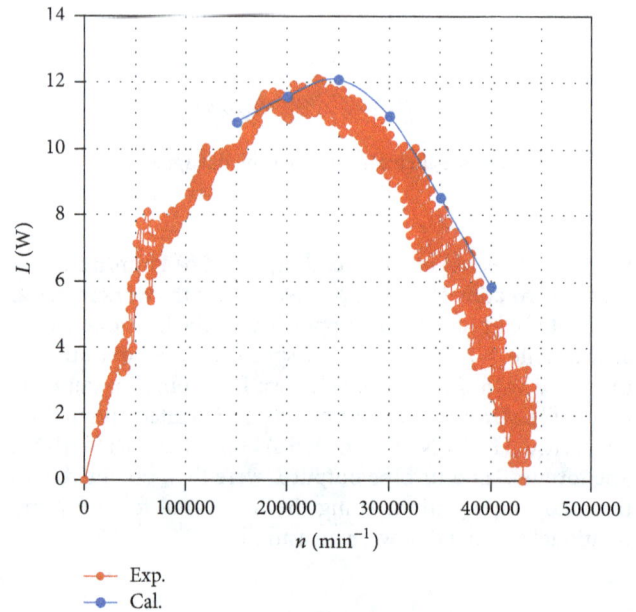

FIGURE 10: Correlation between rotational speed and turbine output.

casing among measurement surfaces. Therefore, in terms of the loss inside of the handpiece, the loss in the rotor region is the most dominant. However, E increased slightly in F1–F4 inlet flow and F6-F7 outlet flow; thus, the loss in the inlet and outlet flows cannot be ignored.

We obtained adiabatic efficiency η_{ad} of the turbine section and the whole handpiece with the following equations:

$$\eta_{ad} = \frac{L}{L_{ad}},$$

$$L_{ad} = \dot{m}\left[\frac{\kappa}{\kappa-1}\frac{P_1}{\rho_1}\left\{1 - \left(\frac{P_2}{P_1}\right)^{(\kappa-1)/\kappa}\right\} + \frac{1}{2}\left(v_1{}^2 - v_2{}^2\right)\right],$$

(2)

where $\dot{m}$ is the mass flow rate and κ is the specific heat ratio (1.4). ρ_1, P_1, and v_1 are fluid density, static pressure, and

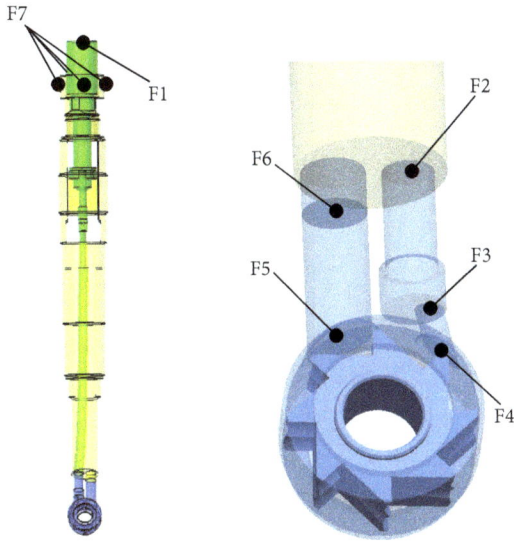

FIGURE 11: Measuring surface.

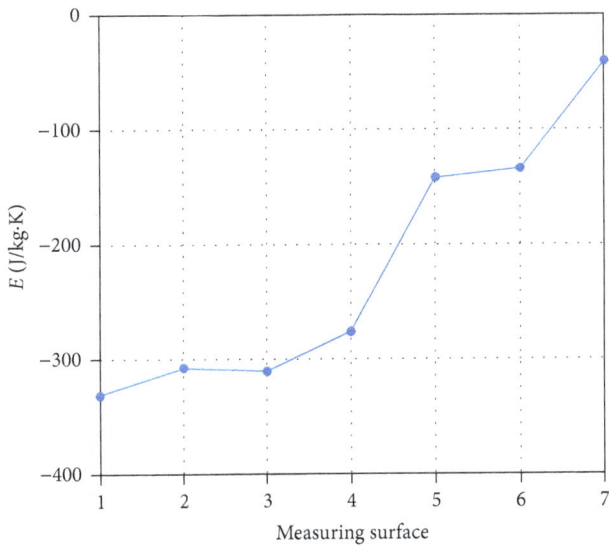

FIGURE 12: Entropy at each measuring surface.

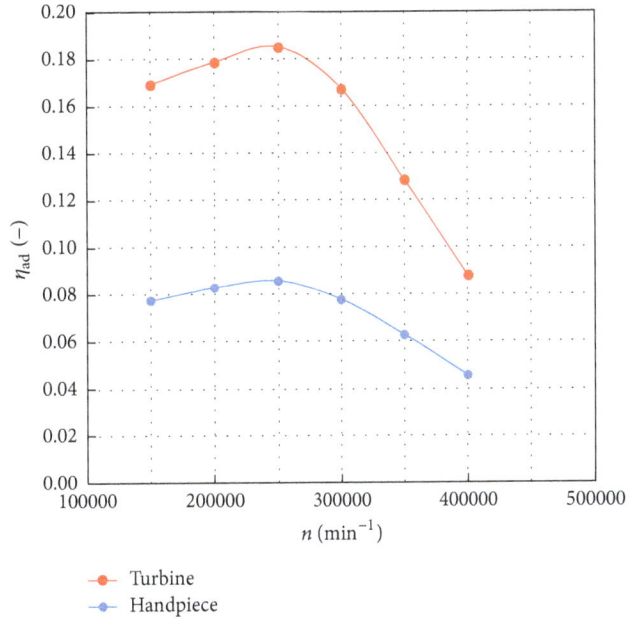

FIGURE 13: Correlation between rotational speed and adiabatic efficiency.

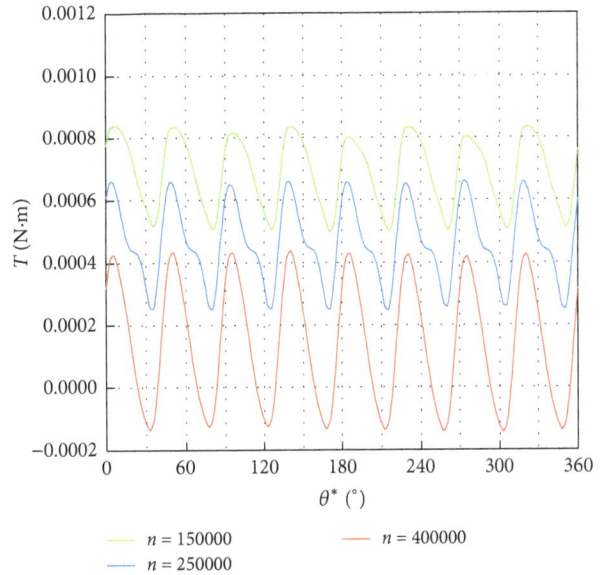

FIGURE 14: Torque fluctuation during one rotation of rotor.

absolute velocity at the inlet, respectively. To calculate η_{ad} of the turbine section, we used the value from the turbine section inlet (F2); to calculate η_{ad} of the whole handpiece, we used the value from the handpiece inlet (F1). P_2 and v_2 are static pressure and absolute velocity at the outlet, respectively, and to calculate η_{ad} of the turbine section we used the value from the turbine section outlet (F6), and to calculate η_{ad} of the whole handpiece, we used the value from the handpiece outlet (F7).

Figure 13 shows the relation between adiabatic efficiency η_{ad} and rotational speed n of the turbine section and the whole handpiece. Figure 13 shows that η_{ad} of the turbine section and the whole handpiece present the maximum value when $n = 250,000\,\text{min}^{-1}$, which is similar to the maximum value of the turbine output, and they show a major drop at high rotational

speed. The maximum η_{ad} of the turbine section was quite low at 0.185. The maximum η_{ad} of the whole handpiece was 0.086, and this is approximately 46% less than the maximum η_{ad} of the turbine section.

4.2. Torque Fluctuation and Internal Flow. Figure 14 shows the torque fluctuation during one rotor rotation for each rotational speed n. Figure 14 shows that T fluctuates greatly during one rotor rotation, showing eight peaks in response to the number of blades. Fluctuation in T increases with increased rotational speed, and especially at $n = 400,000\,\text{min}^{-1}$, the

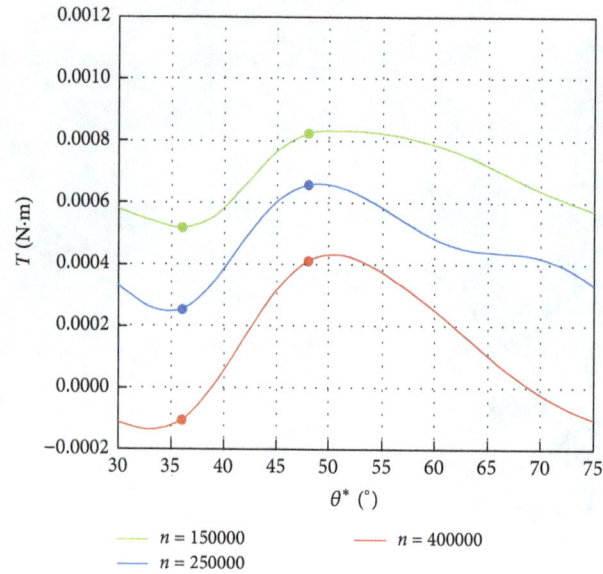

FIGURE 15: Torque fluctuation for one blade pitch.

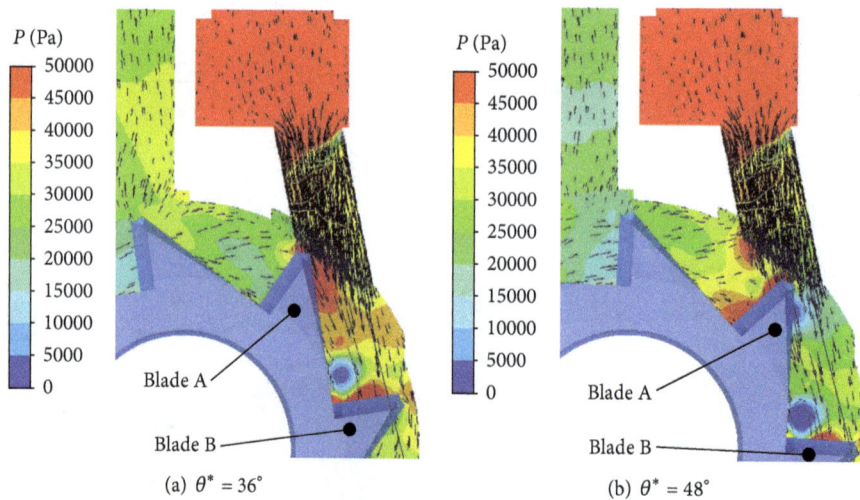

(a) $\theta^* = 36°$

(b) $\theta^* = 48°$

FIGURE 16: Static pressure distributions and absolute velocity vectors ($n = 250{,}000$ min^{-1}).

minimum value presents a negative value. When torque fluctuation increases, it lowers workability of the air turbine headpiece and causes fluid noise; thus, suppression of these issues is a challenge. Since torque fluctuates, periodically to the number of blades, torque fluctuation for one blade pitch is shown in Figure 15. Figure 15 shows that when the blade phase angle is around $\theta^* = 36°$, T becomes a minimum. Subsequently, T increases rapidly and after reaching the maximum around $\theta^* = 48°$, it gradually decreases.

Therefore, we focused on these blade phase angles and examined the internal flow. First, we present static pressure distribution and absolute velocity vectors for the center cross-section of the blade width for each blade phase angle ($\theta^* = 36°$, $48°$) at the maximum output rotational speed of $n = 250{,}000$ min^{-1} (Figures 16(a) and 16(b)). Figures 17–24 show static pressure distribution and absolute velocity vectors of

cross-sections a–a and b–b for each blade (Blade A and Blade B).

At $\theta^* = 36°$, where T becomes minimum, Figure 16(a) shows that with Blade A the jet from the nozzle does not enter the front surface of the blade, but it collides with the back surface of the blade. Therefore, as shown in Figures 17(a) and 17(b), static pressure increases on the back surface of the blade. Especially on the cross-section b–b with a large radius, static pressure increased more than the static pressure on the cross-section a–a. Thus, there was a large negative torque on the Blade A. With Blade B, jet flows facing the front surface of the blade. The larger shift in the flow from entering to exiting the blade causes the torque to also increase more. However, since the distance between the nozzle and blades is large, the jet decreases its speed and spreads in the width direction of the blade. Therefore, as shown in Figures 18(a)

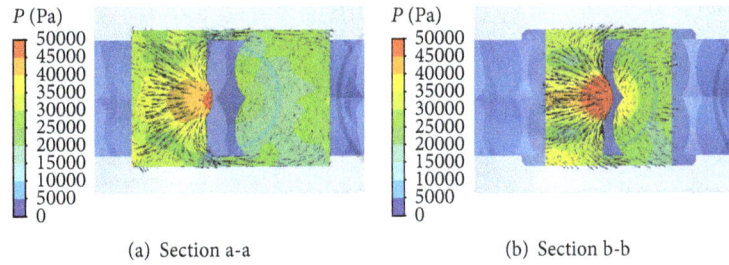

(a) Section a-a

(b) Section b-b

FIGURE 17: Static pressure distributions and absolute velocity vectors of Blade A ($\theta^* = 36°$, $n = 250,000\ \mathrm{min}^{-1}$).

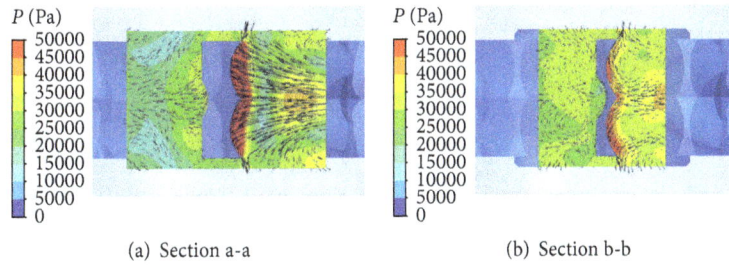

(a) Section a-a

(b) Section b-b

FIGURE 18: Static pressure distributions and absolute velocity vectors of Blade B ($\theta^* = 36°$, $n = 250,000\ \mathrm{min}^{-1}$).

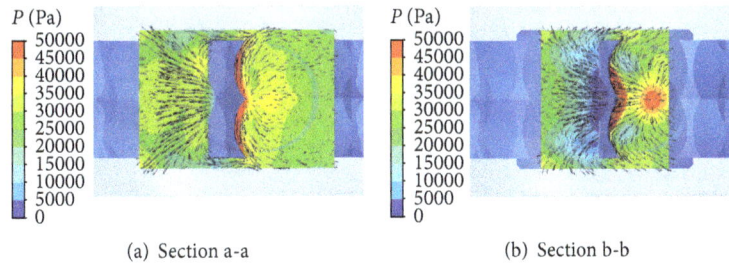

(a) Section a-a

(b) Section b-b

FIGURE 19: Static pressure distributions and absolute velocity vectors of Blade A ($\theta^* = 48°$, $n = 250,000\ \mathrm{min}^{-1}$).

and 18(b), the flow does not exit along the blade in both cross-sections, and the shift in the flow is small. Therefore, although static pressure on the front surface of the blade on the cross-section a–a increases considerably, static pressure on the back surface of the blade is also large; moreover, as it is near the center of the blade with a small radius, it does not sufficiently contribute to the torque. Based on the above, torque near θ^* = 36° presented the minimum value.

In contrast, at $\theta^* = 48°$ where T becomes maximum, as shown in Figure 16(b), the blade tip of Blade A is positioned near the center of the nozzle, and the jet from the nozzle flows to the tip of the Blade A at a close distance where the jet does not lose its speed. Therefore, as shown in Figure 19(b), the flow that entered the front surface of the blade on cross-section b–b flowed out along the blade and static pressure at the front surface of the blade increased, while static pressure on the back surface of the blade decreased. The pressure difference between the front and back surfaces of this blade tip with a large radius is greatly contributing to the torque. As shown in Figure 19(a), since cross-section a–a represents the moment just before Blade A reaches the nozzle, static pressure increase on the front surface of the blade was slightly

lower. With Blade B, the distance between the nozzle and blade is larger, and the jet is prevented by Blade A. Therefore, as shown in Figures 20(a) and 20(b), based on the static pressure distribution on the front and back surfaces of the blade, torque on Blade B is smaller than when θ^* = 36°. As such, near $\theta^* = 48°$, torque at the tip of Blade A becomes dominant, leading to maximum torque.

Furthermore, we examined the internal flow at low and high rotational speed. Figures 21(a), 21(b), 22(a), and 22(b) show static pressure distribution and the absolute velocity vector of central cross-section of the blade width for each blade phase angle ($\theta^* = 36°, 48°$) at $n = 150,000\ \mathrm{min}^{-1}$ and $n = 400,000\ \mathrm{min}^{-1}$. Figures 23–26 and Figures 27–30 show static pressure distribution and absolute velocity vector of cross-sections a–a and b–b for Blades A and B, respectively.

Figures 21(a) and 21(b) show that, at $n = 150,000\ \mathrm{min}^{-1}$, static pressure on the front surface of Blade B at $\theta^* = 36°$ and $\theta^* = 48°$ increased more than at $n = 250,000\ \mathrm{min}^{-1}$. This is because decreased rotational speed increased the relative speed of the jet to the blade. Thus, as shown in Figures 23–26, compared to when $n = 250,000\ \mathrm{min}^{-1}$, the jet does not spread

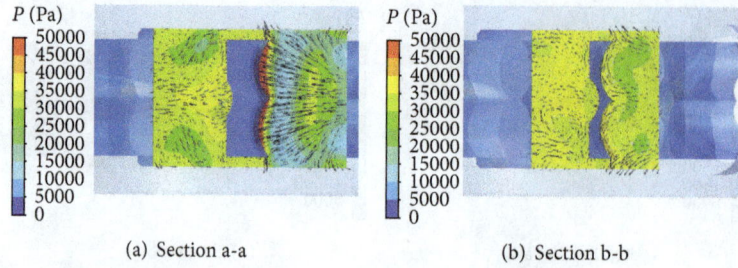

(a) Section a-a

(b) Section b-b

FIGURE 20: Static pressure distributions and absolute velocity vectors of Blade B ($\theta^* = 48°$, $n = 250,000\,\mathrm{min}^{-1}$).

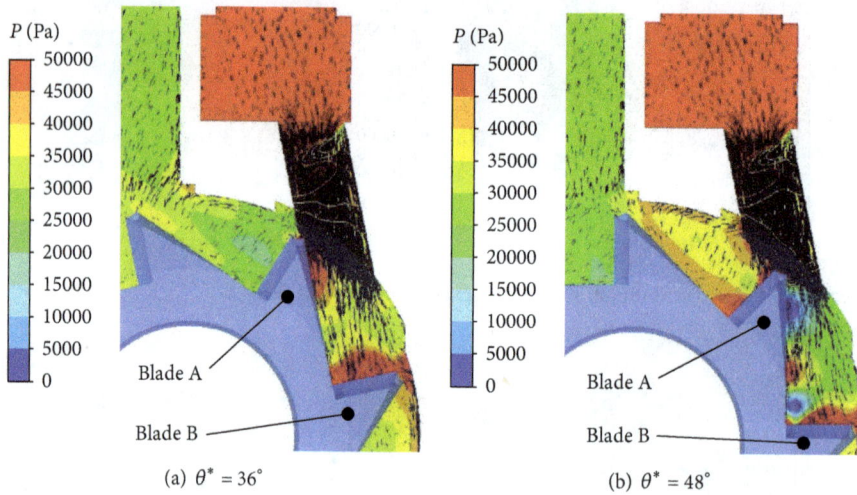

(a) $\theta^* = 36°$

(b) $\theta^* = 48°$

FIGURE 21: Static pressure distributions and absolute velocity vectors ($n = 150,000\,\mathrm{min}^{-1}$).

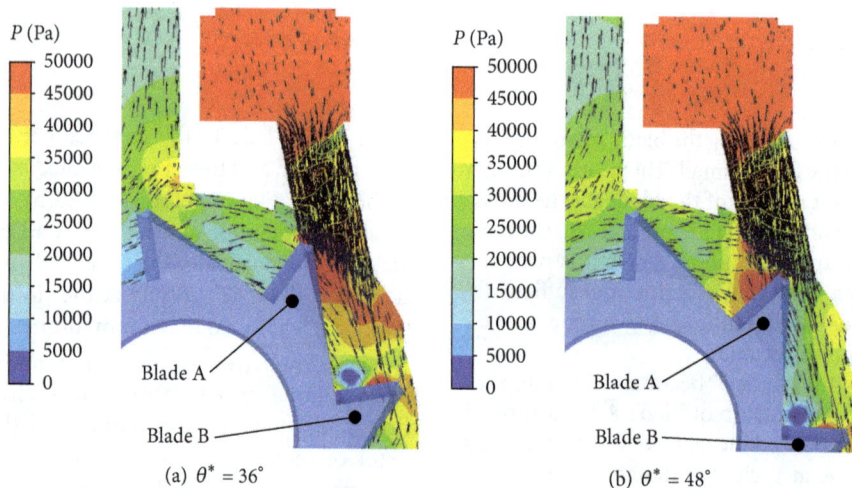

(a) $\theta^* = 36°$

(b) $\theta^* = 48°$

FIGURE 22: Static pressure distributions and absolute velocity vectors ($n = 400,000\,\mathrm{min}^{-1}$).

to the blade width direction, and flow that entered the front surface of the blade flows out along the blade. Therefore, the shift in the flow increases, and as shown in Figure 14, at $n = 150,000\,\mathrm{min}^{-1}$, torque increased.

On the contrary, Figures 22(a) and 22(b) show that, at $n = 400,000\,\mathrm{min}^{-1}$, static pressure on the front surface of Blade B at $\theta^* = 36°$ and $\theta^* = 48°$ decreased compared to when $n = 250,000\,\mathrm{min}^{-1}$. As rotational speed increased, the

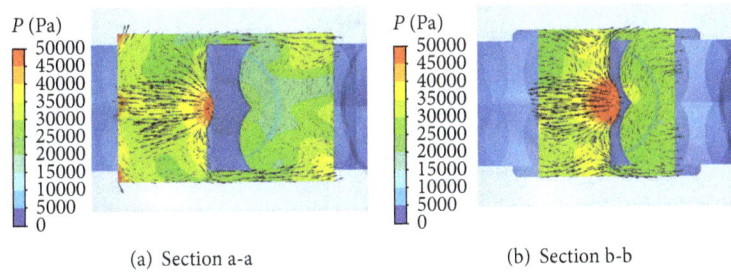

(a) Section a-a　　　　　　　　(b) Section b-b

FIGURE 23: Static pressure distributions and absolute velocity vectors of Blade A ($\theta^* = 36°$, $n = 150{,}000\,\text{min}^{-1}$).

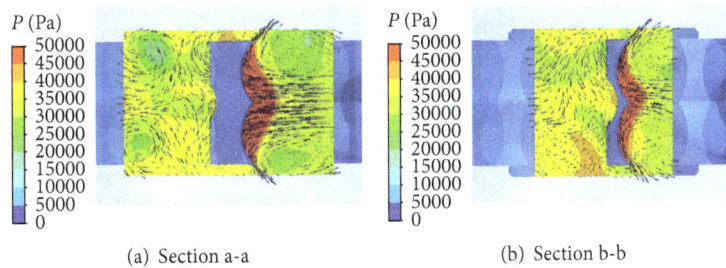

(a) Section a-a　　　　　　　　(b) Section b-b

FIGURE 24: Static pressure distributions and absolute velocity vectors of Blade B ($\theta^* = 36°$, $n = 150{,}000\,\text{min}^{-1}$).

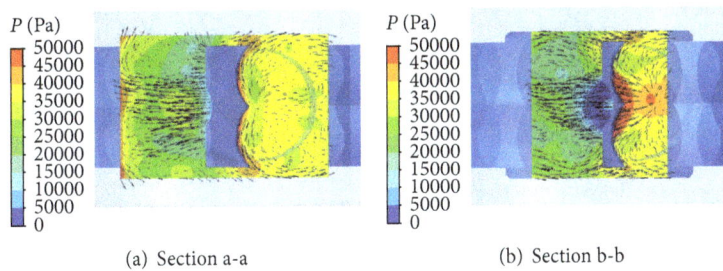

(a) Section a-a　　　　　　　　(b) Section b-b

FIGURE 25: Static pressure distributions and absolute velocity vectors of Blade A ($\theta^* = 48°$, $n = 150{,}000\,\text{min}^{-1}$).

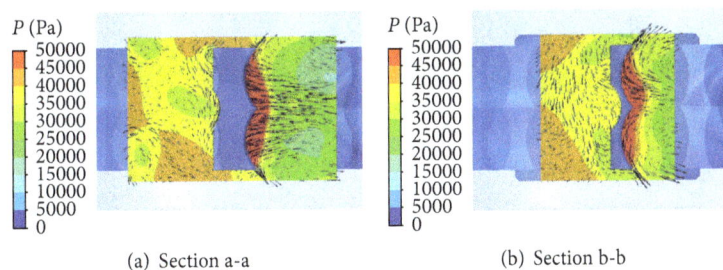

(a) Section a-a　　　　　　　　(b) Section b-b

FIGURE 26: Static pressure distributions and absolute velocity vectors of Blade B ($\theta^* = 48°$, $n = 150{,}000\,\text{min}^{-1}$).

relative speed of the jet to the blade decreased. As shown in Figures 27–30, compared to when $n = 250{,}000\,\text{min}^{-1}$, the jet spreads in the width direction of the blade, and the flow that entered the front surface of the blade does not exit along the blade, reducing the shift in the flow. Especially when θ^* $= 36°$, static pressure at the back surface of Blade A increased more than when $n = 250{,}000\,\text{min}^{-1}$. Therefore, as shown in Figure 14, when $n = 400{,}000\,\text{min}^{-1}$, torque was small and the fluctuation of the torque increased, showing a negative torque at $\theta^* = 36°$.

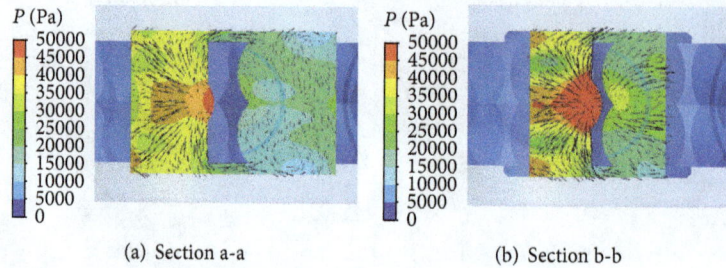

(a) Section a-a   (b) Section b-b

FIGURE 27: Static pressure distributions and absolute velocity vectors of Blade A ($\theta^* = 36°$, $n = 400{,}000$ min^{-1}).

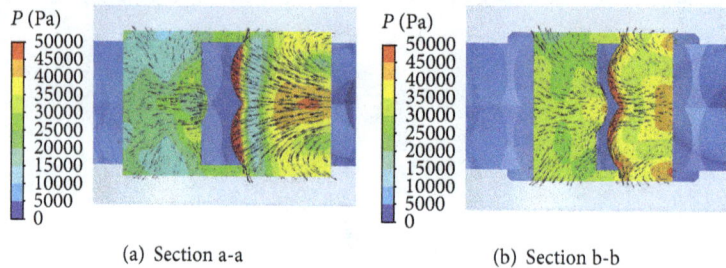

(a) Section a-a   (b) Section b-b

FIGURE 28: Static pressure distributions and absolute velocity vectors of Blade B ($\theta^* = 36°$, $n = 400{,}000$ min^{-1}).

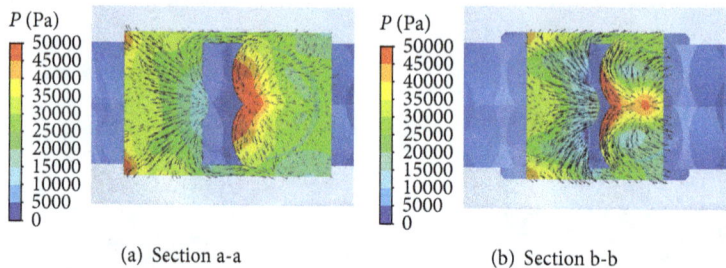

(a) Section a-a   (b) Section b-b

FIGURE 29: Static pressure distributions and absolute velocity vectors of Blade A ($\theta^* = 48°$, $n = 400{,}000$ min^{-1}).

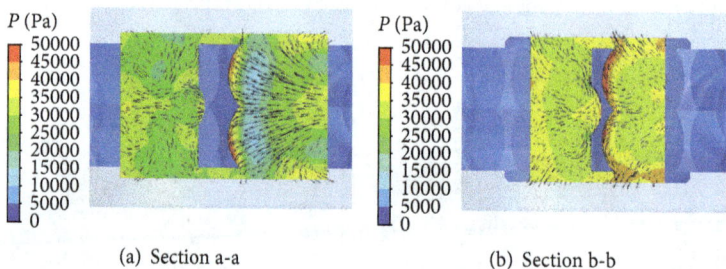

(a) Section a-a   (b) Section b-b

FIGURE 30: Static pressure distributions and absolute velocity vectors of Blade B ($\theta^* = 48°$, $n = 400{,}000$ min^{-1}).

5. Conclusions

We examined performance and internal flow of a dental air turbine handpiece through experiments and numerical analysis and obtained the following findings:

(1) Experimental and calculated values of torque and turbine output were consistent in the constant speed load method with a descending speed of 1 mm/min.

(2) Loss in the rotor region is dominant in the loss inside of the handpiece.

(3) The maximum value of adiabatic efficiency in the turbine section was approximately 0.185, which is quite low. Adiabatic efficiency of the whole handpiece decreased approximately 46% compared to the efficiency in the turbine section.

(4) With an increase in the rotational speed, the torque decreases but its fluctuation increases.

(5) When the blade tip is located near the center of the nozzle, the torque attains its maximum value. This is because

the jet enters the tip of the blade at a close enough distance to not reduce its speed and then it flows out along the blade.

Nomenclature

b: Blade width, mm
B: Casing width, mm
d: Nozzle inner diameter, mm
D_1: Rotor outer diameter, mm
D_2: Rotor inner diameter, mm
D_3: Casing inner diameter, mm
E: Entropy, J/kg·K
F_H: Horizontal force, N
L: Turbine output, W $= 2\pi nT/60$
L_{ad}: Adiabatic air power, W
$\dot{m}$: Mass flow rate, kg/s
n: Rotational speed, min^{-1}
P: Static pressure, Pa
r: Radius of test bar, mm
T: Torque, N·m $= F_H r/1000$
v: Absolute velocity, m/s.

Greek Letters

β: Blade angle, °
η_{ad}: Adiabatic efficiency $= L/L_{ad}$
θ^*: Blade phase angle, °
κ: Specific heat ratio
ρ: Fluid density, kg/m^3.

Conflicts of Interest

The authors declare that there are no conflicts of interest regarding the publication of this paper.

Acknowledgments

The authors extend their appreciation to Mr. Takuma Komuro of Ibaraki University in helping them run the experiments and numerical analysis.

References

[1] H. Miyairi, "Dental air turbine hand pieces," *Journal of the Society of Mechanical Engineers*, vol. 100, no. 949, pp. 1221–1223, 1997 (Japanese).

[2] H. Miyairi, H. Fukuda, and M. Nagai, "Studies on the performance of the dental air turbine handpieces: (part 7) the rotational performance and test method of the dental air turbine handpieces," *Journal of the Japanese Society for Dental Materials and Devices*, vol. 1, no. 4, pp. 328–337, 1982 (Japanese).

[3] M. Taira, K. Wakasa, M. Yamaki, K. Ohmoto, N. Satou, and H. Shintani, "Fundamental Studies on High-speed Rotational Properties of Dental Air-turbine Handpieces," *Journal of the Japanese Society for Dental Materials and Devices*, vol. 13, no. 4, pp. 381–387, 1994 (Japanese).

[4] J. E. Dyson and B. W. Darvell, "Dental air turbine handpiece performance testing," *Australian Dental Journal*, vol. 40, no. 5, pp. 330–338, 1995.

[5] J. E. Dyson and B. W. Darvell, "Flow and free running speed characterization of dental air turbine handpieces," *Journal of Dentistry*, vol. 27, no. 7, pp. 465–477, 1999.

[6] J. E. Dyson and B. W. Darvell, "Torque, power and efficiency characterization of dental air turbine handpieces," *Journal of Dentistry*, vol. 27, no. 8, pp. 573–586, 1999.

[7] H. Kimura and M. Kusano, "Basic studies on development of new type air turbine handpiece for dental use. Part 1. Analysis of noise from air turbine handpiece for dental use," *Shika zairyo, kikai = Journal of the Japanese Society for Dental Materials and Devices*, vol. 7, no. 5, pp. 829–833, 1988.

[8] T. Nomura, M. Itou, M. Uchida, Y. Yajima, and T. Takamata, "Examination of the Stillness Sound Design of an Air Turbine Handpiece for Dentistry : 1. Noise evaluation of different types of turbine wings," *Journal of the Matsumoto Dental University Society*, vol. 24, no. 1, pp. 58–71, 1998 (Japanese).

[9] J. Lee and K. Kim, "Numerical Study on the Effect of Turbine Blade Shape on Performance Characteristics of a Dental Air Turbine Handpiece," *Korean Society of Machine Tool Engineers*, vol. 13, pp. 34–42, 2009 (Korean).

[10] M. Juraeva, K. J. Ryu, and D. J. Song, "Optimum design of a saw-tooth-shaped dental air-turbine using design of experiment," *International Journal of Precision Engineering and Manufacturing*, vol. 15, no. 2, pp. 227–234, 2014.

[11] J. Makhsuda, B. Hwan Park, J. Kyung Ryu, and J. Dong Song, "Computational Approach to Improve the Performance of the Air-Turbine With Different Type Impeller Blades For the Dental Handpiece," *International Journal of Mechanical and Production Engineering*, vol. 4, no. 10, pp. 95–99, 2016.

[12] K. J. Ryu, M. Juraeva, B. H. Park, and D. J. Song, "Computational Approach to Improve the Performance of the Air-Turbine With Gull Type Impeller Blade For the High-Speed Dental Handpiece," *International Journal of Mechanical and Production Engineering*, vol. 4, no. 10, pp. 100–104, 2016.

[13] J. Makhsuda, B. H. Park, K. J. Ryu, and D. J. Song, "Optimum Design of the Air-Turbine for the High-Speed Dental Air-Turbine Hand piece using Design of Experiment," *International Journal of Mechanical and Production Engineering*, vol. 5, no. 4, pp. 34–38, 2017.

[14] M. Juraeva, B. H. Park, K. J. Ryu, and D. J. Song, "Computational Approach to Design the High-Speed Dental Air-Turbine Hand piece," *International Journal of Management and Applied Science*, vol. 5, no. 4, pp. 30–33, 2017.

[15] ANSYS, Inc., "ANSYS CFX-Solver Theoretical guide", pp. 81-85, 2010.

[16] F. R. Menter, "Two-equation eddy-viscosity turbulence models for engineering applications," *AIAA Journal*, vol. 32, no. 8, pp. 1598–1605, 1994.

[17] ANSYS, Inc., "ANSYS CFX-Solver Modeling guide", pp. 144-145, 2010.

2

Spatial Fluctuating Pressure Calculation of Underwater Counter Rotating Propellers under Noncavitating Condition

L. X. Hou ⓘ **and A. K. Hu**

School of Naval Architecture and Ocean Engineering, Dalian Maritime University, No. 1 Linghai Street,
Ganjingzi District, Dalian, Liaoning 116026, China

Correspondence should be addressed to L. X. Hou; 07093129@163.com

Academic Editor: Gerard Bois

The spatial fluctuating pressure field (FPF) of counter rotating propeller (CRP) under noncavitating condition is investigated. The hydrodynamic performance and pressure distributions on the blade surfaces are obtained through low-order potential-based panel method, which is also used to analyze the hydrodynamic interaction between the front and rear propellers of CRP as well as the hydrodynamic interference between any solid surface and propeller. The interaction between the given solid spherical surface and propeller is used to simulate the spatial FPF of propeller, and the fluctuating pressure induced by a propeller over one revolution is analyzed in frequency domain through fast Fourier transform. The method proposed is validated through two given propellers by comparing the calculation results with test data. The FPFs of the front and rear propellers are calculated and compared with that of the corresponding single propeller. The result shows that the CRP produces weaker FPF compared with the single propeller.

1. Introduction

Working in the nonuniform wake field behind ship hull, the propeller is subjected to unsteady surface loadings which lead to fluctuating pressure. It is well known that the fluctuating pressure can cause serious vibration problem on ship hull as well as the appendages behind the vessel. The fluctuating pressure induced by marine propeller can be classified into cavitation and noncavitation fluctuating pressure. The marine propeller cavitation is the most prevalent source of underwater fluctuating pressure in ocean. To the authors' knowledge, a lot of work has been done on the fluctuating pressure analysis of marine propeller under cavitation condition. Breslin et al. [1] studied propeller-induced hull pressures arising from intermittent blade cavitation, loading, and thickness using theoretical method coupled with experiments. Numerical methods have been developed based on surface panel method to compute the fluctuating pressure of cavitating propellers and compared the computational results with experimental data [2]. Seol [3] and Berger et al. [4] addressed the pressure fluctuation induced by a propeller sheet cavitation. The developed time domain

prediction methods provided reasonable results, and these results are in good agreement with the experimental results. Kanemaru and Ando [5] simulated unsteady sheet cavitation patterns, cavity volume evolution, and pressure fluctuations around marine propellers in nonuniform wake using the simple surface panel method SQCM with consideration of viscous effects to improve the calculation accuracy. The CFD technology has been widely used to predict the fluctuating pressure generated by cavitating propeller. Kawamura and Kiyokawa [6] simulated cavitating flows around a propeller rotating in a ship wake. Their results demonstrated that the magnitude of the pressure fluctuations increased greatly during cavitation, though the pressure fluctuations associated with the cavitation were still underestimated and the higher frequency components were not reproduced. Sato et al. [7] predicted the sheet cavitation behavior and pressure fluctuations using CFD software. Their simulations accurately predicted the 1st blade frequency component of the pressure fluctuations, while the high frequency components were severely underestimated due to the inability to simulate the tip vortex cavitation. Ji et al. [8] verified that the acceleration due to the cavity volume changes was the main source of the pressure fluctuations excited by the propeller cavitation by adopting the CFD technique. Lloyd et al. [9] calculated the pressure pulses inside the cavitation tunnel using the computational fluid dynamics code ReFRSCO. In order to predict the pressure pulse, Peralli et al. [10] investigated the wetted and cavitating flow around the INSEAN E779A propeller in a cavitation tunnel using the uRANS and BEM-BEM, respectively, and discussed the pro and cons of these two methods. Bensow and Gustafsson [11] investigated how hull forces and pressure are influenced by small propeller tip clearance by creating a setup where systematic variation of tip clearance could be achieved at similar propeller conditions. The simulations were performed using a scale resolved PANS approach combined with cavitation modelling considering the fluid as a mixture and incorporating mass transfer source terms.

However, submarines and torpedoes are usually operated deep enough under the sea to avoid cavitation [12], and some low speed vessels do not have noticeable cavitation phenomenon. It is of great significance to have investigations about the fluctuating pressure generated by marine propeller under noncavitating condition. Early researches about fluctuating pressure generated by propeller under noncavitating condition mainly obtained some empirical formulas according to a large quantity of accumulated data. Garguet and Lepeix [13], Tsakonas et al. [14], and Hu et al. [15] considered the influences of the hull and the free surface by introducing solid wall correction factor. Chen and Zhou [16] calculated the noncavitating fluctuating pressure of propeller in given wake field through CFD software. Güngör and Bedii Özdemir [17] investigated the performance of an inclined propeller in both noncavitating and cavitating conditions using a finite volume based solver and compared the results with the experimental data. The sliding mesh technique was used to implement the rotations in URANS solver with the renormalization group (RNG) k-ε turbulent model. In the noncavitating case, the amplitudes of the pressure fluctuations were in agreement with the experimental data, but those of pressure pulses in the cavitating condition were underpredicted.

As the world's energy shortage problem gets increasingly serious and the energy efficiency design index (EEDI) for new ship came into effect on January 1, 2013, reducing fuel consumption and building green ship not only relate directly to the operating costs but also help to deal with other risk factors [18]. As the power source of ships, efficient propulsion machinery will reduce fuel consumption and operation costs. The application of the counter rotating propeller (CRP) has been developed significantly in last decades. Compared with single propeller, both of the front and rear propellers of CRP have lower loadings while supplying the same thrust as the single propeller. Therefore, the CRP has better cavitation performance under the same operating condition. Thus, it is significant to investigate the noncavitation fluctuating pressure of CRP theoretically. Nowadays, works concerning the fluctuating pressure field of CRP are hard to find.

In the present paper, the spatial FPF of CRP under noncavitating condition is investigated in detail. The low-order potential based panel method is adopted throughout

this study, and the computation formulas for fluctuating pressure prediction are proposed. The calculation program based surface panel method is validated by comparing the calculation results and test data of DTRC P4118. Two given single propellers' fluctuating pressures are calculated and the results are compared with the test data to validate the fluctuating pressure prediction method. Then the FPFs of a set of CRP and the corresponding single propeller are calculated. The hydrodynamic interaction between the front and rear propellers of CRP is considered. Through analysis, the FPF characteristics of CRP can be obtained, which will provide a basis for proper stern vibration control strategies under noncavitating condition.

2. Methods

The panel method has been proved to be effective for hydrodynamic performance prediction of propeller [19]. Studies by Hsin [20] and Kinnas and Fine [21] have given specific description about the details and fundamentals of panel method. Thus, this paper only gives a brief description. This method derived from Green's theorem and the velocity potential ϕ_p at a fixed point p located anywhere in the flow field can be expressed as follows:

$$4\pi\varsigma\phi_p = \int_S \left[\phi_q \frac{\partial G}{\partial n} - G \frac{\partial \phi_q}{\partial n} \right] dS + \int_{S_w} \Delta\phi \frac{\partial G}{\partial n} dS, \quad (1)$$

where S, S_w represent the propeller surface and the wake surface respectively, $\Delta\varphi$ is the potential jump across the wake sheet, ϕ_q represents the velocity potential at any point q on S, and G denotes Green's function. In the case of unbounded three-dimensional fluid domain, G is given as

$$G = G(p;q) = \frac{1}{R(p;q)} \quad (2)$$

with $R(p;q)$ being the distance between points p and q. ς in (1) has values as follows:

 (i) $\varsigma = 1$, if p lies in the flow field, but not on S.

 (ii) $\varsigma = 1/2$, if p lies on S.

 (iii) $\varsigma = 0$, if p lies outside S.

The propeller surface and wake sheet are discretized with hyperboloidal panels. A constant source and a constant dipole are then distributed on each panel. During unsteady calculation, the integral equation (1) is solved at each time step, and time-dependent terms are updated at the next time step. The specific unsteady treatment is given in literature [20]. The time domain is discretized into equal time intervals Δt, and (1) at each time step $n = t/\Delta t$ can be discretized as follows:

$$\sum_{j=1}^{N_P} a_{i,j}\phi_j(n) + \sum_{m=1}^{N_R} W_{i,m,1}\Delta\phi_{m,1}(n) = \text{RHS}_i(n),$$

$$i = 1, 2, \ldots, N_P,$$

$$\text{RHS}_i(n)$$

$$= \sum_{Z=1}^{N_B} \sum_{j=1}^{N_P} b_{i,j}^Z \sigma_j^Z(n) - \sum_{Z=2}^{N_B} \sum_{j=1}^{N_P} a_{i,j}^Z \phi_j^Z(n)$$

$$- \sum_{Z=2}^{N_B} \sum_{m=1}^{N_R} \sum_{l=1}^{N_W} W_{i,m,l}^Z \Delta\phi_{m,l}^Z(n)$$

$$- \sum_{m=1}^{N_R} \sum_{l=2}^{N_W} W_{i,m,l}^Z \Delta\phi_{m,l}(n), \tag{3}$$

where N_B is the number of blades. For each blade, N_C is the number of chordwise panels, N_R is the number of spanwise panels, $N_P = 2 \times N_C \times N_R$ is the total number of panels, and N_W is the number of chordwise panels in the wake. The influence coefficients $a_{i,j}^Z$ and $b_{i,j}^Z$ are defined as the potentials induced at panel i by unit (constant) strength dipole and source distributions, respectively, located at panel j on blade Z. The wake influence coefficients $W_{i,m,l}^Z$ are defined similarly. The definitions of the influence coefficients are given in the dissertation of Hsin [20].

As the propeller hub is nonlifting, the dipoles on the panels of hub are set to zero, so just a constant source is distributed on each panel of hub. Once the velocity potential is determined, the velocities on the propeller surfaces can be obtained by differentiating the resulting velocity potential. Then the pressure distribution is calculated through Bernoulli's equation.

In this paper, the solid surface S is used for spatial fluctuating pressure field analysis of propeller. The local intensity of the distribution is denoted by $\sigma(q)$, where the source point q now denotes a general point of the surface S, and then the normal velocity boundary condition must be satisfied at any point p on the surface S

$$2\pi\sigma(p) + \iint_S \sigma(q) \frac{\partial}{\partial n} \frac{1}{r(p,q)} dS(q) = -v_{tp} \cdot n_p,$$
$$v_{tp} = V_0 + v_{inp}, \tag{4}$$

where $r(p,q)$ is the distance between p and q, v_{tp} is the inflow velocity of point p, v_{inp} is the velocity induced by the propeller at point p, and can be obtained through panel method.

As the solid surface just has source distribution, the velocity induced by the solid surface at any point d on the propeller surface can be expressed as follows:

$$v_d = \frac{1}{4\pi} \iint_S \sigma(q) \nabla \left(\frac{1}{R(d,q)} \right) dS(q), \tag{5}$$

where $R(d,q)$ is the distance between d and q.

The solid surface is divided into a number of quadrilateral source panels; the dimensions of which are small in comparison with the surface. The solution is constructed in terms of the source strengths on the surface. The integral equation for the source strengths is approximated by a matrix equation on

the assumption of uniform strength on each panel. Equations (4)~(5) can be discretized as follows:

$$2\pi\sigma_i + \sum_{j=1}^{N_0} \sigma_j \iint_{s_j} \frac{\partial}{\partial n_i} \frac{1}{r_{ij}} ds = -v_{ti} \cdot n_i \quad i = 1, \dots, N_0, \tag{6}$$

$$v_{ti} = V_0 + v_{ini}, \tag{7}$$

$$v_d = \frac{1}{4\pi} \cdot \sum_{j=1}^{N_0} \sigma_j \iint_{s_j} \nabla \left(\frac{1}{R_j} \right) ds. \tag{8}$$

Here, N_0 is the panel number of the solid surfaces, v_{ti} is the inflow velocity of the ith panel, and v_{ini} is the velocity induced by the propeller on the ith panel of the solid surface. For (6), the second term is zero when the jth and ith panels are on the same plane; namely, $\sigma_i = -v_{ti} \cdot n_i/2\pi$. R_j is the distance between the point d and the jth panel of the solid surface.

The velocity induced by the solid surface on the propeller surface can be obtained through (8). By adding the velocities induced by the solid surface to the inflow velocities the unsteady hydrodynamic performance of the propeller as well as the velocities induced by the propeller on the solid surface panels are recalculated. Solving (6) can get new source strength σ_i. Through this iterative calculation, the stable source strengths and velocities on the panels of the solid surface can be obtained at each time step. The pressure distribution on the solid surface is obtained through Bernoulli's equation. The hydrodynamic performance coefficients can be defined as follows:

$$J = \frac{V_0}{nD};$$

$$K_T = \frac{T}{\rho n^2 D^4};$$

$$K_Q = \frac{Q}{\rho n^2 D^5}, \tag{9}$$

where J is the advance coefficient, ρ is the ambient fluid density, n and D, respectively, denote the rotational speed and the diameter of the propeller, V_0 is the ship speed, T and Q are the thrust and torque of the propeller, and K_T and K_Q are the thrust and torque coefficients of propeller, respectively.

As the propeller works in the nonuniform flow field, the pressure on the solid surface changes periodically. The Fourier transform is adopted to transform the pressure signals from the time domain to frequency domain, and the fluctuating pressure coefficient can be obtained:

$$K_{Pn} = \frac{P_n}{\rho n^2 D^2} \quad n = 1, 2, 3, \dots, \tag{10}$$

where P_n is the nth blade frequency harmonic component of the fluctuating pressure (zero to peak) and K_{Pn} represents the fluctuating pressure coefficient of the nth blade frequency harmonic component.

3. Method Validation

The surface panel method proposed in this paper is applied to the propeller DTRC 4118 whose specific parameters are given

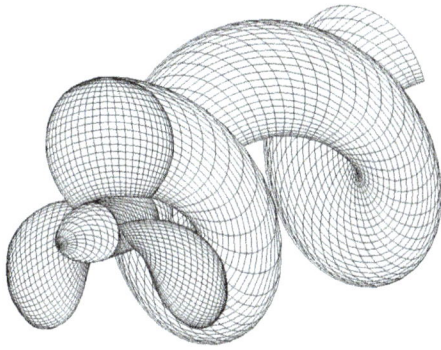

FIGURE 1: Panel arrangement of DTRC 4118 propeller.

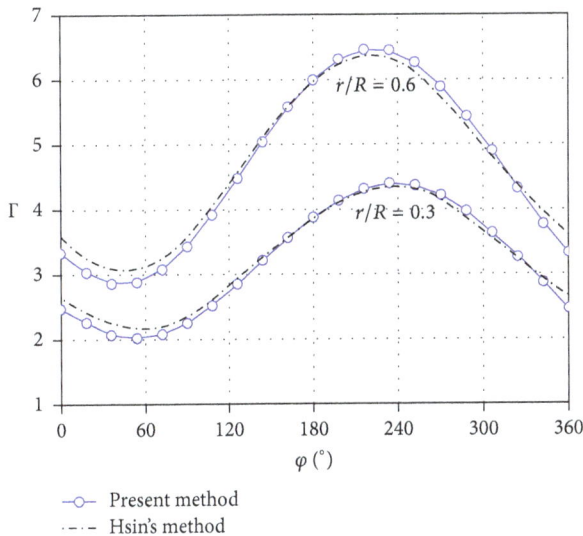

- -o- Present method
- -·-· Hsin's method

FIGURE 2: Comparison with Hsin's method.

TABLE 1: Main parameters of the propeller.

Title	1st model	2nd model
Model diameter (D_m)	0.24 m	0.24 m
Number of blades (Z)	6	6
Expanded blade area ratio (Exar)	0.683	0.81
Pitch ratio$_{(mean)}$	1.039	1.043
Skew	32°	32°
Direction of rotation	Right-handed	Right-handed

which gives the circulation distributions versus blade angle at two propeller radiuses; namely, $r/R = 0.3$ and $r/R = 0.6$. It can be known that the calculation results show excellent agreement with the results of Hsin.

The method proposed in this paper is validated by comparing the theoretical calculation results with the experiment values of two containership propellers. The main parameters of two propellers studied in this paper are shown in Table 1. Figure 3 gives the propeller models used in tests.

The tests were carried out in the cavitation tunnel laboratory of Shanghai Ship & Shipping Research Institute (SSSRI). The pressure fluctuation signals were amplified by the amplifier YD-28 and stored by the wave profile storage. Then the sampled signals were input into computer for fast Fourier transform (FFT) processing. The wake field used for both of propellers is given in Figure 4, the in-plane wake velocities are not considered.

The pressure fluctuation survey was arranged at five measuring points. Five pressure sensors were installed in a plate right above propeller model. The distance between the plate and center line of propeller shaft was 0.188 m (model radius 0.12 m plus tip clearance 0.068 m) in this case. Figure 5 gives the outline of this arrangement.

In the tests, the thrust coefficient was kept the same with that of full scale condition. The thrust coefficient of the first propeller is 0.2079 and that of the second propeller is 0.2231. The rotational speeds of both of propellers were kept fixed to a value of $n = 21.48$ rps throughout the tests. The advance speeds in tests were set to satisfy the required thrusts in the given wake field. The results of pressure fluctuation measurement are presented in form of pressure fluctuating coefficient K_{Pn}. For theoretical calculation, the coordinate is denoted by (X, Y, Z), a right-handed system with the X-axis pointing downstream along the propeller axis; the coordinate origin is located at the center of the propeller disk. The Y-axis points to starboard and Z-axis upward. Figure 6 gives the panel distributions of the 1st propeller surface and a plate large enough as well as the slipstream of the key blade. The 2nd propeller has the same panel distributions. The theoretical calculations adopt the same operating conditions with the model experiments. Table 2 gives the time-averaged hydrodynamic performance coefficients over one revolution. It can be known that the calculation results are in good agreement with the experiment data and that the errors are in acceptable scope.

by Hsin [20]. The panel arrangements of propeller model and slipstream of the key blade are illustrated in Figure 1. The cosine spacing is employed in the chordwise direction and spanwise direction of the propeller blade. The calculation accuracy will be much satisfying when the propeller blades are discretized using more than 14 chordwise panels and 14 spanwise panels [19]. The panel number of the propellers studied in this paper is much more than 14 × 14, and the calculation accuracy can be ensured. What is more, the slipstream model is linear. The linear slipstream model can satisfy the calculation accuracy required well [22] and is also used for other calculations of this paper. The incoming wake is assumed to be axial, with a once per revolution circumferential variation:

$$U_{Sx} = V_0 \left(1 + u_g \cos \theta_S \right) \qquad (11)$$

with u_g being the amplitude of the wake variation, taken equal to 0.2. The advance coefficient is 0.833, and the propeller is operated at 120 rpm with a forward velocity of 1.6 m/s. The density of the undisturbed medium of standard water is 1026 kg/m³. In order to validate the panel method given in this paper, the calculation results are compared with the results of Hsin. The comparison results are shown in Figure 2,

(a)

(b)

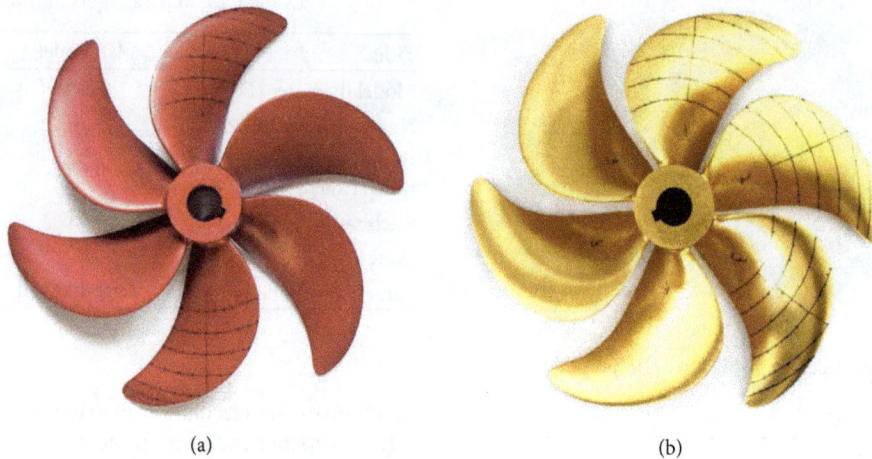

FIGURE 3: Propeller models for experiments: (a) 1st propeller model. (b) 2nd propeller model.

TABLE 2: Time-averaged results of hydrodynamic performance.

	K_T-cal	K_T-exp	K_T-error	$10K_Q$-cal	$10K_Q$-exp	K_Q-error
1st propeller	0.2125	0.2079	2.23%	0.3624	0.3506	3.37%
2nd propeller	0.2279	0.2231	2.18%	0.3786	0.3643	3.93%

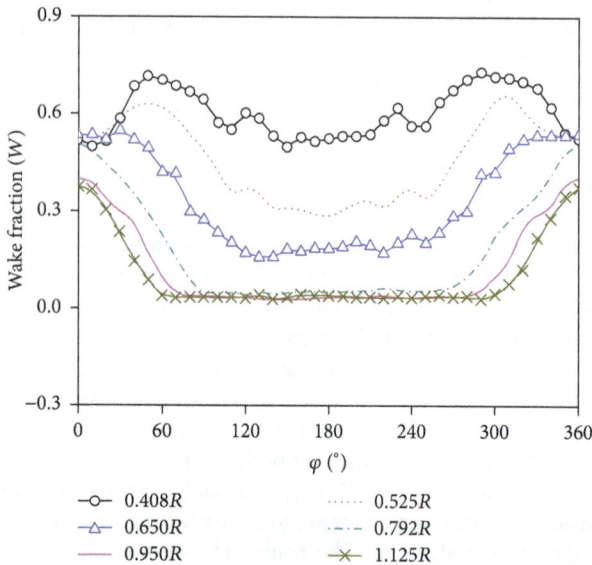

FIGURE 4: Simulated axial wake contours.

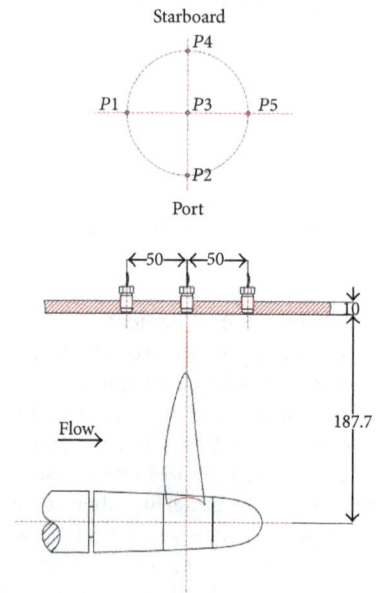

FIGURE 5: Arrangement of sensors for measuring pressure fluctuation.

Figures 7 and 8 give the comparisons between the experiment values and the calculation results of the fluctuating pressure coefficients of the first four blade frequency harmonic components. It can be known that the calculation results are slightly less than the experiment values as a whole; this is mainly because cavitation can not be avoided totally during the experiment in the cavitation tunnel but not be prominent. What is more, somewhat difference exists between the theoretical model and the experimental arrangement. Nevertheless, the result shows that the calculation results agree with the experiment values well and provides a satisfying verification of the theory proposed in this paper.

4. Calculation Results

4.1. Hydrodynamic Performance Prediction. Taking a container ship as the research object, the wake field of the model scale vessel at the propeller plane measured in water tank is shown in Figure 9. The counter rotating propeller HEU96

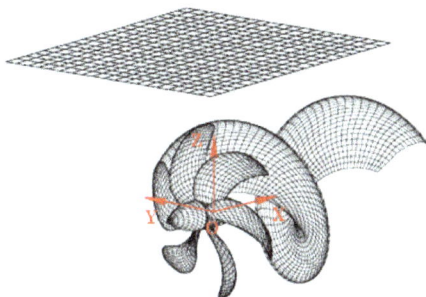

FIGURE 6: Propeller and plate for verification.

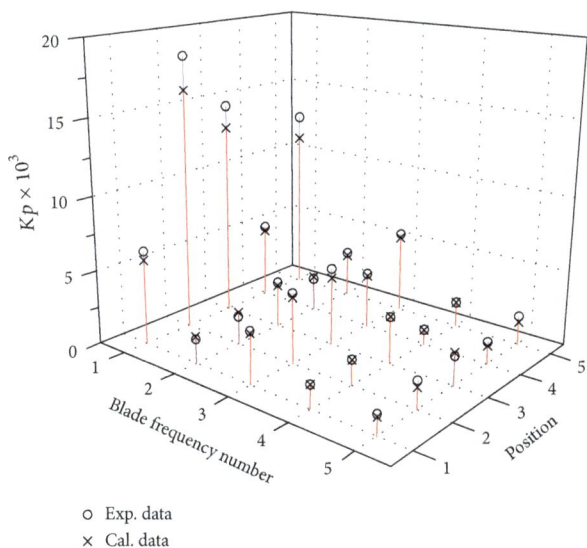

FIGURE 7: Comparisons between the experiment values and calculation results of 1st propeller.

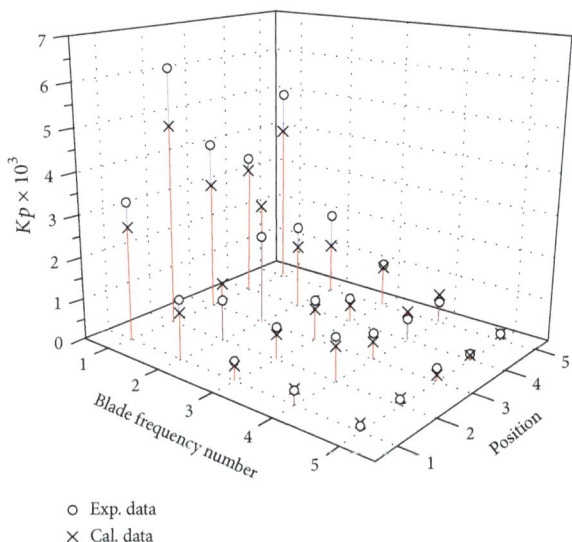

FIGURE 8: Comparisons between the experiment values and calculation results of 2nd propeller.

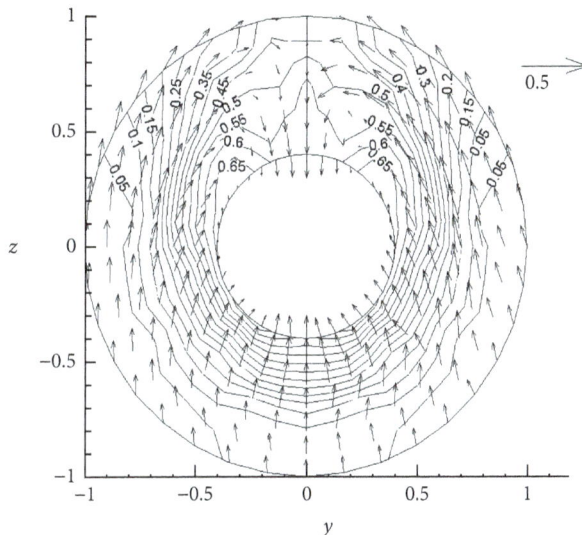

FIGURE 9: Wake field at propeller plane.

CRP and the corresponding single propeller HEU96 propeller studied in this paper are designed by Harbin Engineering University. The 5-bladed HEU96 propeller operates at 8.77 rps with a forward velocity of 1.6 m/s. The front and rear propellers of HEU96 CRP have 4 and 5 blades, respectively, and the front propeller is a right-handed propeller and rear one left-handed. Compared with the single propeller the CRP can have an increase in propulsive efficiency by 12.09%. The CRP has the same advance coefficient and forward velocity with the single propeller. The interference between the front and rear propellers is considered through induced velocities during the calculation iteration.

This paper has unsteady hydrodynamic performance predictions of CRP and single propeller, respectively, and the unsteady hydrodynamic performance of CRP is predicted considering the interactions between the front and rear propellers. As the rear propeller works in the slipstream of forward propeller, the influence coefficients of the slipstream of forward propeller may have singularity when the panel centroids of slipstream get too close to those of rear propeller. It is assumed that the singular influence coefficient appears when the distance between two panel centroids is less than one hundredth of the length of relevant propeller panel diagonal. This paper ignores the singular influence coefficients of slipstream directly as skimming individual singular influence coefficient has little effect on the hydrodynamic performance calculation [22].

Figures 10 and 11 give the thrust and torque coefficients over one revolution. The thrust and torque coefficients of the front and rear propellers of CRP are defined based on the single propeller parameters. The CRP produces roughly the same thrust with the single propeller, whereas the torque of CRP is obviously smaller than that of the single propeller. Therefore, the CRP can raise the propulsive efficiency significantly compared with the single propeller. Both the thrust and the torque show significant periodicities. The fluctuations

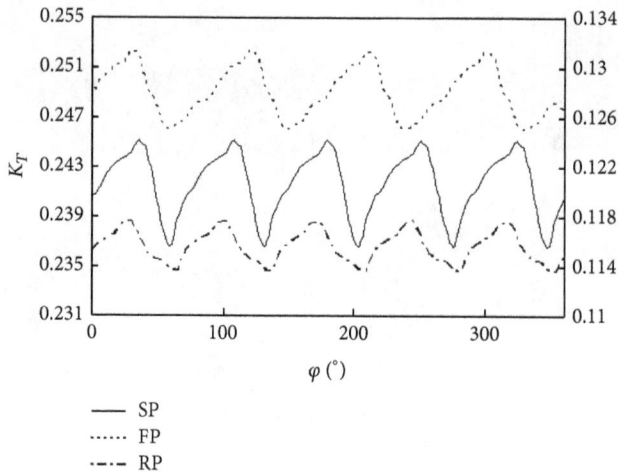

FIGURE 10: Thrusts over one revolution for single propeller (SP) and CRP (FP: front propeller, RP: rear propeller).

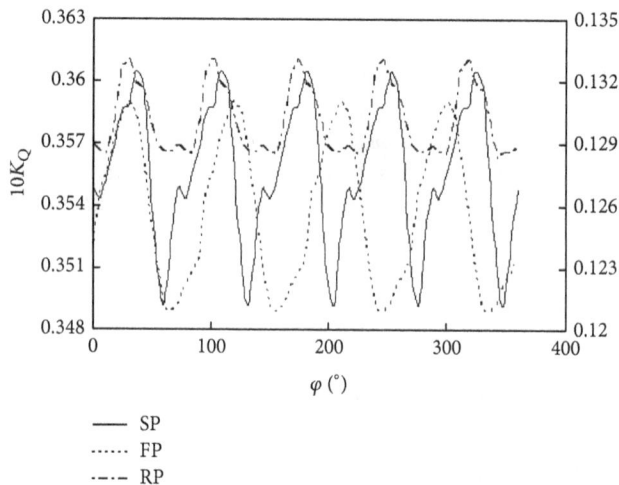

FIGURE 11: Torques over one revolution for single propeller (SP) and CRP (FP: front propeller, RP: rear propeller).

in thrust and torque of front and rear propellers of CRP are somewhat smaller compared with the single propeller.

As the rear propeller blades run through the slipstream of the front propeller, the front propeller has strong influence on the rear propeller and makes the inflow velocities of the rear propeller more circumferentially uniform compared with the wake velocities behind the ship hull. Therefore, the oscillation amplitudes of the thrust and torque of the rear propeller are smaller compared with those of the front propeller, especially; the fluctuation of rear propeller torque is evened out obviously. As the influence of the rear propeller on the front propeller is weak the oscillation amplitudes of the thrust and torque of the front propeller are not improved significantly but still less than those of the single propeller due to lower blade loading.

As the FPF of propeller is closely related to unsteady hydrodynamic forces, the CRP can have better fluctuating pressure performance.

4.2. FPF Calculation. In order to calculate and analyze the spatial FPF of CRP, the spherical surfaces of CRP and the corresponding single propeller are constructed, as shown in Figure 12. The Cartesian coordinate (X, Y, Z) is used, and X points in the downstream direction, Y starboard, and Z upstream direction. The center of the spherical surface is located at the midpoint between the front and rear propeller plane centers. The diameter D_s of spherical surface of CRP is 3.5 times the front propeller diameter. The diameter of the spherical surface for the single propeller is the same as that of the CRP, and the center of the spherical surface is located at the single propeller disk center. We assume that the existence of spherical surface has no effect on the initial inflow velocity of propulsor.

For the single propeller, the FPF calculation is clear and simple. However, the analysis about the FPF of CRP is complex owing to the complicated construction of CRP. For better understanding about the calculation process of the FPF of CRP, Figure 13 gives the iterative process of FPF calculation of CRP. In Step 1 of Figure 13, the induced velocities from rear propeller and spherical surface are not considered for the performance prediction of forward propeller in the first iterative process. In Step 3, the induced velocities from the spherical surface are not considered for the performance prediction of the rear propeller in the first iterative process. In Steps 2 and 4, the induced velocities from the forward propeller and the rear propeller are calculated at each time step during unsteady hydrodynamic performance predictions. Since the relative positions of the forward propeller, rear propellers, and the spherical surface are specified at any time step, the induced velocities between the forward propeller and rear propeller as well as those of the forward and rear propellers at the panel centroids of spherical surface are calculated at each time step. In Step 7, the induced velocities of the spherical surface at the panel centroids of the forward and rear propellers are calculated at each time step. In Step 6, the forces of system refer to the total thrust and the total torque of the forward and rear propellers. Once the total velocities on the spherical surface are determined, the pressure can be calculated through Bernoulli's equation, and the fluctuating pressure induced by the forward and rear propellers can be obtained.

Comparing with the hydrodynamic performance calculations of CRP in Section 4.1, the FPF calculations of CRP need more computation as the interferences between the spherical surface and propellers are considered. The FPF calculations of CRP converge after 8 loop iterations, while the convergence results of hydrodynamic forces given in Figures 10 and 11 just need 5 iterations.

During the calculations, the influence coefficients of the slipstream of propeller may have singularity when the panel centroid of slipstream gets too close to that of spherical surfaces. The singular influence coefficients are also ignored. The calculated three-dimensional FPFs of the single propeller and CRP are shown in Figures 14, 15, and 16. For the sake of constructive analysis, the front hemispherical surfaces are translated 5 times the front propeller diameter along the negative X direction and the rear hemispherical surfaces are translated the same distance along the positive X direction,

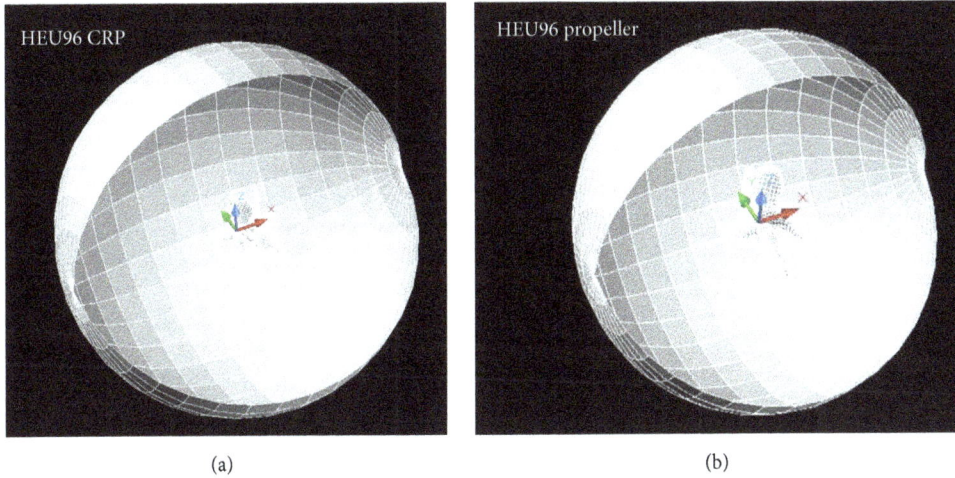

FIGURE 12: Spherical surfaces for calculation: (a) CRP spherical surface system and (b) single propeller spherical surface system.

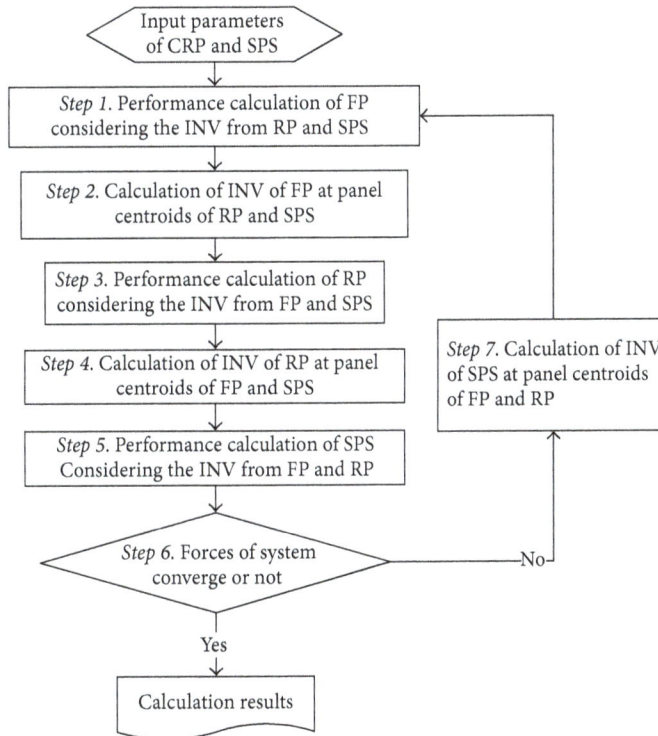

FIGURE 13: Iteration flow diagram of FPF calculations of CRP. INV is the abbreviation of induced velocities. FP and RP represent the forward and rear propellers, and SPS denotes the spherical surface.

and the positions of propellers remain unchanged. Since the single propeller and CRP have different diameters, the fluctuating pressure coefficients of the front and rear propellers of CRP are defined based on the diameter and rotational speed of the single propeller. It can be known from Section 3 that the fluctuating pressure coefficient K_{Pn} decreases rapidly with the increase of blade passage frequency (BPF); thus the first-order fluctuating pressure coefficient K_{P1} is given here.

It can be seen from Figure 14 that the single propeller has strong fluctuating pressure distribution on the solid spherical surface within the longitudinal cylinder of propeller. In other words, the fluctuating pressure induced by the single propeller is obvious ahead of the propeller and within the slipstream. The rotating blades make strong disturbance in the slipstream by inducing a swirling motion of the water. In addition, the suction role of propellers can induce obvious fluctuating pressure. We can know from further analysis about Figure 14 that the FPF at the port side is significantly stronger than that at the starboard. This is mainly because the propeller has stronger disturbance at the port side. What is more, the fluctuating pressure near the propeller plane is weak.

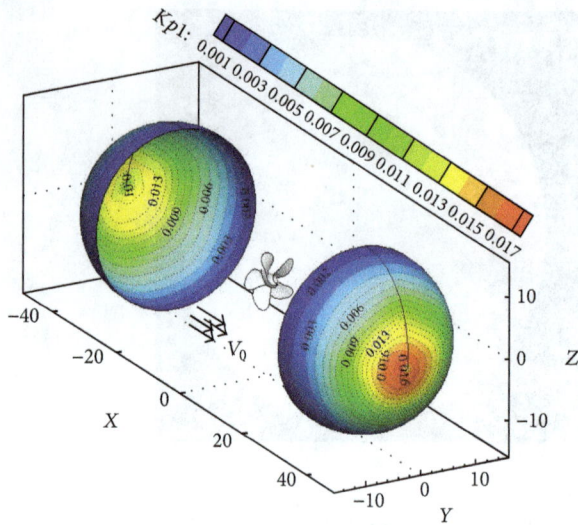

FIGURE 14: 3D contour of single propeller FPF.

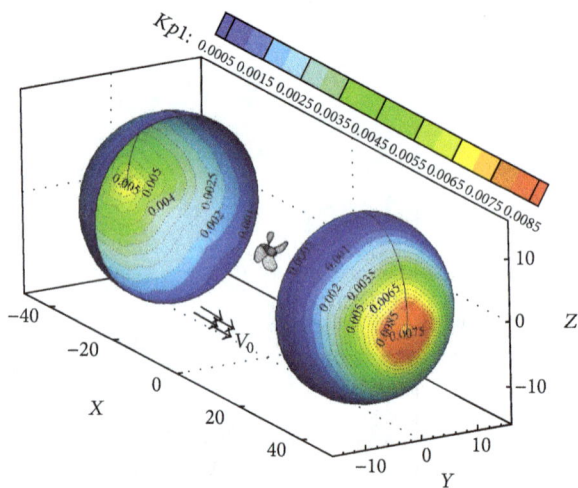

FIGURE 15: 3D contour of front propeller FPF.

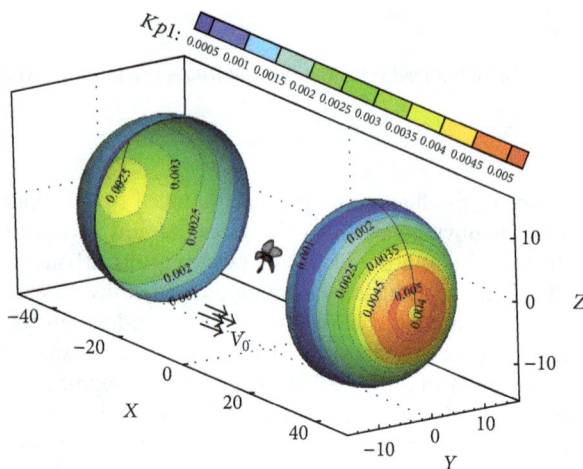

FIGURE 16: 3D contour of rear propeller FPF.

Figures 15 and 16 give the three-dimensional FPFs of the front and rear propellers of CRP. The distribution tendencies of FPFs of the front and rear propellers of CRP are similar to that of the single propeller, the fluctuating pressures downstream are obviously stronger than those upstream, and the weakest fluctuating pressures occur at the propeller rotation plane. The fluctuating pressures of the front and rear propellers of CRP are cyclical, and the cycles have the same numbers of the propeller blades. Compared with the single propeller, the fluctuating pressures induced by the front and rear propellers of CRP are significantly weakened. The first reason is that the load distributions of both the front and rear propellers of CRP are much lower than that of the single propeller with the same thrust being obtained and that the peaks of the fluctuating pressures of the front and rear propellers of CRP are significantly decreased within the slipstream. As the hydrodynamic force amplitudes of the rear propeller are smaller than those of the front propeller, the fluctuating pressure value of the rear propeller is slightly smaller than that of the front propeller. What is more, the interference between the front and rear propellers has an influence on the fluctuating pressure distributions of both propellers. As the rear propeller blades run through the slipstream of the front propeller, the front propeller has a strong influence on the hydrodynamic performance of the rear propeller and makes the inflow velocities of the rear propeller more circumferentially uniform compared with the wake velocities behind the ship hull. Therefore, the pulsation amplitude of unsteady loading on the blade surfaces is smaller, and the FPF of the rear propeller is more uniform and smaller than that of the front propeller.

The total FPFs of CRP are given in Figure 17 in which the fluctuating pressure coefficients are the sums of the fluctuating pressure coefficients of the forward propeller and those of the rear propeller. It can be known from Figures 14 and 17 that the maximum fluctuating pressure of CRP is obviously smaller than that of the single propeller.

In order to have further analysis and better comparison about FPFs between the single propeller and CRP, this paper gives the two-dimensional distribution of the fluctuating pressure of propeller in the longitudinal section in center plane of ship. The positions are given in terms of the angle θ and the distance d. The axial angle θ is measured from the downstream propeller axis and the distance d is $D_s/2$, as shown in Figure 18.

Figure 19 gives the two-dimensional distribution of the fluctuating pressure of the single propeller. The spectra values are plotted as $Kp1$ (first-order fluctuating pressure coefficient) versus position angle, in degree. The single propeller has a strong fluctuating pressure radiation near the hub axis, and local minimum values appear at the hub axis. The fluctuating pressures downstream are obviously greater than that upstream. The weakest fluctuating pressure occurs at the blade rotation plane, and the fluctuating pressure has a strong radiation tendency towards the positions on the hub axis. The resulting fluctuating pressure directivities of the front and rear propellers are shown in Figure 20. The fluctuating pressure directivities of the front and rear propellers have similar tendency to that of single propeller; the fluctuating

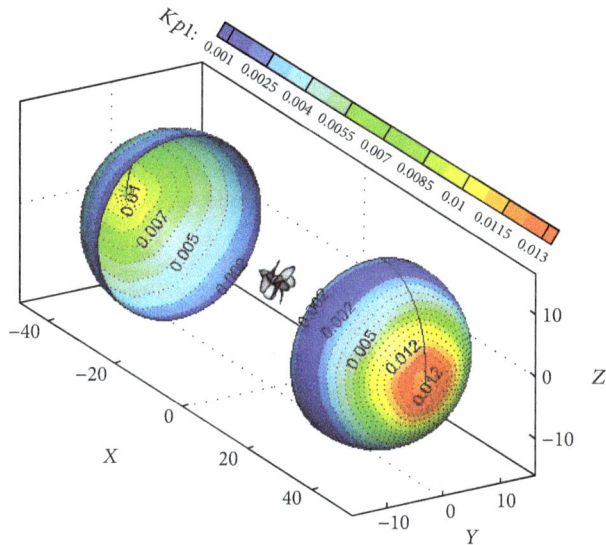

FIGURE 17: 3D contour of the total FPF of CRP.

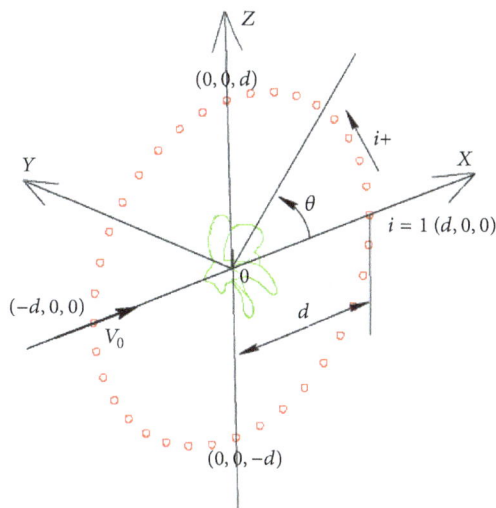

FIGURE 18: Schematic of calculation positions.

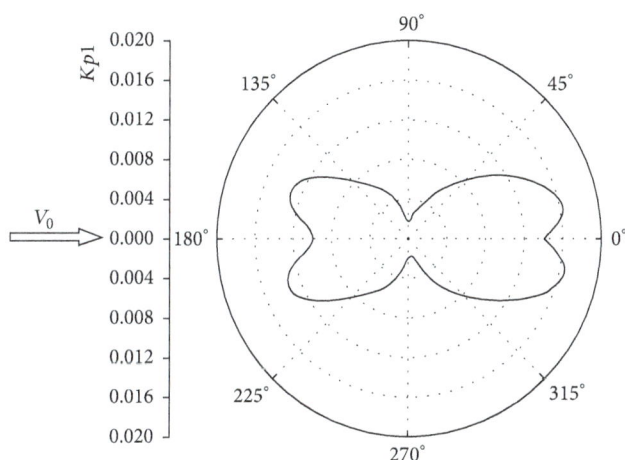

FIGURE 19: Two-dimensional distribution of the fluctuating pressure of single propeller.

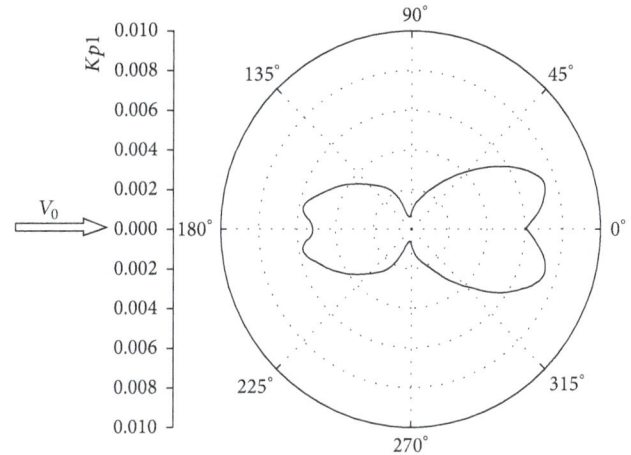

(a)

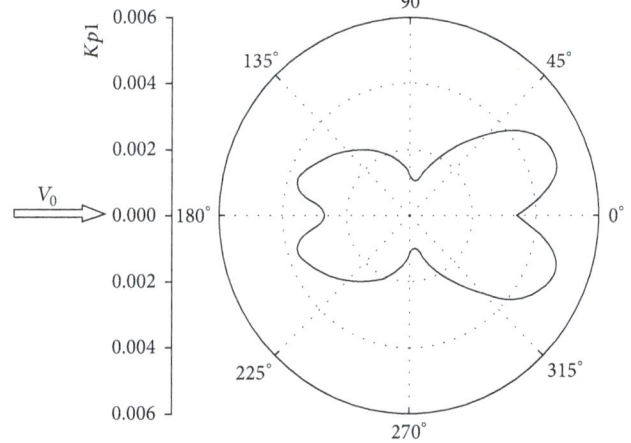

(b)

FIGURE 20: Two-dimensional distributions of the fluctuating pressures of the forward and rear propellers of CRP. (a) Forward propeller. (b) Rear propeller.

pressures downstream are obviously stronger than those upstream.

In order to have further analysis and better comparison between the FPFs of the single propeller and the CRP system, the overall two-dimensional distribution of the first-order fluctuating pressure of CRP is obtained through superposing the first-order fluctuating pressure of the front and rear propellers, as shown in Figure 21. The fluctuating pressure of the CRP is significantly smaller than that of the single propeller in the upstream as well as at the positions near the hub axis in the downstream. However, the fluctuating pressures of CRP and the single propeller are close near the Y-Z plane in the downstream, and the fluctuating pressure of CRP is even greater than that of the single propeller on the Y-Z plane. It can be known that the CRP and the single propeller have close value of fluctuating pressures in the downstream near the Y-Z plane under the same condition.

5. Conclusions

A theoretical method for the analysis of the fluctuating pressure characteristics of CRP is proposed based on the

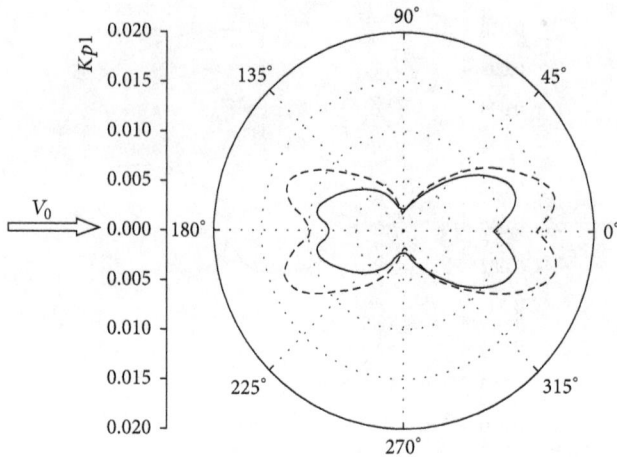

FIGURE 21: Comparison between the overall two-dimensional distribution of the fluctuating pressure of CRP (solid line) and that of single propeller (dash line).

φ:	Angular position
n:	Rotational speed of propeller
D:	Diameters of propeller
V_0:	Inflow velocity
v_{in}:	Induced velocity of propeller
v_t:	Total velocity on solid surface
v_d:	Induced velocity of solid surface
T:	Thrust of propeller
Q:	Torque of propeller
$J = V_0/nD$:	Advance coefficient of propeller
$K = T/\rho n^2 D^4$:	Thrust coefficient of propeller
$K = Q/\rho n^2 D^5$:	Torque coefficient of propeller
P_n:	nth fluctuating pressure
K_{Pn}:	nth fluctuating pressure coefficient
u_g:	Amplitude of the wake variation.

Conflicts of Interest

The authors declare that there are no conflicts of interest regarding the publication of this paper.

low-order potential based panel method. Through systematic research, the following conclusions can be obtained.

The verification using two given propellers proves the feasibility of the method proposed in this paper. The agreement between the calculation results and the experiment data is excellent. The single propeller has strong fluctuating pressure distribution ahead of the propeller and within the slipstream and may bring serious damage to the vessel hull and the appendages behind the propeller. The FPFs of the front and rear propellers of CRP have similar distribution tendency with the single propeller. The propellers have a strong fluctuating pressure radiation near the hub axis, and local minimum values appear at the hub axis. The fluctuating pressure downstream is greater than that upstream. The weakest fluctuating pressure occurs at the blade rotation planes, and the fluctuating pressure has a strong radiation tendency towards the positions on the hub axis.

Through analysis, the overall fluctuating pressure of CRP is significantly smaller than that of the single propeller in the upstream as well as at the positions near the hub axis in the downstream. The fluctuating pressures of CRP and the single propeller are close near the Y-Z plane in the downstream, and the fluctuating pressure of CRP is even greater than that of the single propeller on the Y-Z plane. Overall, the CRP can improve the stern vibration and cause less damage to the vessel hull as well as appendages behind the thruster.

As the surface panel method adopted in this paper is based on the inviscid flows, the real flow field of single propeller and the CRP cannot be described accurately. Therefore, the calculation results may deviate slightly from the actual situation. The surface panel method will be further improved and more methods as well as model tests will be applied for this topic in the future.

Nomenclature

ρ: Ambient fluid density
ϕ: Velocity potential

References

[1] J. P. Breslin, R. J. Van Houten, and J. E. Kerwin, "Theoretical and experimental propeller-induced hull pressures arising from intermittent blade cavitation, loading and thickness," *Transactions-Society of Naval Architects and Marine Engineers*, vol. 90, pp. 111–151, 1982.

[2] Y.-Z. Kehr, C.-Y. Hsin, and Y.-C. Sun, "Calculations of pressure fluctuations on the ship hull induced by intermittently cavitating propellers," *Proceedings of the National Science Council, Republic of China - Part A: Physical Science and Engineering*, vol. 22, no. 5, pp. 642–653, 1998.

[3] H. Seol, "Time domain method for the prediction of pressure fluctuation induced by propeller sheet cavitation: Numerical simulations and experimental validation," *Ocean Engineering*, vol. 72, pp. 287–296, 2013.

[4] S. Berger, M. Baue, and M. Druckenbrod, "An efficient viscous/inviscid coupling method for the prognosis of propeller-induced hull pressure fluctuations," in *Proceedings of 7th International Conference on Ships and Shipping Research*, 2012.

[5] T. T. Kanemaru and J. Ando, "Numerical analysis of pressure fluctuation on ship stern induced by cavitating propeller using a simple surface panel method "SQCM"," in *Proceedings of the 2nd International Symposium on Marine Propulsors*, Hamburg, Germany, 2011.

[6] T. Kawamura and T. Kiyokawa, "Numerical prediction of hull surface pressure fluctuation due to propeller cavitation," in *Proceedings of the Japan Society of Naval Architects and Ocean Engineers*, Nagasaki, Japan, 2008.

[7] K. Sato, A. Oshima, and H. Egashira, "Numerical prediction of cavitation and pressure fluctuation around marine propeller," in *Proceedings of the 7th International Symposium on Cavitation*, 2009.

[8] B. Ji, X. Luo, X. Peng, Y. Wu, and H. Xu, "Numerical analysis of cavitation evolution and excited pressure fluctuation around a propeller in non-uniform wake," *International Journal of Multiphase Flow*, vol. 43, no. 1, pp. 13–21, 2012.

[9] T. Lloyd, G. Vaz, D. Rijpkema, and B. Schuiling, "The potsdam propeller test case in oblique flow: prediction of propeller performance, cavitation patterns and pressure pulses,"

in *Proceedings of the Second Workshop on Cavitating Propeller Performance*, 2015.

[10] P. Peralli, T. Lloyd, and G. Vaz, "Comparison of uRANS and BEM-BEM for propeller pressure pulse prediction: E779A propeller in a cavitation tunnel," in *Proceedings of the 19th Numerical Towing Tank Symposium*, Nantes, France, 2016.

[11] R. E. Bensow and R. Gustafsson, "Effect of propeller tip clearance on hull pressure pulses," in *Proceedings of the 5th International Symposium on Marine Propulsors*, Espoo, Finland, 2017.

[12] D. Ross, *Mechanics of Underwater Noise*, Pergamon Press, Oxford, UK, 1976.

[13] M. Garguet and R. Lepeix, "The problem of influence of solid boundaries on propeller-induced hydrodynamic forces," in *Proceedings of Symposium on High Powered Propulsion of Large Ships*, Wageningen, Netherlands, 1974.

[14] S. Tsakonas, J. P. Breslin, and J. Teeters, *Correlation of Theoretical Predictions of Propeller-Induced Hull Pressures with Available Data*, Stevens Institute of Technology, Hoboken, NJ, USA, 1980.

[15] J. Hu, Y. M. Su, and S. Huang, "Numerical simulation of propeller-induced vibrating pressure on stern," *Journal of Harbin Engineering University*, vol. 26, no. 3, pp. 292–296, 2005.

[16] M. Chen and Q. Zhou, "Numerical simulation of fluctuating propeller forces and comparison with experimental data," *Applied Mechanics and Materials*, vol. 105-107, pp. 518–522, 2012.

[17] E. Güngör and İ. Bedii Özdemir, "Simulation of oblique propeller flow including cavitation and pressure pulses," *Underwater Technology*, vol. 33, no. 4, pp. 203–213, 2016.

[18] M. Osborne, "Design constraints limit options for EEDI compliance," *Naval Architecture*, vol. 25, no. 1, pp. 4–27, 2012.

[19] L. X. Hou, A. K. Hu, and C. H. Wang, "Investigation about the spatial pressure field by the propeller," *Journal of Wuhan University of Technology (Transportation Science Engineering)*, vol. 39, no. 4, pp. 743–746, 2015.

[20] C. Y. Hsin, *Development and analysis of panel methods for propellers in unsteady flow, [Ph.D. thesis]*, Department of Ocean Engineering, Massachusetts Institute of Technology, Cambridge, Mass, USA, 1990.

[21] S. A. Kinnas and N. E. Fine, "A nonlinear boundary element method for the analysis of unsteady propeller sheet cavitation," in *Proceedings of the 19th Symposium on Naval Hydrodynamics*, 1992.

[22] L. Hou, C. Wang, X. Chang, and S. Huang, "Hydrodynamic performance analysis of propeller-rudder system with the rudder parameters changing," *Journal of Marine Science and Application*, vol. 12, no. 4, pp. 406–412, 2013.

3

Static Mechanical Properties and Modal Analysis of a Kind of Lift-Drag Combined-Type Vertical Axis Wind Turbine

Fang Feng ⓘ,[1,2] Chunming Qu,[3] Shouyang Zhao,[3] Yuedi Bai,[3] Wenfeng Guo,[3] and Yan Li ⓘ[2,3]

[1]*College of Science, Northeast Agricultural University, Harbin, China*
[2]*Heilongjiang Provincial Key Laboratory of Technology and Equipment for Utilization of Agricultural Renewable Resources in Cold Region, Harbin 150030, China*
[3]*College of Engineering, Northeast Agricultural University, Harbin, China*

Correspondence should be addressed to Fang Feng; fengfang@neau.edu.cn and Yan Li; liyanneau@163.com

Academic Editor: Tareq S. Z. Salameh

In order to explore a set of methods to analyze the structure of Lift-Drag Combined-Type Vertical Axis Wind Turbine (LD-VAWT), a small LD-VAWT was designed according to the corresponding Standards and General Design Requirements for small vertical axis wind turbines. The finite element method was used to calculate and analyze the static mechanical properties and modalities of main parts of a kind of small-scale LD-VAWT. The contours of corresponding stress and displacement were obtained, and first six-order mode vibration profiles of main parts were also obtained. The results show that the main structure parts of LD-VAWT meet the design requirements in the working condition of the rated speed. Furthermore, the resonances of all main parts did not occur during operation in the simulations. The prototype LD-VAWT was made based on the analysis and simulation results in this study and operated steadily. The methods used in this study can be used as a reference for the static mechanical properties and modal analysis of vertical axis wind turbine.

1. Introduction

The vertical axis wind turbine (VAWT) has a simple structure and does not need special device to catch the wind. In addition, it is environmentally friendly; therefore, it has a rapid development in recent years. Among them, the Straight-Bladed Vertical Axis Wind Turbine (SB-VAWT) has been studied more deeply due to better power characteristics and higher transfer efficiency of wind energy. However, the starting characteristic is not well, which is one of the important factors restricting the development of SB-VAWT [1]. Therefore, improving the starting characteristics of SB-VAWT has become the research focus for many scholars [2]. Tang Jing et al. [3] have installed wind hood at the top and bottom of the SB-VAWT to increase the flow speed which can improve the start-up performance of wind turbine; Wu Zhicheng et al. [4] have changed symmetric wind rotor into eccentric wind rotor in order to improve the starting torque of the wind turbine. The theoretical calculations and model tests for the aerodynamic characteristics are numerous; however, the analysis of static and dynamic mechanical properties of wind turbine structure for designing prototype is little. M. SaqibHameed et al. [5] have shown that larger centrifugal load mainly causes bending deformation of blade and used finite element method to compare the mechanical properties of blades made in aluminum and glass fibre reinforced plastic (FRP). The results show that FRP is more suitable to blade material; Lin Wang [6] has used the finite element analysis and genetic algorithm to optimize the structure and weight of blade in the requirement of strength; Zhang Tingting et al. [7] have analyzed the dynamics of the main axis of Darrieus wind turbine, calculated the range of wind speed which can avoid resonance, and obtained the optimal thickness of tube wall of main axis; Wang Jianyu [8] has analyzed the influence of blade shedding vortex on the dynamics of tower and main axis. The research shows that shedding vortex can induce

TABLE 1: Basic structural parameters of LD-VAWT.

Name	Symbol	Value
Rated power [kW]	P	3
Rated wind speed [m/s]	V	10
Wind rotor diameter [mm]	d	4000
Wind rotor height [mm]	H	5200
Number of main blades	N	3
Chord length of main blade [mm]	c	400
Airfoil of main blade	—	NACA0018
Attack angle of main blade [°]	β	0
Drag rotor diameter [mm]	d	700
Drag rotor height [mm]	h	2600
Mounting position of drag rotor [mm]	S	250

resonance; Nidal H. Abu-Hamdeh [9] used ANSYS to model the majority of the structural components of a collapsible vertical axis wind turbine, and data from the mathematical models were used to verify the structure of the turbine and shafts were within acceptable stress and strain limits, the result of the experiments verified the mathematical simulation analysis; Yu Tang [10] used ANSYS Workbench static and modal analysis module to make load analysis of wind turbine internal maintenance lifting platform and obtained the maximum stress of platform bridge structure and place and form of deformation; E. Verkinderen [11] analyzed the coupled structure through a multidegree of freedom system, as well as numerically through the finite element (FE) method of H-Darrieus vertical axis wind turbines; Zheng Li [12] presented a method to simulate wind turbine gearbox system with the multibody drivetrain dynamic analysis software, and the modal analysis of wind turbine gearbox can be carried out on the basis of the multibody dynamic theory. The above researches are only focused on the analysis of common vertical axis wind turbine. However the analysis on the structure of LD-VAWT is little and it does not build a perfect set of designing plan and methods yet. Therefore, this paper will search on the static and dynamic mechanical properties of structure based on a small-scale lift-type vertical axis wind turbine [13–21] and propose a set of suitable research plans and methods as references to other kinds of LD-VAWTs and vertical axis wind turbines.

2. Design of Wind Turbine

2.1. Wind Turbine Model. The model of LD-VAWT designed is shown in Figure 1, and the basic structural parameters are shown in Table 1.

2.2. Structure Design of Wind Turbine. (1) Wind Rotor: The wind rotor of LD-VAWT is an important part, which can convert wind energy into mechanical energy. It is composed of main blade, drag rotor, beam, main axis, and so on. The main blade is made of FRP which has characteristics of being light, having high strength, having corrosion resistance, and being manufactured easily. The main blade is hollow and

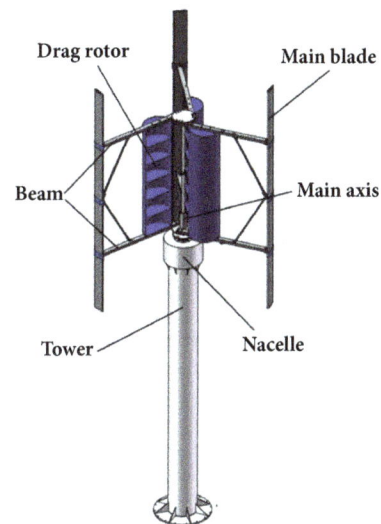

FIGURE 1: Model of LD-VAWT.

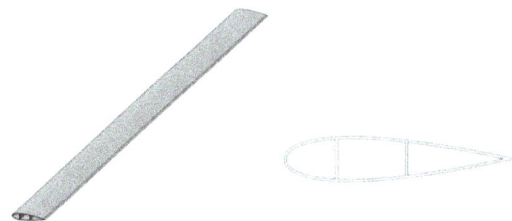

FIGURE 2: The structure of main blade.

stiffened by two ribs, which can reduce the weight of blade. The main blade is shown in Figure 2.

The shape of drag rotor is a semicylindrical surface with thin wall thickness. In order to reduce the weight in the premise of strength requirement, the aluminum alloy material is selected. The thickness of aluminum plate is 3 mm. The structure of drag rotor is shown in Figure 3.

In Figure 1 the beams support the main blades and transmit the torque generated by blade to the main axis. In

FIGURE 3: The structure of drag rotor.

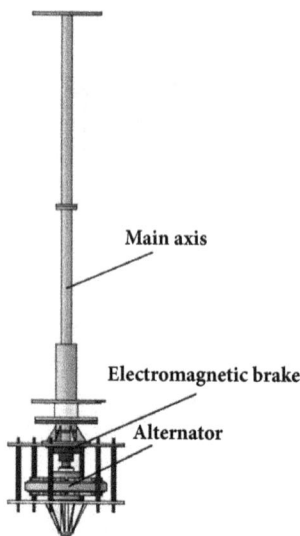

FIGURE 4: The structure of nacelle.

TABLE 2: The parameters of tower.

Name	Value
Height [m]	6
Outside diameter [mm]	630
Inside diameter [mm]	616
Wall thickness [mm]	8

TABLE 3: Material properties of main blade.

Name	Value
Density [kg/m^3]	2×10^3
Elastic Modulus [MPa]	7.2×10^4
Yield limit [MPa]	450
Poisson ratio	0.22
Allowable stress [MPa]	320

the structure is shown in Figure 1, and the configuration parameters are shown in Table 2.

3. Static Mechanical Property Analysis Structure

3.1. Main Blade. The loads of main blade during operation mainly include self-gravity G_L, centrifugal force load F_{rL} caused by the rotation, and aerodynamic load F_{pL} from wind. According to theoretical calculation, the self-gravity G_L is 312 N, the centrifugal force F_{rL} is 6843 N, and the wind load F_{pL} is 95.6 N.

The static mechanical property of main blade is analyzed by finite element method (FEM). The tetrahedral element is selected as mesh type of main blade and the element type is Solid186. Finally, the finite element model of blade has 616368 elements and 212397 nodes. The material of main blade is FRP and the material properties are shown in Table 3.

In order to simulate the connected relation between main blade and beam, a fixed constraint is added at the connection point. Then the wind load is applied on the windward surface of main blade by pressure, the main blade weight is calculated by mass, and gravity acceleration and the centrifugal force are calculated by the rotational inertia load. The loads above are applied on the model of main blade. Finally, the contours of stress and displacement under the rated operation conditions can be obtained as shown in Figures 5 and 6.

From Figure 5, the maximum stress of blade is 45.4 MPa which appears at the connection between main blade and beam. The limited stress of FRP is 320 MPa and the safety factor is 1.5 in this design. Then the ultimate allowable tensile stress of FRP is 213 MPa. According to simulation results, the structural strength of main blade meets design requirements [22]. From Figure 6, the maximum node displacement of main blade appears at the tip of the blade and the value is 15.1 mm, which is larger than the deformation of middle part. It shows that the deformation has less influence on the dynamic characteristics of the wind turbine, which means the structure meets the design requirements [23, 24].

order to enhance bending strength, the square steel is selected as structure of beam. The material of square steel is Q235. The size of cross section is 60×60 mm and thickness of wall is 3 mm.

(2) Nacelle: The nacelle consists of alternator, electromagnetic brake, main axis, and support bars as shown in Figure 4. The alternator is disc-type permanent magnet synchronous generator. The dynamic friction torque of brake is 400 N·m.

The main axis in the nacelle is an important part in designing process. The diameter of axis is designed based on the analog method and empirical method. The minimum diameter of main axis is 40 mm and a pair of angular contact ball bearings with model number 7214 is used.

(3) Tower: The role of the tower is to support and fix the wind rotor and nacelle. The material of tower is Q235,

A: Static Structural
Equivalent Stress
Type: Equivalent (von-Mises) Stress
Unit: MPa
Time: 1

45.449 Max
40.4
35.35
30.301
25.251
20.201
15.152
10.102
5.053
0.003461 Min

0.00   500.00   1000.00 (mm)
250.00   750.00

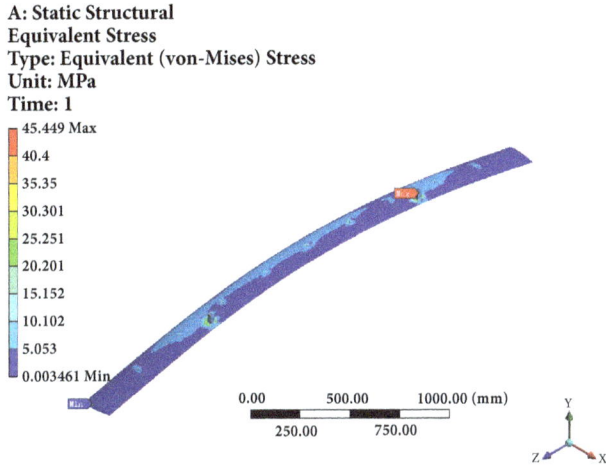

FIGURE 5: Equivalent stress contour of main blade.

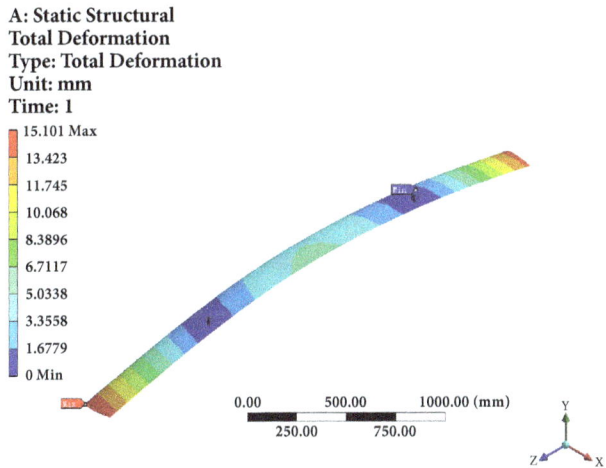

A: Static Structural
Total Deformation
Type: Total Deformation
Unit: mm
Time: 1

15.101 Max
13.423
11.745
10.068
8.3896
6.7117
5.0338
3.3558
1.6779
0 Min

0.00   500.00   1000.00 (mm)
250.00   750.00

FIGURE 6: Displacement variation contour of main blade.

3.2. Drag Rotor. The calculating method of mechanical property of drag rotor is the same as main blade. The self-gravity G_D is 210 N, the centrifugal force F_{rD} is 1381.75 N, and the wind load F_{pD} is 56.77 N.

The solid 185 element is used to mesh and the number of elements and nodes are 457865 and 6956782, respectively. The material of drag rotor is aluminum alloy and the material properties are shown in Table 4.

In the analysis process, the nodes on the upside surface and downside surface of drag rotor are restrained. After calculation, the contours of stress and displacement under the rated condition are obtained as shown in Figures 7 and 8, respectively.

From Figure 7, the maximum stress of drag rotor is 161.4 MPa which is lower than the limit stress of aluminum alloy. Figure 8 shows that the drag rotor has a little displacement, which satisfies the design requirements.

3.3. Beam. The maximum load of beam happens at the rated speed 100 r/min of wind rotor. Therefore, analyses of the stress and deformation of beam are processed under the rated speed

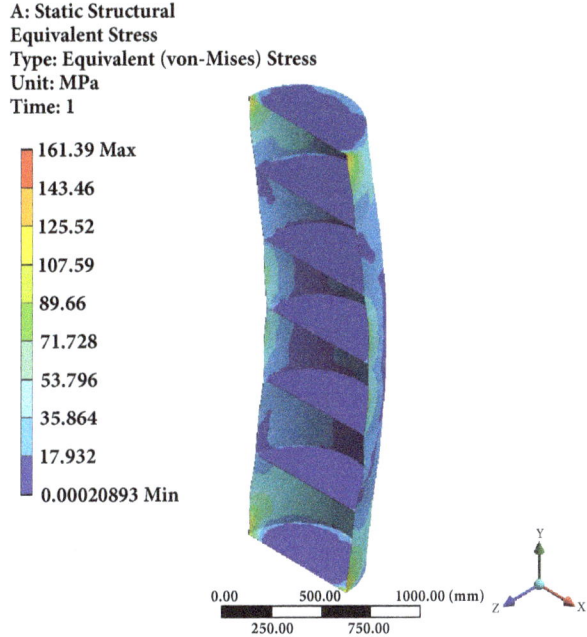

A: Static Structural
Equivalent Stress
Type: Equivalent (von-Mises) Stress
Unit: MPa
Time: 1

161.39 Max
143.46
125.52
107.59
89.66
71.728
53.796
35.864
17.932
0.00020893 Min

0.00   500.00   1000.00 (mm)
250.00   750.00

FIGURE 7: Equivalent stress contour of drag rotor.

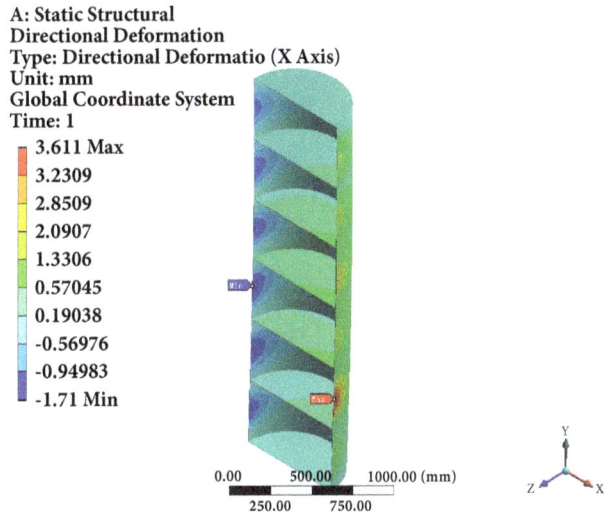

A: Static Structural
Directional Deformation
Type: Directional Deformatio (X Axis)
Unit: mm
Global Coordinate System
Time: 1

3.611 Max
3.2309
2.8509
2.0907
1.3306
0.57045
0.19038
-0.56976
-0.94983
-1.71 Min

0.00   500.00   1000.00 (mm)
250.00   750.00

FIGURE 8: Displacement variation contour of drag rotor.

TABLE 4: Material properties of drag rotor.

Name	Value
Density [kg/m³]	2.7×10^3
Elastic Modulus [MPa]	6.9×10^4
Yield limit [MPa]	276
Poisson ratio	0.33
Tensile strength [MPa]	350

condition. Force and torque can be calculated as shown in Table 5. The material of beam is Q235-A (16Mn) and the properties are shown in Table 6.

TABLE 5: Load distribution of beam.

Name	Value
Self-gravity [N]	115
Gravity of main blade [N]	156
Gravity of drag rotor [N]	105
Torque of main blade [N·m]	119.2
Torque of drag rotor [N·m]	645
Self-centrifugal force [N]	120.4
Centrifugal force of main blade [N]	3421.4
Centrifugal force of drag rotor [N]	808.5

TABLE 6: Material properties of beam.

Name	Value
Density [kg/m^3]	7.86×10^3
Elastic Modulus [GPa]	211
Yield limit [MPa]	196
Poisson ratio	0.3
Tensile strength [MPa]	235

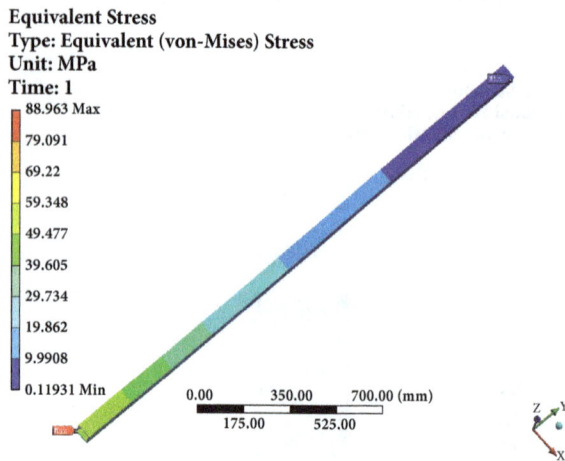

FIGURE 9: Equivalent stress contour of beam.

The Degrees of Freedom (DOF) are constrained on the displacements of X, Y, and Z directions at the end of connection position of beam and main axis. Then the gravity load, centrifugal load, and torque load are applied on the model, respectively. The contours of stress and displacement of beam under the rated condition are obtained by calculation as shown in Figures 9 and 10, respectively.

From Figure 9, the maximum stress is 89 MPa, which appears at the end of connection position between beam and main axis. Therefore the junction should be strengthened. From Figure 10, the maximum displacement happens at the tip of beam where the lift and drag force is fixed and the maximum deformation is 2.4 mm. The strength of beam needs to meet the checking formula (1)

$$\delta_{\max} < \frac{[\delta]}{S} \tag{1}$$

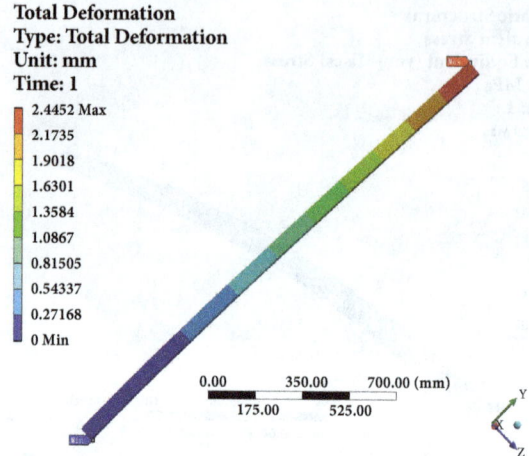

FIGURE 10: Displacement variation contour of beam.

TABLE 7: Received force of main axis.

Name	Value
Gravity of main axis [N]	630
Gravity of wind rotor [N]	3250
Torque of wind rotor [N·m]	1045
Centrifugal force of wind rotor [N]	0

TABLE 8: Material parameters of main axis.

Name	Value
Density [kg/m^3]	7.86×10^3
Elastic Modulus [MPa]	2×10^5
Yield limit [MPa]	400
Poisson ratio	0.3
Tensile strength [MPa]	980

where $\delta_{\max}$ is the maximum stress, $[\delta]$ is yield limit stress of material, in this paper $[\delta]$ is 235 MPa, and $[S]$ is safety factor, which is selected as 1.5.

From the calculation, $\delta_{\max}$ is lower than allowable stress. The stiffness checking formula is shown as follows:

$$\frac{\omega_{\max}}{l} < \frac{[\omega_{\max}/l]}{S} \tag{2}$$

where $\omega_{\max}$ is the maximum displacement of beam, 2.4 mm, l is the length of beam, 2 m, and $[\omega_{\max}/l]$ is the allowable deflection of simply supported beam, $l/500$.

After calculating, $\omega_{\max}/l = 1.2 \times 10^{-3}$ m $< [\omega_{\max}/l]/S = 2.6 \times 10^{-3}$ m, the stiffness of beam under the rated speed meets the design requirements.

3.4. Main Axis. The main axis is mainly subjected to gravity load, centrifugal load, and aerodynamic load. The values of loads are shown in Table 7.

In the static mechanical analysis of main axis, tetrahedral element is used to mesh grids. The numbers of elements and nodes are 87536 and 159853, respectively. The material of main axis is 40Cr, and the properties are shown in Table 8.

Equivalent Stress
Type: Equivalent (von-Mises) Stress
Unit: MPa
Time: 1

26.374 Max
23.443
20.513
17.582
14.652
11.722
8.7912
5.8608
2.9304
7.9636e-6 Min

0.00   500.00   1000.00 (mm)
250.00   750.00

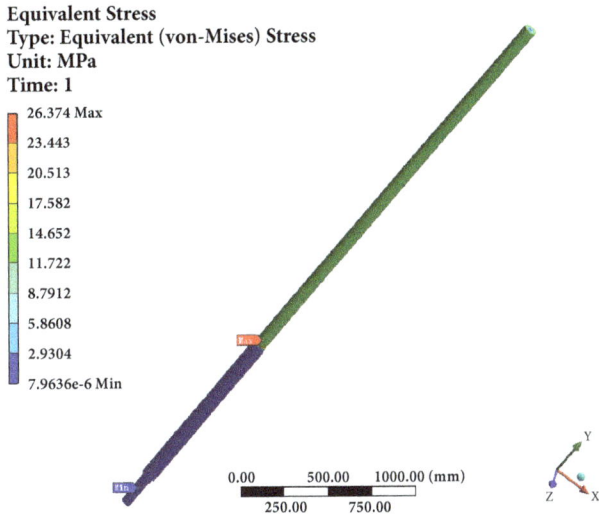

FIGURE 11: Equivalent stress contour of main axis.

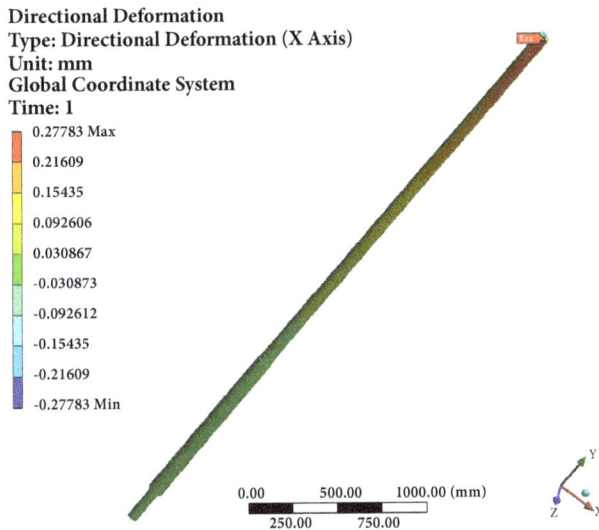

Directional Deformation
Type: Directional Deformation (X Axis)
Unit: mm
Global Coordinate System
Time: 1

0.27783 Max
0.21609
0.15435
0.092606
0.030867
-0.030873
-0.092612
-0.15435
-0.21609
-0.27783 Min

0.00   500.00   1000.00 (mm)
250.00   750.00

FIGURE 12: Displacement variation contour of axis.

According to the assembly relation, the end of main axis connected with generator is constrained. The self-gravity load of main axis is applied with gravity acceleration, the gravity of wind rotor is applied at mounting position of flange, and the torque of wind rotor is also applied at mounting position of flange. The simulation results are shown in Figures 11 and 12, respectively.

From Figure 11, the maximum stress of main axis is 26.3 MPa which is at the connection position between beam and main axis. From Figure 12, the maximum displacement is at the top of main axis which is 0.27 mm.

According to formula (1), maximum stress of main axis δ_{max} is 26.3 MPa, limit stress $[\delta]$ is 980 MPa, and safety factor S is 3. The maximum stress δ_{max} is lower than allowable stress.

TABLE 9: Load distribution of tower.

Name	Value
Horizontal thrust [N]	1132.4
Gravity of wind rotor and cabin [N]	8330
Self-gravity of tower [N]	8291
Torque of cabin [N·m]	286.5
Wind pressure [N/m]	27.1

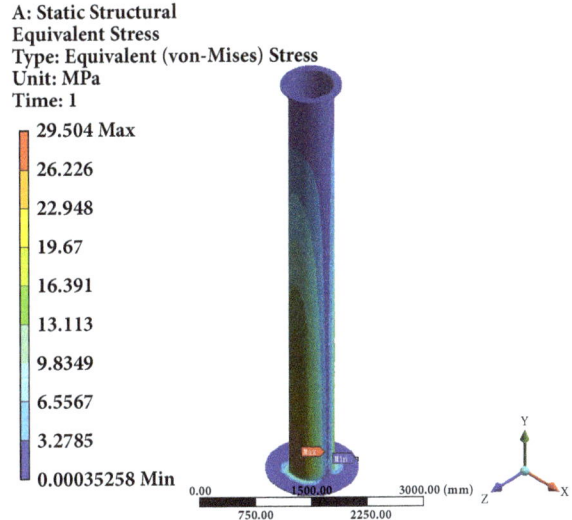

A: Static Structural
Equivalent Stress
Type: Equivalent (von-Mises) Stress
Unit: MPa
Time: 1

29.504 Max
26.226
22.948
19.67
16.391
13.113
9.8349
6.5567
3.2785
0.00035258 Min

0.00   1500.00   3000.00 (mm)
750.00   2250.00

FIGURE 13: Equivalent stress contour of tower.

Similarly the stiffness of main axis needs to meet the stiffness checking formula as follows:

$$y_{max} \leq 0.0005l \qquad (3)$$

where y_{max} is the maximum deformation of main axis and l is the length of main axis, 3400 mm.

The calculation result shows that the maximum deformation of main axis is 0.27 mm.

3.5. Tower. The tower is mainly subjected to horizontal thrust of the wind rotor, the gravity of wind rotor and nacelle, self-gravity, the torque of wind rotor, and the wind pressure acting on the tower. The values of loads distributing on the tower are shown in Table 9.

The material of tower is Q235 and the solid 185 is selected as element type. The numbers of elements and nodes are 15696 and 30864, respectively.

The bottom of tower is constrained. The above loads are applied on the model of tower and the contours of stress and deformation of tower are shown in Figures 13 and 14, respectively.

From Figure 13, the maximum stress of tower is 29.5 MPa which appears at the bottom of tower. From Figure 14, the maximum deformation of tower is 5.3 mm. According to the engineering experience of tower designing [25], the maximum deformation of tower should be less than 0.5~ 0.8% of its height for tower. The limit stress of Q235 is

A: Static Structural
Total Deformation
Type: Total Deformation
Unit: mm
Time: 1

5.2985 Max
4.7098
4.1211
3.5323
2.9436
2.3549
1.7662
1.1774
0.58872
0 Min

0.00   1500.00   3000.00 (mm)
    750.00   2250.00

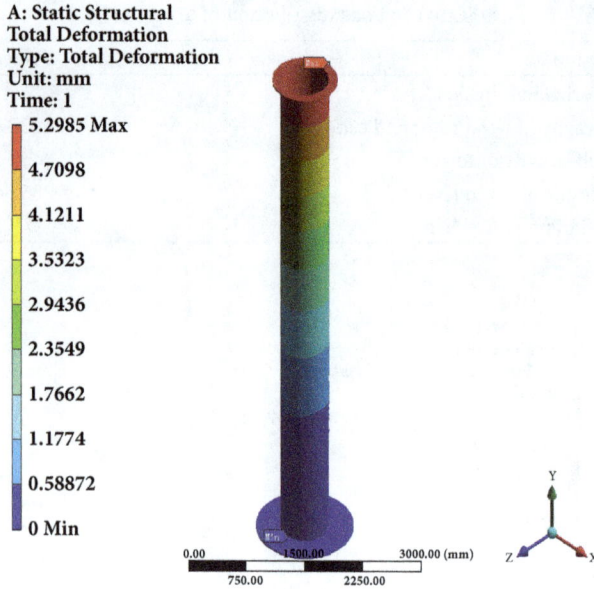

FIGURE 14: Total deformation contour of tower.

TABLE 11: The first six-order mode natural frequency of drag rotor.

Order	Value
First order [Hz]	50.2653
Second order [Hz]	50.3117
Third order [Hz]	50.3735
Fourth order [Hz]	50.4092
Fifth order [Hz]	50.4428
Sixth order [Hz]	76.6961

TABLE 12: The first six-order mode natural frequency of main axis.

Order	Value
First order [Hz]	8.35907
Second order [Hz]	8.35938
Third order [Hz]	52.2301
Fourth order [Hz]	52.2319
Fifth order [Hz]	145.56
Sixth order [Hz]	145.565

TABLE 10: The first six-order mode natural frequency of main blade.

Order	Value
First order [Hz]	18.0535
Second order [Hz]	18.0543
Third order [Hz]	28.2657
Fourth order [Hz]	75.4023
Fifth order [Hz]	80.6801
Sixth order [Hz]	80.7842

156.7 MPa which is higher than the maximum stress 29.5 MPa in Figure 14.

4. Modal Analysis

When the wind turbine works in natural environment, the load is complex and changeable. The power of air, inertia force, and elasticity force applied on the blades of wind turbine can make blade and tower deform and oscillate. If the frequency of exciting force approaches the natural frequency of the structure, the resonance may lead to damage of wind turbine. In order to avoid resonance, the natural frequency of wind turbine should be different from the one of wild exciting force. Therefore, the modal analysis should be carried out during the structural design of wind turbine.

4.1. Main Blade. The model of main blade used in the modal analysis is the same as statistic analysis. The low-order mode of main blade has a great influence on stability and fatigue of blade, and the first six-order modes and the natural frequencies are calculated which are shown in Table 10. The vibration modes are shown in Figure 15.

From Figure 15, the frequency of first order is 18.0535 Hz and the first-order critical rotational speed of main blade is calculated as following formula:

$$n_0 = 60 f \qquad (4)$$

The first-order critical rotational speed n_0 is 1083 r/min, which is higher than rotational speed of main blade. It means that the resonance of main blade will not occur during operation.

4.2. Drag Rotor. The model of drag rotor used in the modal analysis is the same as the static analysis. Similarly the frequencies of first six-order modes are shown in Table 11 and the vibration modes are shown in Figure 16.

From Figure 16, the natural frequency of first-order is 50.2653 Hz, and the first critical rotational speed of drag rotor calculated by the formula (4) is 3016 r/min. The rotational speed of drag rotor is far lower than the critical rotational speed, which means that the resonance of blade will not occur during the operation.

4.3. Main Axis. The main axis is one of the important parts of wind rotor and nacelle, which not only needs to check the strength and stiffness but also avoid resonance phenomenon. Therefore, based on the model of static mechanical property analysis, the natural frequencies of first six-order modes of main axis are shown in Table 12 and the vibration mode of main axis are shown in Figure 17.

From Figure 17, the frequency of first-order mode is 8.35907 Hz. When the wind rotor works at the rotational speed 100 r/min, the exciting frequency subjected by main axis from wind rotor is 1.667 Hz. However, the natural frequency of first-order is 8.35907 Hz and it is higher than

(a) First-order mode vibration profile

(b) Second-order mode vibration profile

(c) Third-order mode vibration profile

(d) Fourth-order mode vibration profile

(e) Fifth-order mode vibration profile

(f) Sixth-order mode vibration profile

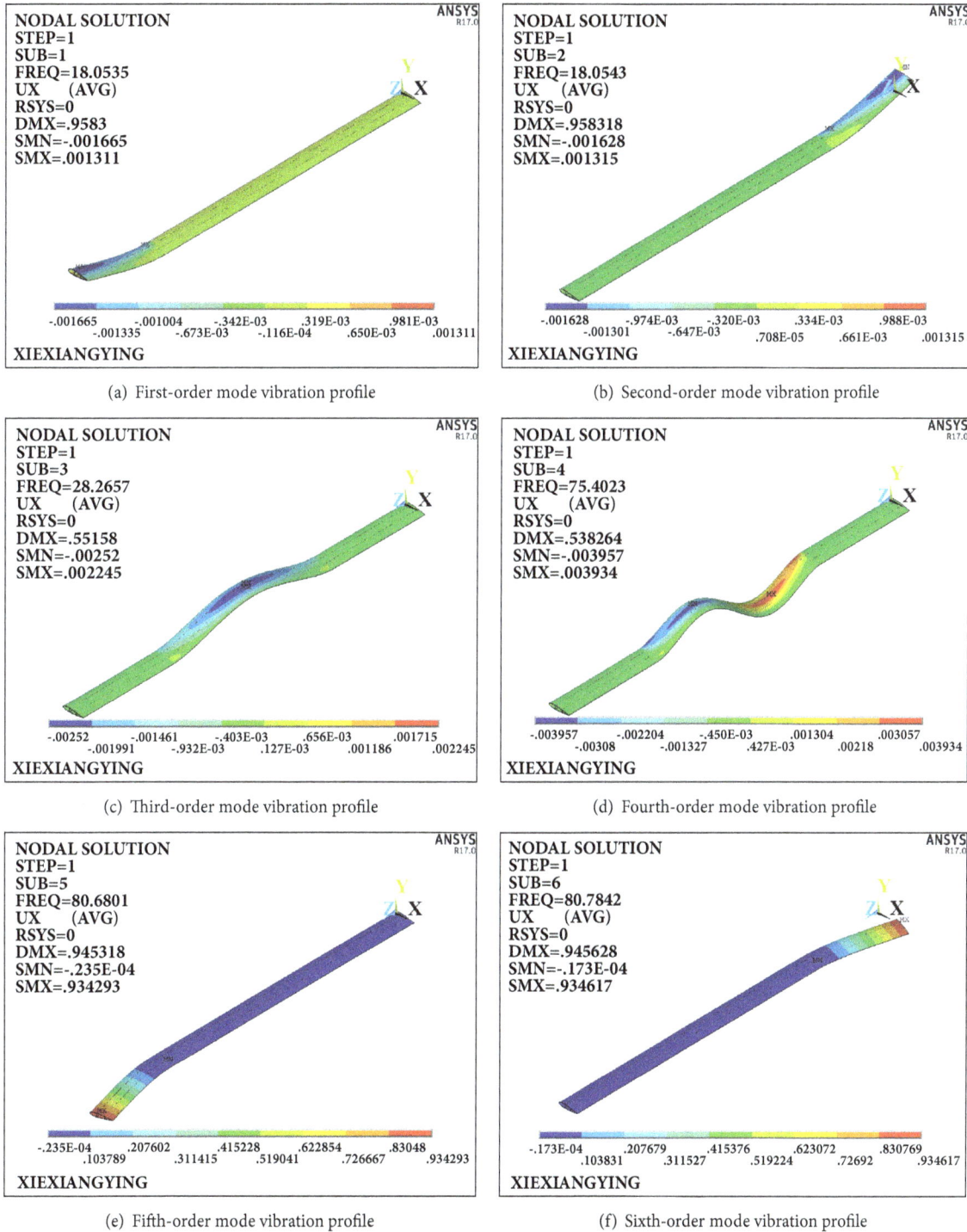

FIGURE 15: The first six-order mode vibration profile of main blade.

the work frequency, which means that the resonance will not occur during the operation.

4.4. Tower. Similarly the FEM model of tower in the modal analysis is the same as the static mechanical property analysis and the contact surfaces between tower and ground are

constrained. The natural frequencies of first six-order mode of tower are shown in Table 13, and the vibration modes of tower are shown in Figure 18.

From Figure 18, the frequency of first-order mode is 15.2166 Hz. According to what has been mentioned previously the frequency f_1 of wind rotor under the rated speed is

(a) First-order mode vibration profile

(b) Second-order mode vibration profile

(c) Third-order mode vibration profile

(d) Fourth-order mode vibration profile

(e) Fifth-order mode vibration profile

(f) Sixth-order mode vibration profile

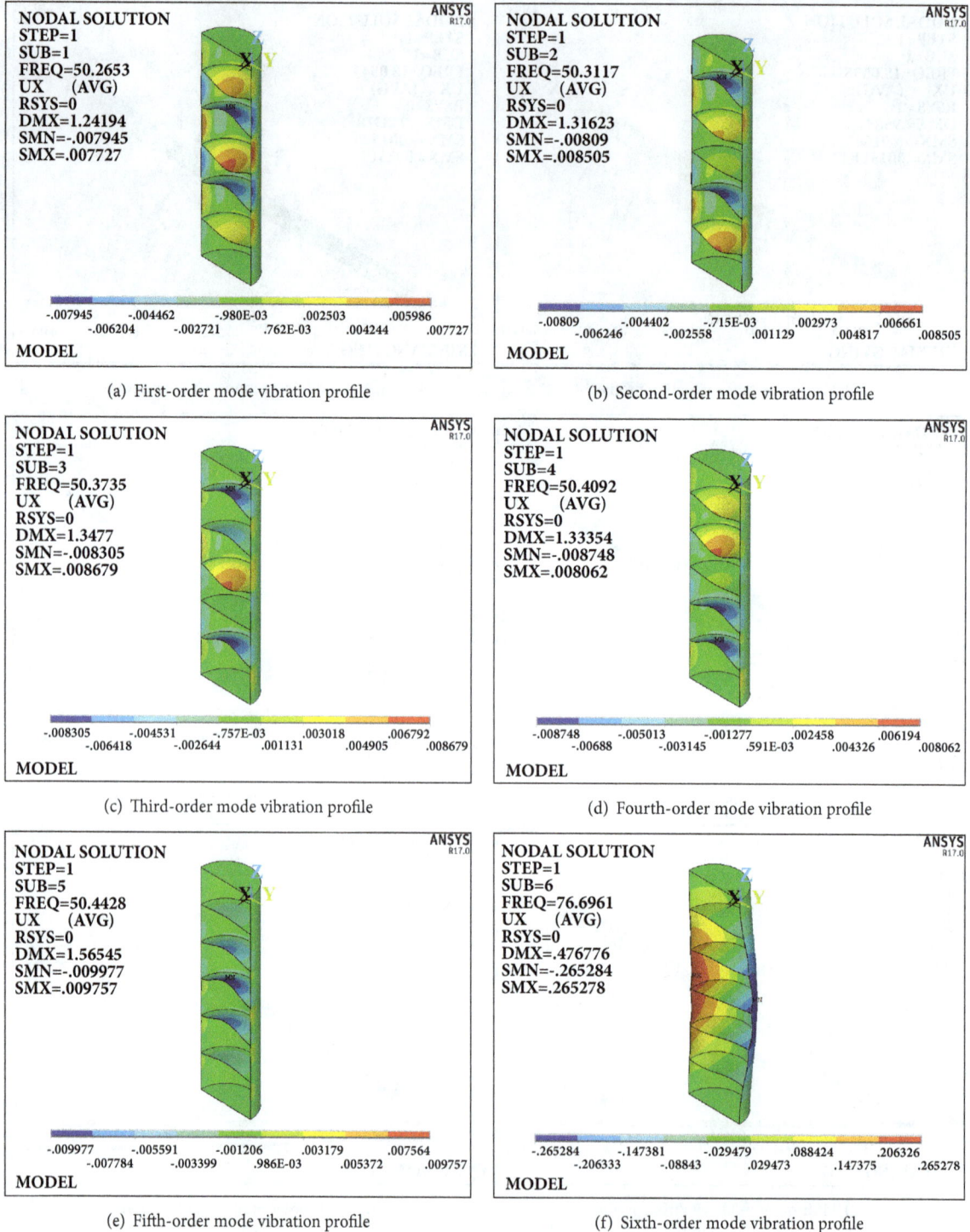

FIGURE 16: The first six-order mode vibration profile of drag rotor.

TABLE 13: The first six-order mode natural frequency of tower.

Order	Value
First order [Hz]	15.2166
Second order [Hz]	15.2166
Third order [Hz]	88.6749
Fourth order [Hz]	88.6749
Fifth order [Hz]	120.362
Sixth order [Hz]	168.738

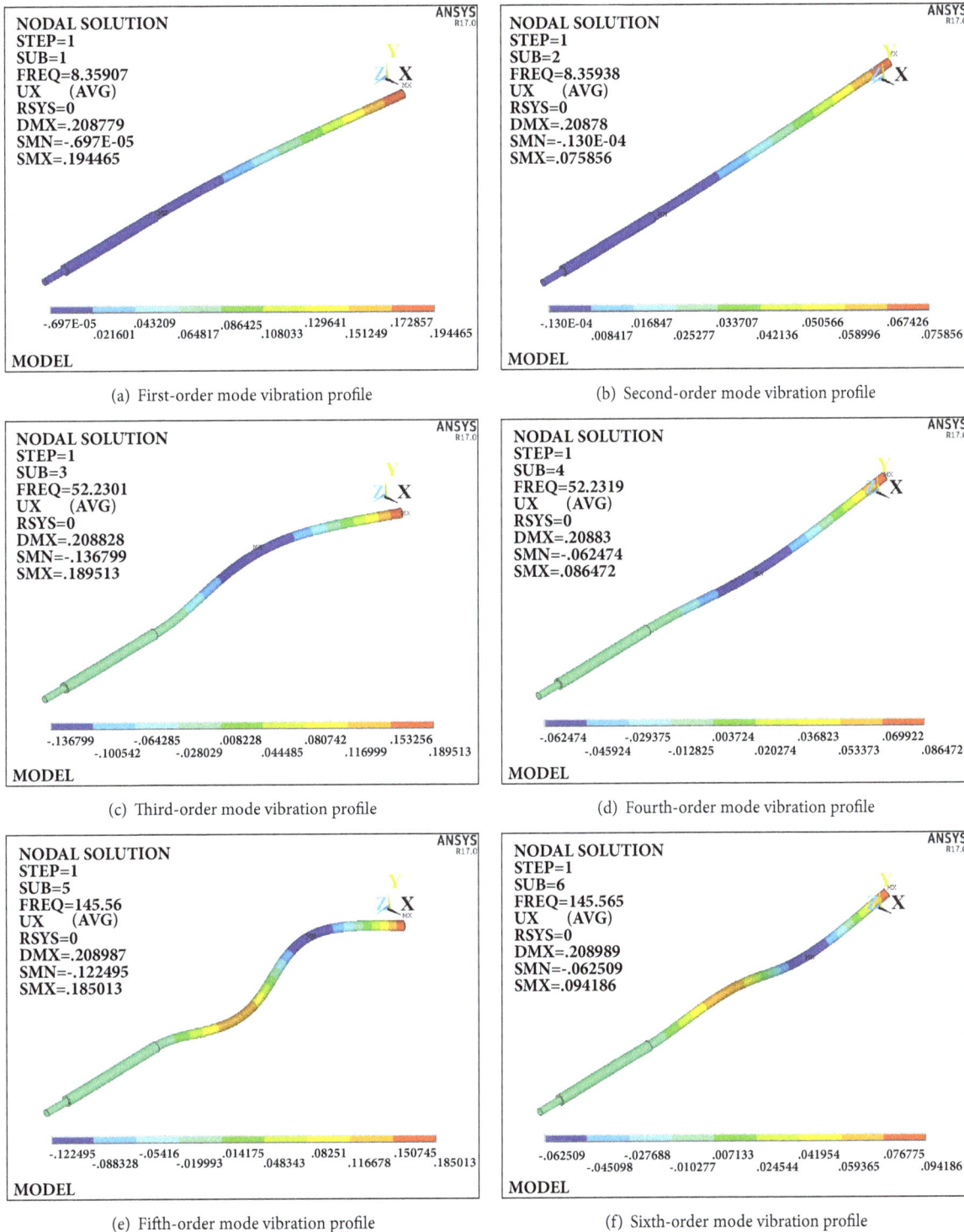

(a) First-order mode vibration profile

(b) Second-order mode vibration profile

(c) Third-order mode vibration profile

(d) Fourth-order mode vibration profile

(e) Fifth-order mode vibration profile

(f) Sixth-order mode vibration profile

FIGURE 17: The first six-order mode vibration profile of main axis.

1.67 Hz. The wind rotor has three blades; therefore the passage frequency f_2 of the main blades is 5.01 Hz. According to the engineering experience [26], the first-order frequency f_{01} of tower must be higher than the passage frequency f_2 of the blade and meet the formula:

$$\frac{(f_{01} - f_{02})}{f_2} \geq 10\% \tag{5}$$

The calculated results meet above conditions, which means that the excitation of wind rotor will not cause the tower to resonate.

5. Prototype of LD-VAWT

According to the design of LD-VAWT with the static mechanical property and modal analysis, the results show

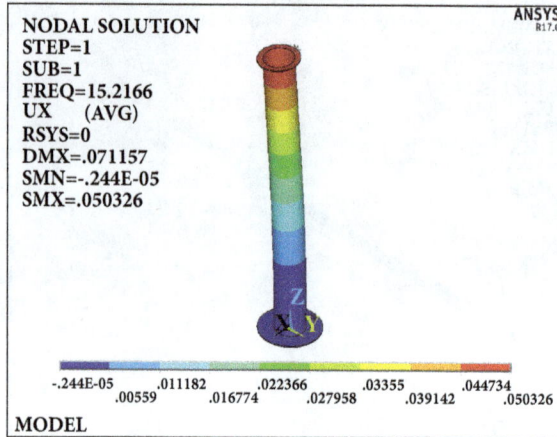
(a) First-order mode vibration profile

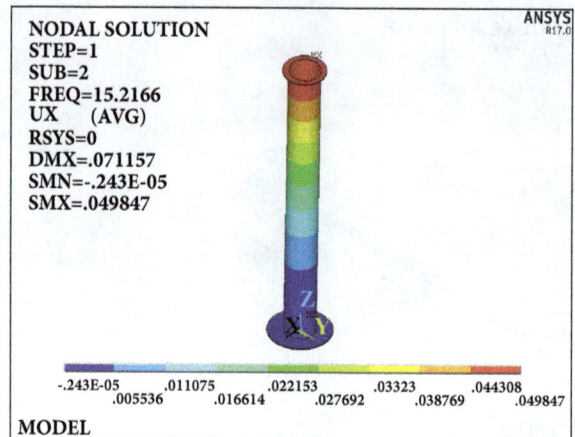
(b) Second-order mode vibration profile

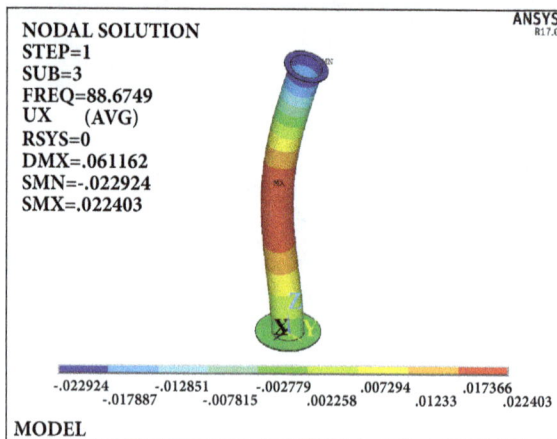
(c) Third-order mode vibration profile

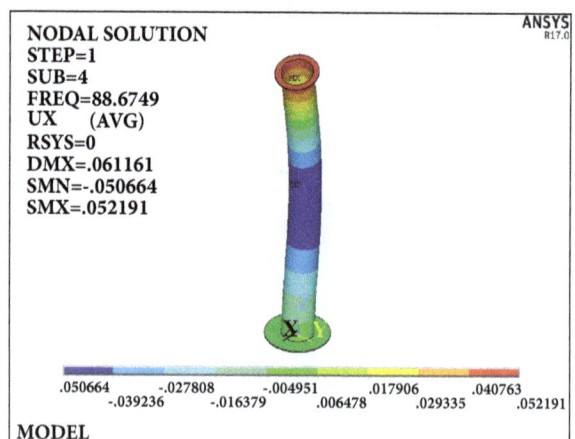
(d) Fourth-order mode vibration profile

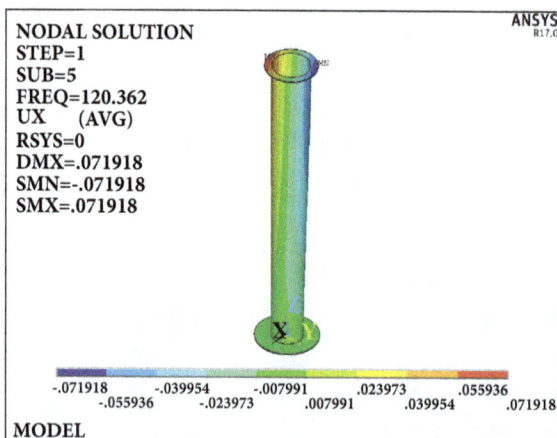
(e) Fifth-order mode vibration profile

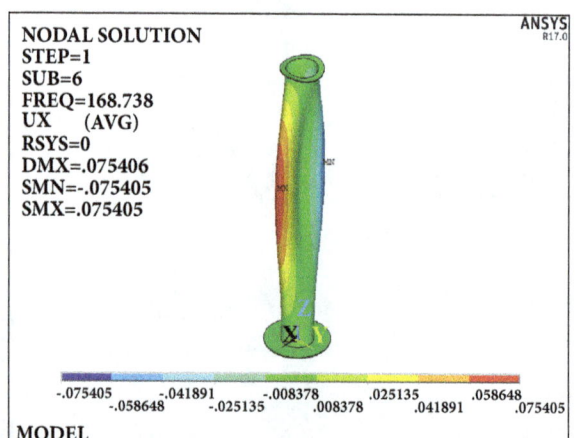
(f) Sixth-order mode vibration profile

FIGURE 18: The first six-order mode vibration profile of tower.

that the design of wind turbine structure is reasonable. A prototype of LD-VAWT was designed and made. It was tested in a farm of Northeast Agricultural University of China which is shown in Figure 19.

Based on the observation of its operation situation for a period of time, the wind turbine can work safely and stably according to the design goal, which shows that the design scheme is practicable and proves that the ideas and methods

for LD-VAWT are correct. The paper provides references to analyze the structure of the LD-VAWT.

6. Conclusions

In order to explore a set of methods about designing and analyzing the structure of LD-VAWT, the paper took a small-scale LD-VAWT as an example and analyzes the static

FIGURE 19: Actual machine of wind turbine.

mechanical property and modal analysis by finite element method; the conclusions are as follows.

The corresponding contours of stress and deformation were obtained by using ANSYS to analyze the static mechanical property of main parts of wind turbine, which concludes that the structure of wind turbine meets the design requirements.

The first six-order mode vibration profiles of main parts were also obtained based on the modal analysis, which concludes that the resonance of each main part will not resonant during the operation.

The prototype LD-VAWT was made based on the analysis and simulation results in this study and operated steadily. The methods used in this study can be used as a reference for the static mechanical properties and modal analysis of vertical axis wind turbine.

Conflicts of Interest

The authors declare that they have no conflicts of interest.

Acknowledgments

This research is sponsored by the Project 2017MS02 supported by the Foundation of Key Laboratory of Wind Energy and Solar Energy Technology, Ministry of Education. The authors thank the supporter.

References

[1] Y. Li, Y. Zheng, S. Zhao et al., "A review on aerodynamic characteristics of straight-bladed vertical axis wind turbine," *Acta Aerodynamica Sinica*, vol. 35, no. 6, pp. 368–382, 2017.

[2] G. Dai, Z. Xu, K. Huangfu, and Y.-J. Zhong, "Research progress in the vertical axis wind turbine," *Fluid Machinery*, vol. 38, no. 10, pp. 39–43, 2010.

[3] Y. Li, J. Tang, K. Tagawa, and F. Feng, "Effect of frustum-shaped wind collection pattern to starting performance of VAWT,"
Journal of Northeast Agricultural University, vol. 47, no. 4, pp. 95–101, 2016.

[4] Z.-C. Wu, *Research on Aerodynamic Characteristics of Vertical Axis Wind Turbine with Eccentric Rotor Structure*, Harbin, Northeast Agricultural University, 2017.

[5] M. S. Hameed, S. K. Afaq, and F. Shahid, "Finite element analysis of a composite VAWT blade," *Ocean Engineering*, vol. 109, pp. 669–676, 2015.

[6] L. Wang, A. Kolios, T. Nishino, P.-L. Delafin, and T. Bird, "Structural optimisation of vertical-axis wind turbine composite blades based on finite element analysis and genetic algorithm," *Composite Structures*, vol. 153, pp. 123–138, 2016.

[7] T.-T. Zhang, H.-X. Wang, and Z.-B. Dai, "Research on vertical-axis wind turbines structure vibration characteristics," *East China Electric Power*, vol. 37, no. 3, pp. 452–455, 2009.

[8] J.-Y. Wang, *Study of the Effect of Vortex Shedding on a 5 KW H-Type Vertical Axis Wind Turbine*, Harbin: Harbin Institute of Technology, 2016.

[9] N. H. Abu-Hamdeh and K. H. Almitani, "Construction and numerical analysis of a collapsible vertical axis wind turbine," *Energy Conversion and Management*, vol. 151, pp. 400–413, 2017.

[10] Y. Tang, Y. H. Wu, K. Zhang, J. Sun, and E. W. Song, "Finite element analysis of wind turbine lifting platform bridge structure," *Applied Mechanics and Materials*, vol. 687-691, pp. 398–401, 2014.

[11] E. Verkinderen and B. Imam, "A simplified dynamic model for mast design of H-Darrieus vertical axis wind turbines (VAWTs)," *Engineering Structures*, vol. 100, pp. 564–576, 2015.

[12] Z. Li and Y. Chen, "Dynamic research with drivetrain simulation and modal analysis of wind turbine gearbox," *Advanced Materials Research*, vol. 952, pp. 161–164, 2014.

[13] I. Paraschivoiu, *Wind Turbine Design with Emphasis on Darrieus Concept*, Shanghai: Shanghai Scientific & Technical Publishers, 2013, translated by C. Li.

[14] Y.-H. Zhu and Z.-H. Liu, "Structure and performance analysis for hybrid vertical axis wind turbines with lift and resistance leaves," *East China Electric Power*, vol. 36, no. 7, pp. 99–101, 2008.

[15] F. Feng, Y. Li, L. Chen, W. Tian, and Y. Zhang, "A simulation and experimental research on aerodynamic characteristics of combined type vertical axis wind turbine," *Acta Energiae Solaris Sinica*, vol. 35, no. 5, pp. 855–860, 2014.

[16] Q.-B. He, *Study on Calculation of Structure And Aerodynamic Characteristics for Vertical Axis Wind Turbine with Double-Layer Retractile Blades*, Harbin: Northeast Agricultural University, 2015.

[17] K. Niklas, "Strength analysis of a large-size supporting structure for an offshore wind turbine," *Polish Maritime Research*, vol. 24, no. 1, pp. 156–165, 2017.

[18] J.-F. Ji, Z.-Y. Deng, L. Jiang, and D.-G. Huang, "5kW masking the optimization design of the lift vertical axis wind turbine," *Journal of Engineering Thermal Physics*, vol. 33, no. 7, pp. 560–564, 2013.

[19] W. Kou, B. Yuan, Q. Li, and L.-T. Fan, "The structural design of a type of vertical shaft wind generator," *Electrical and Electronic Engineering*, vol. 27, no. 5, pp. 25–28, 2011.

[20] Z. Xu, Y.-L. Huo, Y. Chen, H.-W. Yang, and H.-F. Tan, "Optimum design and study on the properties of a new combined type vertical axis wind turbine," *Journal of Zhejiang University of Technology*, vol. 43, no. 3, pp. 261–264, 2015.

[21] J.-J. Qu, M.-W. Xu, Z.-J. Li, and C. Zhi, "A kind of lift and drag hybrid vertical axis wind turbine," *Renewable Energy Resources*, vol. 28, no. 1, pp. 101–104, 2010.

[22] K. Kong, X. L. Zhou, and M.-Z. Cheng, "Structural modelling analysis and testing of wind turbine rotor blade," *Mechanical Electrical Engineering Technology*, vol. 47, no. 5, pp. 45–48, 2018.

[23] GB/T 29494-2013 Small vertical axis wind turbines [S].

[24] GB/T 13981- Design general requirements for small wind turbine[S],.

[25] X. Sun, Y. Chen, Y. Cao, G. Wu, Z. Zheng, and D. Huang, "Research on the aerodynamic characteristics of a lift drag hybrid vertical axis wind turbine," *Advances in Mechanical Engineering*, vol. 8, no. 1, 2016.

[26] X. Song and J.-X. Dai, "Mechanical modeling and ANSYS simulation analysis of horizontally axial wind turbine tower," *Journal of Gansu Sciences*, vol. 23, no. 1, pp. 91–95, 2011.

4

Investigations on the Effect of Radius Rotor in Combined Darrieus-Savonius Wind Turbine

Kaprawi Sahim ⓘ, Dyos Santoso, and Dewi Puspitasari

Mechanical Engineering Department, Sriwijaya University, Sumatera Selatan, Indonesia

Correspondence should be addressed to Kaprawi Sahim; kaprawis@yahoo.com

Academic Editor: Tariq Iqbal

Renewable sources of energy, abundant in availability, are needed to be exploited with adaptable technology. For wind energy, the wind turbine is very well adapted to generate electricity. Among the different typologies, small scale Vertical Axis Wind Turbines (VAWT) present the greatest potential for off-grid power generation at low wind speeds. The combined Darrieus-Savonius wind turbine is intended to enhance the performance of the Darrieus rotor in low speed. In combined turbine, the Savonius buckets are always attached at the rotor shaft and the Darrieus blades are installed far from the shaft which have arm attaching to the shaft. A simple combined turbine offers two rotors on the same shaft. The combined turbine that consists of two Darrieus and Savonius blades was tested in wind tunnel test section with constant wind velocity and its performance was assessed in terms of power and torque coefficients. The study gives the effect of the radius ratio between Savonius and Darrieus rotor on the performance of the turbine. The results show that there is a significant influence on the turbine performance if the radius ratio was changed.

1. Introduction

Darrieus and Savonius wind turbines are the most common VAWTs. The Savonius wind turbine is an aerodynamically drag based, self-starting turbine with low cut in speed, but its inefficiency curtails it to fewer applications, whereas Darrieus wind turbines are aerodynamically lifting based turbines having higher cut in speed with higher coefficients of performance. Thus, it can rotate faster than the wind velocity. Combined rotors are the combination of Darrieus and Savonius rotors mounted on the same shaft. Mostly combined wind turbines are available in the vertical axis configuration. Figure 1 gives a performance comparison of the various types of all conventional turbines that have been constructed [1]. As can be seen in the figure, turbines with horizontal axis have high power coefficient among other turbines which reach 49% in $\lambda \approx 7$. All turbines with vertical shaft have low power coefficient and the smallest power coefficient is Savonius turbine having maximum power coefficient 0.15 on $\lambda \approx 0.8$.

1.1. Darrieus Turbine. Various types of Darrieus turbine rotor configuration exist; among them are: egg beater type, straight

blade type, and helical blade rotor. To increase the Darrieus turbine performance, Takao et al. [2] placed a guide vane row upstream of the rotor in order to enhance its torque, so the power coefficient of the turbine is 1.5 times higher than that of the original turbine which has no guide vane. The micro Darrieus turbine consists of usually two or three pairs of airfoils of NACA type and each pair has a single blade. Configuration of each pair consists of two blades, where one of them is auxiliary blade and gives a higher static torque [3]. The number of Darrieus blades influences the turbine performance. The fewer the number of blades, the higher the performance of the rotor [4]. The use of double blades of solo Darrieus rotor makes the turbine lower in performance [5].

The effect of an upstream flat deflector on the power output of two counterrotating straight-bladed Darrieus wind turbines was investigated experimentally [6, 7]. The result shows that the power output of a turbine increases significantly. The effect of solidity will increase the torque of H-Darrieus turbine, but increasing solidity alone does not improve the performance of the rotor since there is an effect of blockage and interaction [8]. The new model of Darrieus blade is introduced by Zamani et al. [9] in using J-shaped profile; the wind turbine performance and self-starting of

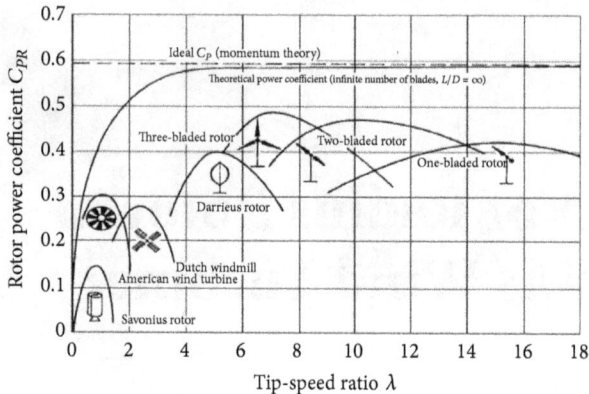

FIGURE 1: Typical performance of wind power machines.

turbine are improved. It has been experimented that, by the combination of both Darrieus and Savonius rotor, higher coefficient of performance could be achieved at moderate wind speed.

1.2. Savonius Turbine. A vertical axis wind turbine (VAWT) is summarized as accepting wind from any direction and self-starting at low cut in speeds. Two semicircle blades were usually chosen for the Savonius rotor because of their excellent start-up performance. The double-step Savonius bucket was proposed since the starting torque is never negative [10]. The combination of two geometries of the surface curve of Savonius bucket was studied by Sanusi et al. [11]. The concave side is a half-circle and the other side is in a semicircular form. They show experimentally that the combined blade improved the performance of Savonius turbine. Torque variation on the angular positions of the rotor blade and lower efficiency are the main drawbacks of Savonius turbine. To overcome the problem, Thiyagaraj et al. [12] placed a set of additional blades called secondary blades in front of the concave side of main rotor blades. Thus the efficiency is improved and the torque values increased along the angular positions. To increase the power coefficient of Savonius turbine, Al-Kayiem et al. [13] proposed installing a curtain in upstream flow of turbine and the rotor without shaft. By these configurations, the coefficient of power increase significantly.

A considerable improvement in the performance of Savonius rotor can be achieved by using multiple quarter blades instead of single blade as shown by S. Sharma and R. K. Sharma [14]. The twisted Savonius bucket was tested to know the optimum twisted angle of the blade. The optimum angle is found to be 15° [15] and for blade arc angle the range is 110°–135° [16]. The performance of the newly developed two-blade Savonius wind turbine is reported by Roy and Saha [17] by modifying the blade called Bach, Benesh, and semielliptical blades. New model of Savonius blade having arm is given by Roy and Ducoin [18]. The performance is better than the conventional turbine. The use of ventilation in the Savonius blade improved the performance of the conventional Savonius windmill. This improvement in power coefficient achieves 25% of low wind speed [19]. The

power coefficient can be increased by placing two deflectors upstream of rotor [20].

The performance and shape characteristics of a helical Savonius wind turbine at various helical angles have an influence on the turbine performance [21]. Type of the blade fullness of Savonius was introduced by Tian et al. [22] which gives the optimum performance of Savonius wind turbine. The aspect ratio defined as height to diameter of Savonius rotor influences torque of the rotor. The torque of a rotor can be increased by decreasing aspect ratio, and the speed can be increased by increasing aspect ratio [23].

1.3. Combined Darrieus and Savonius Turbine. The Darrieus-Savonius rotor combines the advantage of Savonius rotor, that is, high starting torque, and the advantage of Darrieus rotor, that is, a high power coefficient, into a single combination of these rotors. A combined Savonius-Darrieus type vertical axis wind rotor has many advantages over individual Savonius or individual Darrieus wind rotor, such as better efficiency than the Savonius rotor and high starting torque than the Darrieus rotor. However to make the rotor completely self-starting, the rotor is incorporated into a hybrid system with Savonius rotor as its starter. It was found that the hybrid design fully exhibits the self-starting capability at all azimuthal positions, signified by the positive static torque coefficient values, and the power performance of the hybrid rotor increases to 0.34 [24].

There are some possible combined configurations of turbines. Savonius is placed in the middle of the Darrieus rotor, Darrieus rotor is placed above the Savonius rotor, and Savonius was placed at the top of Darrieus rotor. The first configuration gives the highest coefficient of power [25]. The efficiency of Darrieus and Savonius turbine depends on upstream fluid velocity. The higher the velocity, the higher the efficiency of the turbine [26].

The overlap between Savonius blades of combined Darrieus-Savonius rotor increases the performance of the combined turbine [27]. Combined Darrieus-Savonius turbine is tested in water current flow by using a flat deflector upstream of the turbine. The deflector increases the performance of the combined turbine [28]. A combined three-bladed Darrieus and three-bucket Savonius rotor is explored by Ghosh et al. [29]. They show that the power coefficient is obtained at 0.53 for Darrieus rotor mounted on the Savonius rotor which is better than rotor in which Savonius is mounted on Darrieus rotor. The optimum value of attachment angle of Savonius buckets to the Darrieus rotor is perpendicular to the Darrieus blade arm because at this condition it gives greater torque at low speed as shown experimentally by Kyozuka [30]. For the different angle of attachment, the performances have small differences. Xiaoting et al. [31] compute the optimum angle of attachment. The optimum angle of attachment of Savonius rotor is 0° and they also show that the radius ratio, KD, between Savonius and Darrieus radius is better if KD < 0.5.

From the literature surveys described above, for a combined Darrieus and Savonius wind turbine, the dimensionless radius ratio between Savonius and Darrieus rotor is not yet studied experimentally. It seems that the gap distance between Savonius and Darrieus blades is an important

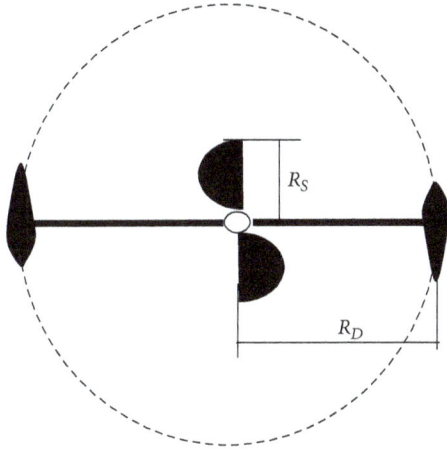

FIGURE 2: Combined Darrieus and Savonius turbine.

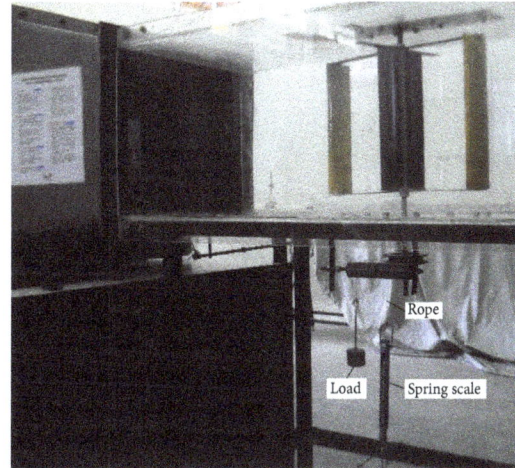

FIGURE 3: Photograph of turbine in wind tunnel.

parameter which can influence the combined turbine performance. It is the reason that the objective of the present study is to explore experimentally the performance of a combined Darrieus and Savonius wind turbine.

2. Experimental Study

The combined wind turbine was tested in a subsonic wind tunnel laboratory with test cross section of $400 \times 400 \, \text{mm}^2$. The design should fit in the cross section; thus a micro combined Darrieus and Savonius wind turbine is designed and fabricated.

2.1. Model Test. The schematic figure of the combined rotor is shown by Figure 2 where the Savonius rotor is placed perpendicular to the Darrieus rotor. In the present investigations, two main configurations in the study were investigated since there are two ways to make the variation of dimensionless radius ratios as follows.

(a) Radius of Savonius bucket R_S is kept fixed and radius of Darrieus rotor R_D is varied (case (a)).

(b) Radius of Darrieus rotor is kept fixed, the radius of Savonius is varied. It means that the diameter of semicircular Savonius bucket became larger (case (b)).

The combined rotor has two Darrieus blades and two Savonius buckets (two semicircle buckets). The Savonius rotor is installed in the middle of the Darrieus rotor with the same shaft. The Darrieus airfoil is a NACA 0020 with chord length 60 mm and span length 300 mm. The blade was made from wood and Darrieus rotor has 340 mm in diameter. The semicircular Savonius buckets have the same length as the Darrieus blade and were made from polyvinyl chloride (PVC) material. The radius ratio is written as RL $= R_S/R_D$, where R_S is radius of Savonius rotor and R_D is radius of Darrieus rotor. The above Darrieus blade dimensions give the solidity of rotor as $\sigma = 0.11$.

2.2. Procedures. The combined rotor was installed in wind tunnel section with wind velocity set at 8 m/s which is

constant along the test. The photograph of turbine in the wind tunnel is shown in Figure 3. Two ball bearings were used in the base to support the shaft. The shaft was made up of mild-steel of 12 mm diameters. The structure of Darrieus arm was made of strip of cast iron to ensure high strength to weight ratio and durability. Flexible digital anemometer and digital laser tachometer were used to measure the wind speed and revolutions per minute of the rotor. Torque meter which is installed on rotor shaft of a wind turbine can determine power and torque coefficients absorbed from a wind turbine. Type of rope brake dynamometer meter was used to measure torque. Torque meter consists of a rope, a brake pulley, a tube spring scale, and a pan. The weighing pan, brake pulley, and tube spring scale are connected by a nylon string of 2 mm in diameter.

The first step of measurements was carried out without load. Then, while the rotor was running, a load was added; then force on tube spring scale and rotation was measured. The same measurements were done for the other loads. At certain load, the rotor was stopped since the load was maximized.

The performance of a wind turbine can be expressed in the form of torque coefficient (C_T) and the coefficient of power (C_P). Tip speed ratio or TSR (λ) is a parameter related to rated wind speed and rotor diameter. λ is the ratio between speed of tip blade and wind speed through the blade; TSR is determined as follows.

$$\lambda = \frac{U}{V} = \frac{\omega \cdot r}{V}, \tag{1}$$

where U is the tip speed or the peripheral velocity of the rotor, in the case of combined turbine the speed of Darrieus rotor, V is the wind speed, ω is the angular velocity of the rotor, and r is the radius of the rotor.

The coefficient of torque or C_T is defined as the ratio between the actual torque developed by the rotor (T) and the theoretical torque available in the wind as expressed by

$$C_T = \frac{T}{T_W} = \frac{4 \cdot T}{\rho \cdot A \cdot d \cdot V^2}$$

$$\text{or } C_T = \frac{C_P}{\lambda}, \tag{2}$$

where ρ is the density of air, T is the torque, and A is the swept area of blades = the rotor height × the rotor diameter. The force acting on the rotor shaft is obtained in (N) by

$$F = (s - m)\, g, \tag{3}$$

where s is the spring balance reading in kg, m is the mass loaded on the pan in kg, and g is the gravitational acceleration. The torque is calculated from the following relation.

$$T = (r_b + r_r) \cdot F, \tag{4}$$

where the r_b is radius of brake drum and the r_r is radius of the rope.

The coefficient of power of a wind turbine (C_P) is the ratio between the maximum power obtained from the wind (P_T) and the total power available from the wind (P_a):

$$C_P = \frac{P_T}{P_a} = \frac{P_T}{(1/2)\, \rho A V^3}, \tag{5}$$

where P_T is the power of turbine that is given by

$$P_T = T \cdot \omega. \tag{6}$$

3. Results and Discussion

Initial experiments were carried out for solo Savonius and Darrieus separately and then the two turbines were combined together. In this study, two important parameters of turbine performance in terms of torque and power coefficient were presented and all parameters of turbine performance were calculated from (1) to (6).

Figure 4 shows the variation of the coefficient of torque with tip speed ratio. The coefficient attains its highest value of the tip speed ratio $\lambda = 1.8$ for solo Darrieus rotor and it decreases very slowly with lower speed. These characteristics give it less self-starting capability, while the torque coefficient decreases to zero or no load condition as speed ratio is larger or at $\lambda = 2.7$. The solo Savonius rotor has a higher coefficient for low speed compared to solo Darrieus rotor and the torque coefficient decreases rapidly with λ in the narrow range of $\lambda < 0.6$. We note that the solo Savonius has rotor speed not greater than the upstream wind speed. When these two rotors are combined, it seems that the torque coefficients of all RL values increase significantly for low speeds. The higher the RL, the greater the torque coefficient. As stated by Xiaoting et al. [31] the larger the RL ratio, the more the inference of the flow field by the Savonius rotor to the Darrieus rotor.

The power coefficient characteristics of the turbine rotors are presented in Figure 5. The solo Darrieus turbine has the

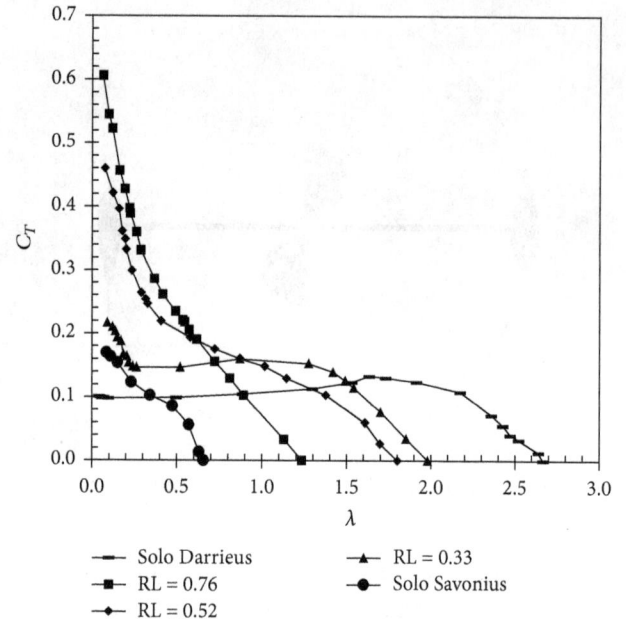

FIGURE 4: Coefficient of torque (case (a)).

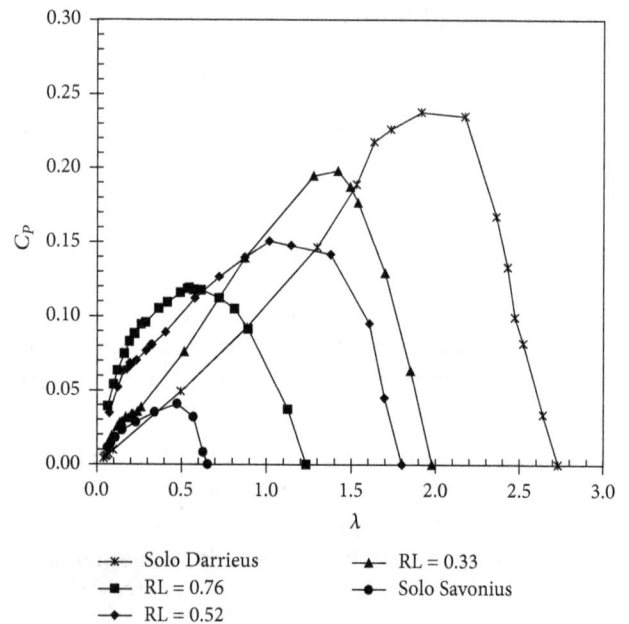

FIGURE 5: Coefficient of power (case (a)).

highest power coefficient that attains maximum values $C_P = 0.238$ at about $\lambda = 2$. The solo Darrieus turbine has also a wide range of operation speed from $\lambda = 0$ to 2.7. It is much larger than that of Savonius turbine. It is observed that the lowest power coefficient is found in solo Savonius turbine that has a maximum value of $C_P = 0.041$ at about $\lambda = 0.5$. For the combined turbine in which the Darrieus blades are at the nearest from the Savonius rotor or RL = 0.76, the maximum power coefficient is found to be $C_P = 0.12$. When RL decreases from RL = 0.76 to 0.33, the value of C_P increases considerably. The lower value of RL will improve the

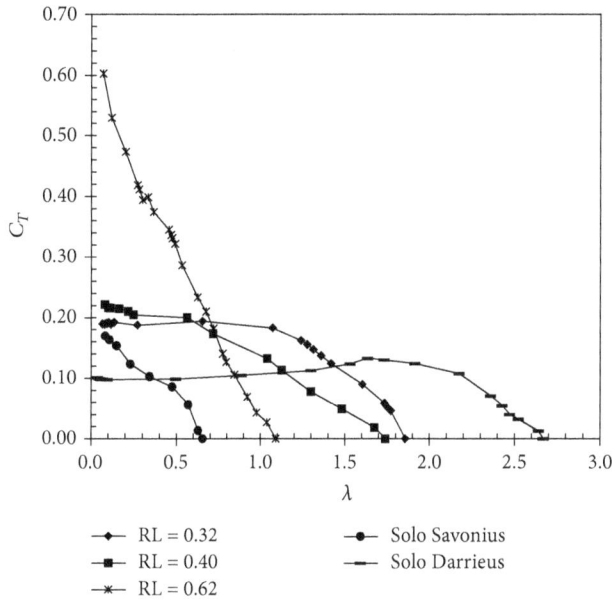

FIGURE 6: Torque coefficient variation (case (b)).

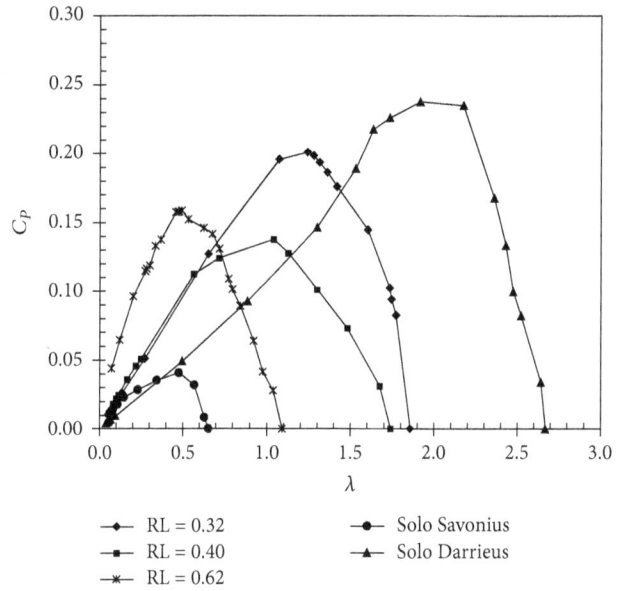

FIGURE 7: Coefficient of power (case (b)).

power coefficient of combined turbine since the maximum power coefficient increases and the curves shifted to the right which makes the operation speed become larger. The higher speed of the combined turbine is due to the Darrieus being far from Savonius and no interference occurs. As we noted the dimensionless parameter RL represents the ratio between the frontal area of Savonius rotor and the frontal area of the Darrieus rotor since both rotors have the same height. In this study, the Savonius rotor has ratio $H/D = 2.7$, where H is the height of the rotor and D is the rotor diameter. The power coefficient of a turbine is influenced by the amount of energy loss due mechanical structure. In this case, a friction exists between wind and blades, and mechanical friction exists in the bearing. However, the friction in the system is difficult to quantify.

In the case of the radius of Darrieus rotor being kept fixed and the radius of Savonius bucket being varied, the torque coefficient of combined rotor is shown in Figure 6. It is observed that, at low speed, the torque increases in the increase in RL. The torque variation is considerably large for RL = 0.62; it decreases rapidly with λ, while for the other two RL the torque variation is not so large. RL = 0.62 means that the Savonius bucket is larger and/or it absorbed much energy from wind to rotate the rotor. Smaller blades will absorb less wind energy due to the decrease of frontal area. As the diameter of Savonius was increased, the inertia of the rotor increased thereby reducing its speed. For RL = 0.32 and 0.40, the two curves seem to be similar except at very low speed since these two values of RL are not much different. However, it can be concluded that the variations in RL are sensitive to the torque variations and the higher the RL, the higher the torque coefficient, but the speed reduced.

The variations in RL value show also the variation in an aspect ratio of the Savonius rotor (AR = H/D) in combined rotor. A low value of RL indicates high AR. The power

coefficient of combined rotor for case (b) is shown in Figure 7. As seen in the figure, the maximum power coefficients have been obtained around 0.198 for RL = 0.32, 0.138 for RL = 0.40, and 0.158 for RL = 0.62. The curves show that the higher power coefficient is found for low RL, but when RL increases, the power coefficient decreases. For low speed, the power coefficient of rotor with RL = 0.62 is higher than that of the rotor with RL = 0.40 since the torque profile is much higher at lower speed for RL = 0.62. It is concluded that the higher the dimension of Savonius bucket is, the higher the losses become; thereby the performance decreases.

The torque coefficient for all configurations (cases (a) and (b)) of the combined turbine has higher values than that of solo Darrieus and solo Savonius for low speed. The maximum ranges of power coefficient of all configurations and all RL are similar, which vary from the range of 0.12 to 0.20 (Figures 5 and 7), but the speed ranges are higher for first configuration (case (a)). It is observed further that, for the low value of λ and for RL = 0.52 (Figure 4), the torque coefficient is quite high compared to that of RL = 0.33 but it is lower than that for RL = 0.76. The operational speed range of RL = 0.76 is lower compared to the two other values of RL. The other part (Figure 5), the maximum power coefficient for RL = 0.52, is not too low. So it seems that the optimum condition occurs to RL = 0.52. For the better condition in the design of a combined wind turbine, the rotor should have a self-starting point at low speed, a high torque operation, a high speed, and a high power coefficient. So it is suggested to use the radius ratio $0.30 \leq RL \leq 0.50$ in the design of a combined turbine. For very low RL can produce a negative torque [31]

4. Conclusions

In this study, an experimental study in a wind tunnel has been carried out in order to explore the influence of rotor radius of combined Darrieus-Savonius turbine on performance. The dimensionless radius is called RL, which is the ratio of the Savonius rotor to the Darrieus rotor. The results conclude that the variation in Darrieus rotor radius in a combined turbine has an important effect on the turbine performance. The higher the RL, the lower the power coefficient and the higher the torque coefficient. Similarly, in case of increasing the Savonius rotor radius, the power coefficient is lower but torque coefficient is higher.

The combined Darrieus-Savonius wind turbine is simple in construction, low cost, and self-starting at low wind speed. This makes it suitable for generating electrical energy, such as for household application at rural areas and for a sailboat.

Conflicts of Interest

The authors declare that there are no conflicts of interest regarding the publication of this paper.

Acknowledgments

The authors acknowledge Sriwijaya University for the research grant of scheme Unggulan Profesi.

References

[1] R. E. Wilson and P. B. S. Lissaman, "Applied aerodynamics of wind machines," *Research Applied to National Needs*, vol. GI-41840, 1974.

[2] M. Takao, T. Maeda, Y. Kamada, M. Oki, and H. Kuma, "A straight-bladed vertical axis wind turbine with a directed guide vane row," *Journal of Fluid Science and Technology*, vol. 3, no. 3, pp. 379–386, 2008.

[3] M. Scungio, F. Arpino, V. Focanti, M. Profili, and M. Rotondi, "Wind tunnel testing of scaled models of a newly developed Darrieus-style vertical axis wind turbine with auxiliary straight blades," *Energy Conversion and Management*, vol. 130, pp. 60–70, 2016.

[4] Q. Li, T. Maeda, Y. Kamada, J. Murata, K. Furukawa, and M. Yamamoto, "Effect of number of blades on aerodynamic forces on a straight-bladed Vertical Axis Wind Turbine," *Energy*, vol. 90, pp. 784–795, 2015.

[5] Y. Hara, T. Kawamura, H. Akimoto, K. Tanaka, T. Nakamura, and K. Mizumukai, "Predicting double-blade vertical axis wind turbine performance by a quadruple-multiple streamtube model," *The International Journal of Fluid Machinery and Systems*, vol. 7, no. 1, pp. 16–27, 2014.

[6] D. Kim and M. Gharib, "Efficiency improvement of straight-bladed vertical-axis wind turbines with an upstream deflector," *Journal of Wind Engineering & Industrial Aerodynamics*, vol. 115, pp. 48–52, 2013.

[7] C. Stout, S. Arnott, S. Islam et al., "Efficiency improvement of vertical axis wind turbines with an upstream deflector efficiency improvement of vertical axis wind turbines with an upstream deflector," *Energy Procedia*, vol. 118, p. 9, 2017.

[8] S. Joo, H. Choi, and J. Lee, "Aerodynamic characteristics of two-bladed H-Darrieus at various solidities and rotating speeds," *Energy*, vol. 90, part 1, pp. 439–451, 2015.

[9] M. Zamani, S. Nazari, S. A. Moshizi, and M. J. Maghrebi, "Three dimensional simulation of J-shaped Darrieus vertical axis wind turbine," *Energy*, vol. 116, pp. 1243–1255, 2016.

[10] J.-L. Menet, "A double-step Savonius rotor for local production of electricity: A design study," *Journal of Renewable Energy*, vol. 29, no. 11, pp. 1843–1862, 2004.

[11] A. Sanusi, S. Soeparman, S. Wahyudi, and L. Yuliati, "Experimental study of combined blade savonius wind turbine," *International Journal of Renewable Energy Research*, vol. 6, no. 2, pp. 615–619, 2016.

[12] J. Thiyagaraj, I. Rahamathullah, and P. S. Prabu, "Experimental investigations on the performance characteristics of a modified four bladed Savonius hydro-kinetic turbine," *International Journal of Renewable Energy Research*, vol. 6, no. 4, pp. 1530–1536, 2016.

[13] H. H. Al-Kayiem, B. A. Bhayo, and M. Assadi, "Comparative critique on the design parameters and their effect on the performance of S-rotors," *Journal of Renewable Energy*, vol. 99, pp. 1306–1317, 2016.

[14] S. Sharma and R. K. Sharma, "Performance improvement of Savonius rotor using multiple quarter blades – A CFD investigation," *Energy Conversion and Management*, vol. 127, pp. 43–54, 2016.

[15] U. K. Saha and M. J. Rajkumar, "On the performance analysis of Savonius rotor with twisted blades," *Journal of Renewable Energy*, vol. 31, no. 11, pp. 1776–1788, 2006.

[16] A. Kumar and R. P. Saini, "Performance analysis of a single stage modified Savonius hydrokinetic turbine having twisted blades," *Journal of Renewable Energy*, vol. 113, pp. 461–478, 2017.

[17] S. Roy and U. K. Saha, "Wind tunnel experiments of a newly developed two-bladed Savonius-style wind turbine," *Applied Energy*, vol. 137, pp. 117–125, 2015.

[18] S. Roy and A. Ducoin, "Unsteady analysis on the instantaneous forces and moment arms acting on a novel Savonius-style wind turbine," *Energy Conversion and Management*, vol. 121, pp. 281–296, 2016.

[19] R. Hariyanto, S. Soeparman, W. Denny, and N. S. Mega, "Experimental study on improvement the performance of savonius windmill with ventilated blade," *International Journal of Renewable Energy Research*, vol. 6, no. 4, pp. 1403–1407, 2016.

[20] G. Kailash, T. I. Eldho, and S. V. Prabhu, "Performance study of modified savonius water turbine with two deflector plates," *International Journal of Rotating Machinery*, vol. 2012, Article ID 679247, 12 pages, 2012.

[21] J.-H. Lee, Y.-T. Lee, and H.-C. Lim, "Effect of twist angle on the performance of Savonius wind turbine," *Journal of Renewable Energy*, vol. 89, pp. 231–244, 2016.

[22] W. Tian, B. Song, J. H. Van Zwieten, and P. Pyakurel, "Computational fluid dynamics prediction of a modified savonius wind turbine with novel blade shapes," *Energies*, vol. 8, no. 8, pp. 7915–7929, 2015.

[23] W. Vance, "Vertical Axis Wind Rotors - Status and Potential," in *Advanced Concepts Division Science Applications*, Inc. La Jolla, California, Calif, USA, 1973.

[24] S. Bhuyan and A. Biswas, "Investigations on self-starting and performance characteristics of simple H and hybrid H-Savonius vertical axis wind rotors," *Energy Conversion and Management*, vol. 87, pp. 859–867, 2014.

[25] A. Siddiqui, A. Hameed, S. N. Mian, and R. Khatoon, "Experimental Investigations of Hybrid Vertical Axis Wind Turbine," in *Proceeding of the 4th International Conference on Energy, Environment and Sustainable Development (EESD '16)*, 2016.

[26] N. H. Mahmoud, A. A. El-Haroun, E. Wahba, and M. H. Nasef, "An experimental study on improvement of Savonius rotor

performance," *Alexandria Engineering Journal*, vol. 51, no. 1, pp. 19–25, 2012.

[27] R. Gupta, A. Biswas, and K. K. Sharma, "Comparative study of a three-bucket Savonius rotor with a combined three-bucket Savonius-three-bladed Darrieus rotor," *Journal of Renewable Energy*, vol. 33, no. 9, pp. 1974–1981, 2008.

[28] S. Kaprawi, D. Santoso, and R. Sipahutar, "Performance of combined water turbine darrieus-savonius with two stage savonius buckets and single deflector," *International Journal of Renewable Energy Research*, vol. 5, no. 1, pp. 217–221, 2015.

[29] A. Ghosh, A. Biswas, K. K. Sharma, and R. Gupta, "Computational analysis of flow physics of a combined three bladed Darrieus Savonius wind rotor," *Journal of the Energy Institute*, vol. 88, no. 4, pp. 425–437, 2015.

[30] Y. Kyozuka, "An experimental study on the darrieus-savonius turbine for the tidal current power generation," *Journal of Fluid Science and Technology*, vol. 3, no. 3, pp. 439–449, 2008.

[31] L. Xiaoting, F. Sauchung, O. Baoxing, W. Chili, Y. C. Christopher, and P. Kaihong, "A Computational Study of the Effects of the Radius Ratio and Attachment Angle on the Performance of a Darrieus-Savonius Combined Wind Turbine," *Renewable Energy*, vol. 113, pp. 329–334, 2017.

5

Design, Modeling, and CFD Analysis of a Micro Hydro Pelton Turbine Runner: For the Case of Selected Site in Ethiopia

Tilahun Nigussie,[1] **Abraham Engeda,**[2] **and Edessa Dribssa**[1]

[1]*School of Mechanical and Industrial Engineering, Addis Ababa Institute of Technology, Addis Ababa, Ethiopia*
[2]*Department of Mechanical Engineering, Michigan State University, East Lansing, USA*

Correspondence should be addressed to Tilahun Nigussie; tilahun.nigussie@aait.edu.et

Academic Editor: Rafat Al-Waked

This paper addresses the design, modeling, and performance analysis of a Pelton turbine using CFD for one of the selected micro hydro potential sites in Ethiopia to meet the requirements of the energy demands. The site has a net head of 47.5 m and flow rate of 0.14 m³/s. The design process starts with the design of initial dimensions for the runner based on different literatures and directed towards the modeling of bucket using CATIA V5. The performance of the runner has been analyzed in ANSYS CFX (CFD) under given loading conditions of the turbine. Consequently, the present study has also the ambition to reduce the size of the runner to have a cost effective runner design. The case study described in this paper provides an example of how the size of turbine can affect the efficiency of the turbine. These were discussed in detail which helps in understanding of the underlying fluid dynamic design problem as an aid for improving the efficiency and lowering the manufacturing cost for future study. The result showed that the model is highly dependent on the size and this was verified and discussed properly using flow visualization of the computed flow field and published result.

1. Introduction

Despite the fact that many rural communities have good access to plenty of water resources, the most serious problem faced by a country like Ethiopia is that of rural electrification. One of the most important and achievable methods to produce electricity is to introduce a standalone electric power generation, using renewable resources. Rural Electrification Fund (REF), which is operating under the Ministry of Water, Irrigation and Energy (MWIE), is working to control the energy crisis in the country [1–3]. In its effort, it has identified some potential micro hydro sites in the country. This potential for small-scale hydro power is estimated to be 10% of the total potential (1,500–3,000 MW) [3, 4]. So far, out of the total potentials for micro hydro power (MHPs) in the country (over 1000 MW) only a minute portion of it (less than 1%) is developed [4]. If these water resources were properly harnessed, it will help Ethiopia to meet its power demand and maintain her economic growth for the next decade. Due to the existence of these numerous sites in Ethiopia, suitable for

micro hydro turbine installations, the need for development of micro hydro turbines using locally available materials and with local manufacturing capability has been identified for those sites which have been evaluated and proved to be viable, the aim being to cut the equipment cost which is imported from various countries from Europe and Asia. As a result, future of micro hydro developments in the country would need a manufacturer to provide turbines and parts with new design.

Figure 1 shows some MHP areas. These areas are mainly in the Western and South-Western part of the country and they are characterized by high mean annual rainfall ranging from 300 mm to over 900 mm. For many of these sites, a Pelton turbine is the only option. This is due to higher mountains providing higher heads and seasonal variation in flow rates appropriate for the choice of Pelton turbines for hydro power projects in the country [4–6].

Depending on water flow and design, Pelton wheels operate best with heads from 15 meters to 1,800 meters, although there is no theoretical limit. In this turbine, water

Mean annual
Water surplus

▨ More than 900 mm	▨ 300 to 500
▨ 700 to 900	▨ 100 to 300
▨ 500 to 700	▨ Less than 100 mm

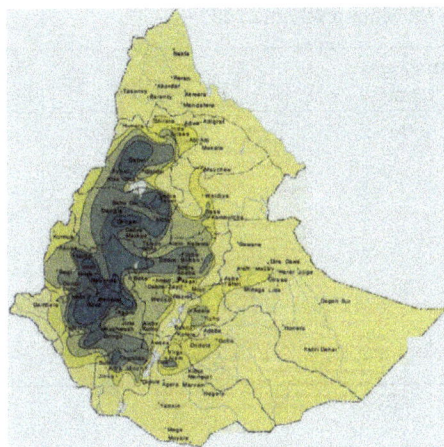

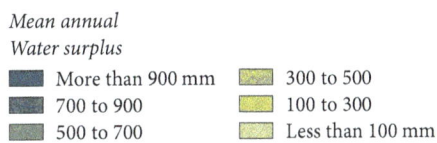

FIGURE 1: Distribution of MHP potential sites in Ethiopia [4].

is brought down through the penstock pipe to a nozzle, and it comes out into the turbine casing. The jet is then directed at a wheel, or runner, which has a number of buckets around its edge. The force of the jet on this wheel makes it turn and gives the output power [7–9]. However, investigation reveals that there is no company or institution engaged in supplying this micro Pelton turbine locally in the country. As a result, the necessity and the possibilities to design and manufacture Pelton turbines locally are increasing. More often the material and the skilled labour as well as technical staff are available but what is missing is the information and knows-how [4, 6].

More precisely, in the new context where harvesting small hydro potentials can become economically viable, there is also a need to provide solutions to reduce the design cycle time and cost for Pelton runners. It is known that, with increasing demand, the performance analysis of turbine such as efficiency and dynamic behavior is also an important aspect to analyze its suitability under different operating conditions [7]. Additionally, it is used by the turbine producer to guarantee the hydraulic performance of a turbine to the customer. However, it is known that design of Pelton turbine is mainly conducted from know-how and extensive experimental testing, which provide an empirical understanding of factors that are important to turbine design. But, in today's highly competitive market of turbine, the performance is often difficult to determine in the short term with this traditional practice. Therefore, the incorporation of computational fluid dynamics (CFD) in the design of micro hydro turbines appears to be necessary in order to improve their efficiency and cost-effectiveness beyond the traditional design practices [7, 10–12].

The main topic of investigations by the CFD method has focused on the interactions between the jet and the rotating buckets as well as the relative flows within the buckets. These are flows that are so far not easily accessible by experimental

measurements [11]. CFD simulations are therefore likely considered as an available way for investigating complex flows in Pelton turbines, provided that they are reliable and able to reveal the possibility of improving the system efficiency and reducing the manufacturing cost. In addition, CFD provide deeper understanding of the flow mechanisms that govern performance. The most detailed CFD analysis of rotating Pelton turbine was done by Perrig et al. [13, 14] by considering five buckets and the computed results were compared with experimental results at best efficiency point (BEP).

As explained by Zhang [7] in Pelton turbine book, hydraulic design of a Pelton turbine, the related practical experiences have thus always played a major role besides applying general design rules. Even the optimum bucket number and the size of a Pelton wheel, for instance, were determined only by experience or model tests without relying on any hydro mechanical background. The main reasons for this were the complex flow conditions in both the high-speed jet and the unsteady interaction between the high-speed jet and the rotating Pelton buckets [7, 11]. But nowadays, the largest amount of publications on modeling of Pelton turbines uses commercial code ANSYS CFX [7, 10, 14]. The capability of solving complex impulse turbine related problems that include multiphase flow with free surfaces has been demonstrated by a number of studies and is becoming increasingly significant [10–18]. However, large computational cost of the simulations is also the main factor why there is a lack of publications on CFD usage for Pelton turbines [11]. As a result, various authors made different simplifying assumptions in order to reduce this cost as much as possible and make Pelton simulation for performance predictions possible. Most CFD simulations reviewed in the literatures were assuming symmetry in the flow as simplifying assumption and therefore they model only half of a runner or a bucket. Because of the periodic behavior assumption, the majority of simulations used also only a fraction of a runner with the number of buckets in the section modeled being 2, 3, 5, 7, or even 10 [11]. Many authors used only 3 consecutive buckets where the torque was measured only on the bucket in the middle. This torque measured on a single middle bucket was then used to construct the torque on the runner assuming that every bucket would undergo identical loading [10, 11]. The first bucket was required to produce the back-splashing water that impacts the middle bucket. The third bucket was required to realistically cut the jet when it is impacting the second bucket. Even though, it was shown that it is possible to model the complete runner; this could be seen as unnecessary usage of computational resources. For instance, using the same computational resources and a reduced complexity simulation with only 3 buckets would allow simulations with better discretized grids (therefore improving accuracy) or analyzing more operating points or design variations and enable the optimization of Pelton turbine (see [11, 13, 14]).

So far, few investigators have only reported diverse values of maximum efficiencies as shown in Table 1. Because of commerciality of the turbine, many of the investigators normalized their results in their publications. Most papers report that the shape of the efficiency curve is well captured, while actual differences between measured and numerically

TABLE 1: Some Pelton turbine studies and maximum efficiency levels attained.

Investigators	Net head (m)	Flow rate (m³/sec)	Runner speed (rpm)	PCD (mm)	Maximum efficiency, %
Panthee et al. [10]	53.9	0.05	600	400	82.5
Panagiotopoulos et al. [19]	100	135% of BEP	1000	400	86.7
Solemslie and Dahlhaug [20]	70	—	—	513	77.75
Pudasaini et al. [21]	80.85	0.09218	600	490	87.71

TABLE 2: Proposed site data for MHP development [22].

Region	Zone	Wereda	Kebel	River name	Head (m)	Flow rate (L/s)
Oromia	W/showa	Tokikutay	Melkey Hera	Indris	>50	140

predicted efficiency remain unreported. An exception is [10, 19–21], where predicted efficiency was quite close to the measurements. Fewer still go deeper into the design and performance analysis of the turbine.

Due to the natural limitations concerning publication of results concerning turbines from commercial companies, one of the goals of this research paper has been to design and simulate using a Pelton turbine and compare the obtained result with the reference from Table 1. The results in Table 1 may have also depended on the laboratory settings and other controls of design parameters; the results still indicate room for further performance improvement. Even though this is a well-established turbine technology, there are many unanswered questions regarding design and optimization. Thus, further development is still relevant today. As a main feature, previous studies have mainly focused on experimental studies of turbine efficiency as a function of different geometrical parameters of the buckets. No paper that consists of design analysis, modeling, numerical performance analysis of rotating Pelton turbine runner, and its comparison with experimental data to reduce the size of the turbine has been published as far as we know. Therefore, a design study that takes into account the links between the size of the runner and the flow field characteristics is needed. This paper addresses this issue for one of the selected potential sites (Indris River) in Ethiopia to meet the requirements of the energy demands. The design process consists of sequential stages: parametric design of turbine, and CFD simulation with the baseline design and a smaller size of the turbine from the baseline design. The result of this study will be an input to the optimization of the runner for future study. In this way, a study was carried on Pelton turbine runner specifically designed and the result of this research was compared with that of Table 1 (published results). In the second part of this article, optimization of the geometry of this turbine, local manufacturing of the optimal geometry and testing of this turbine in the lab, and finally comparison of the results obtained will be presented. The result of this research will also ensure availability of documented procedure by contributing knowledge for designing micro hydro Pelton turbine in the

country. Most importantly, a general awareness and technical understanding of successful micro hydro turbine technology will be developed and fostered at the local and regional levels so that rural electrification projects can be implemented effectively.

2. Method and Methodology

2.1. Problem Description. After surveying different villages in the South-West District of Ethiopia, we narrowed our choices down to one: Melkey Herra Village. It is a rural community, Keble, roughly 149 kilometers from Addis Ababa, and is well renowned for tourism. Its geographical coordinates are 08° 51′ 40″ North and 37° 45′ 10″ East which does not have the access to electricity. The selected water resource for hydroelectric generation is "Indris River" for the communities living in the village called "Melkey Herra" [22]. The micro hydroelectric project is a main priority to the community. Due to the recent crises of electricity in the community, it is utmost need for the utilization of micro hydropower. The data (Table 2) has been found from primary and secondary data collection.

2.2. Steps Involved in Design of the Turbine. To start the initial design, calculations will be conducted to size turbine parts. The theory behind these is mainly taken from the "Micro Hydro Pelton Turbine Manual, by Thake [6]". Different assumptions are made in the design process using design guides and literature.

2.2.1. Calculation of the Net Head (H_n). The net head at the nozzle exits can be expressed by the following formula [6]:

$$H_n = H_g - H_l, \tag{1}$$

where H_g is the gross head and H_l is total head losses due to the open channel, trash rack, intake, penstock, and gate or valve. These losses approximately equal 5% of gross head [6]. This makes the net head available at the end of penstock as 50 m − 2.5 m = 47.5 m.

2.2.2. Selection of Turbine. For the proposed site data head, 50 m, and flow rate, 140 liter/sec combination (from Table 2),

TABLE 3: Calculation summary to determine the turbine speed (N) and PCD.

Name	Symbol	Unit	Value			
Number of jets	n_{jet}	—	1	2	3	4
Jet diameter	d_{jet}	mm	77.0	54.4	44.4	38.5
Runner PCD	PCD	m	699.667	494.740	403.953	349.834
Available PCD	PCD	mm	700	500	425	350
Turbine speed	N	rpm	371	520	611	742
Gear ratio	X	—	4.04	2.89	2.45	2.02

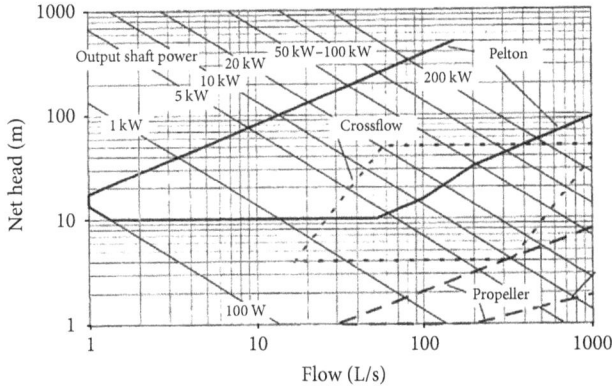

FIGURE 2: Application ranges for different types of turbine [6].

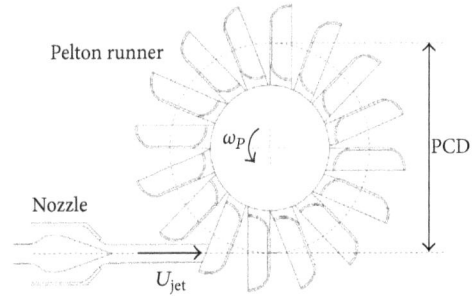

FIGURE 3: Diagram of a Pelton runner showing PCD [6].

principally a choice was necessary between Pelton and cross flow situations as shown in Figure 2. The figure shows the approximate application ranges of turbines for micro hydro. Therefore, this chart can be used for selection of the turbine type. The area highlighted in solid line shows an indicative range of operation for the Pelton runner used in this research paper.

2.2.3. Calculation of Jet Diameter (d_{Jet}). The pressure at the bottom of the penstock creates a jet of water with velocity, V_{jet}

$$V_{\text{jet}} = K_N \sqrt{2gH_n}, \qquad (2)$$

where V_{jet} is jet velocity (m/s), K_N is nozzle velocity coefficient (normally around **0.95** to **0.99**), and H_n is net head at the nozzle. The flow rate (Q) is then given by this velocity multiplied by the cross-sectional area of the jets:

$$Q = A_{\text{jet}} \times V_{\text{jet}} \times n_{\text{jet}}$$
$$= \Pi \frac{d_{\text{jet}}^2}{4} \cdot V_{\text{jet}} \cdot n_{\text{jet}}, \qquad (3)$$

where n_{jet} is number of jets and d_{jet} is diameter of jets (m)

Combining (2) and (3) and using an average value of 0.97 for K_N, then solving for d_{jet} becomes

$$d_{\text{jet}} = \frac{0.54}{H_n^{1/4}} \cdot \sqrt{\frac{Q}{n_{\text{jet}}}}. \qquad (4)$$

The next step in the turbine design process is to determine the pitch circle diameter (PCD) of the turbine.

2.2.4. Calculation of the Runner Circle Diameter (PCD). Figure 3 shows a schematic of a Pelton runner with a pitch circle diameter D (=$2R$) rotating at angular velocity ω_p.

Beginning with the derived formula to determine the turbine speed which can be expressed as

$$2\pi \frac{N}{60} \cdot \frac{D}{2} = x \cdot V_{\text{jet}}, \qquad (5)$$

where (D) or PCD is pitch circle diameter (m) and x is ratio of runner velocity to jet velocity ($x = 0.46$ is used to produce the maximum power out of the turbine) [6]. N is rotational velocity of runner (rpm). Substituting for V_{jet}, from (2), and using $x = 0.46$, (2) becomes

$$D = 37.7 \times \frac{\sqrt{H_n}}{N}. \qquad (6)$$

A spreadsheet is prepared as shown in Table 3 to determine the turbine speed (N) and PCD.

From Table 3 all the gear ratios are possible with belt drives. From the result in Table 3, runner pitch diameters (D) of 350 mm and 425 mm with 4-jet and 3-jet turbine, respectively, have smaller pitch diameter. However, for a turbine of these powers making is very complex and needs considerable expertise, so these are probably not an option [6]. The single jet solution is possible, but a 700 mm PCD runner makes a very big turbine. The best solution is the 500 mm, 2-jet turbine. Therefore, we can take the PCD to be 500 mm in our design analysis. To estimate the final system power output, efficiency values for the nozzles, turbine, and generator can be assumed. Thake [6] provides reasonable values for each. Table 4 shows the values of assumed efficiency in this research.

The equations used to determine the different design parameters of wheel are collected in Table 5.

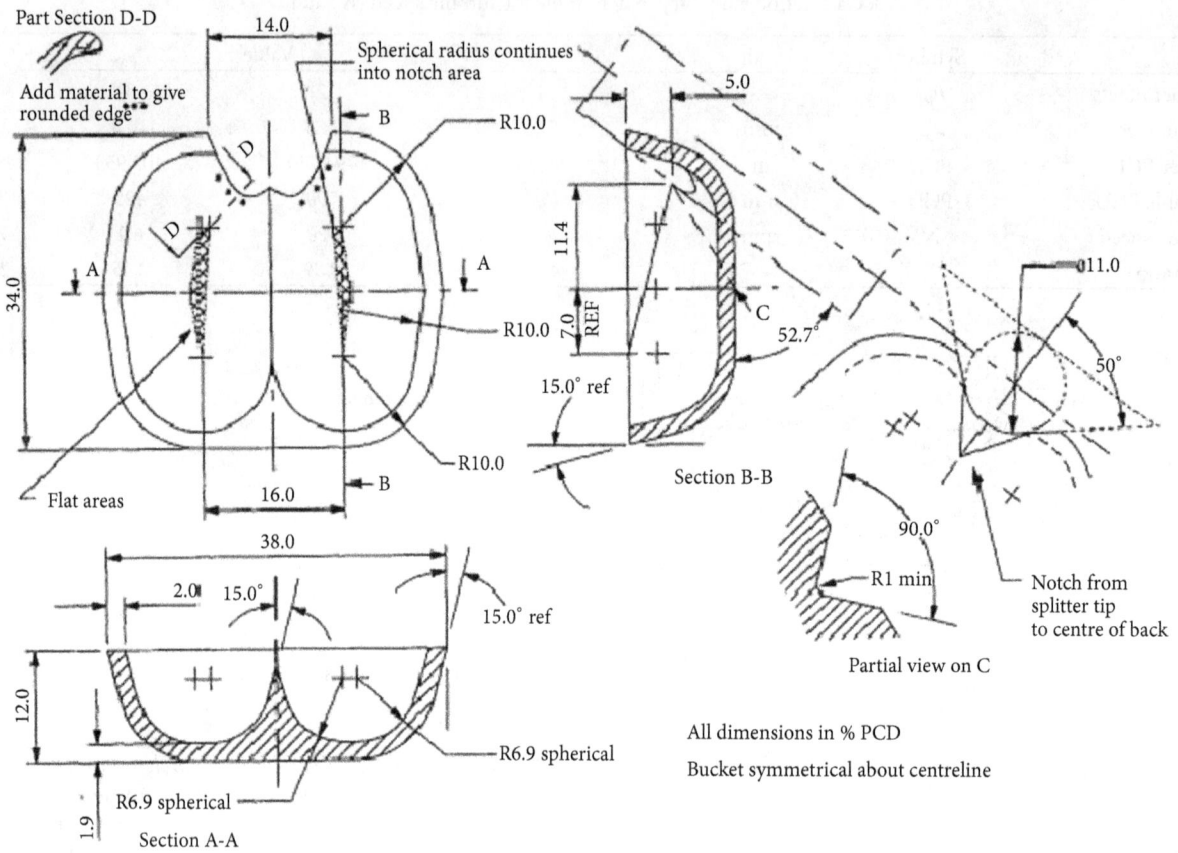

FIGURE 4: A scalable Pelton bucket. All dimensions are in % of PCD [6].

TABLE 4: Pelton turbine parts' assumed efficiency [6].

Part	Symbol	Assumed efficiency
Penstock	η_p	0.95
Manifold	η_m	0.98
Nozzle	η_n	0.94
Runner	η_r	0.8
Drive	η_d	1
Generator	η_g	0.8
Overall efficiency	η_o	0.56

2.2.5. Detail Bucket Geometry Design. Figure 4 shows the dimensions of the bucket as percentage values of the PCD of the turbine. Like the basic bucket model changing the PCD value within the model allows it to be scaled [6]. The physical dimensions of the bucket for the selected site data based on empirical relations for 500 mm PCD are shown in Table 6 which is used in this report for modeling of the bucket.

A basic stem for machining, used for bolting or clamping the buckets to the hub, is shown in Figure 5. Bolted buckets are an ideal solution for this research and are chosen as discussed in [6].

From all the above calculated parameters in Table 6 and some standard parameters using Figures 4 and 5, the

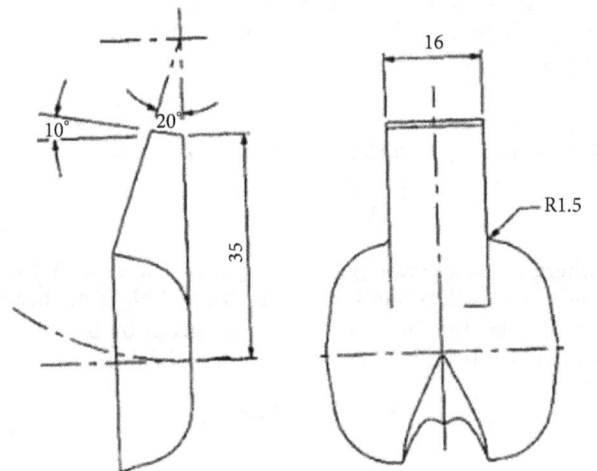

FIGURE 5: A basic bucket stem design for bolted fixing. All dimensions are in % PCD [6].

modeling of the bucket is done using CATIA V5 software as shown in Figure 6(a). The bucket design was specified by defining the outline of the bucket using the dimensions given in Figures 4 and 5.

The basic bucket model was adapted to form an entire bucket by the use of patterning. Two-disc plate was used to

TABLE 5: Basic design calculation summary of Pelton turbine.

Descriptions	Data	Unit	Design guidelines Thake 2000 [6]
$H_{\text{-net turbine}}$	46.5	mm	H-net turbine = $H_{\text{-net}} \times$ manifold efficiency
K_N	0.97	—	Ranges (between 0.95–0.99).
$V_{\text{-jet}}$	29.6	m/s	Velocity of the jet (for $n_{\text{jet}} = 2$)
$d_{\text{-jet}}$	54.4	mm	Diameter of the jet (for $n_{\text{jet}} = 2$) $d_{\text{-jet}} = 0.11 \times$ PCD
$x(U/V\text{-jet})$	0.46	—	For maximum power output (blade speed/V-jet) ratio
Bucket speed	13.6	m/s	$U = x \cdot V_{\text{jet}} = 0.46 \times 29.6$
N	1500	Rpm	Standard generator RPM
N-r	1170	Rpm	Runaway speed = 1.8, turbine optimum speed
η	0.56		Total system efficiency
P	36.57	kW	Estimated electrical power = $\rho g H_{\text{-net}} \times Q \times \eta_0$
P-turbine	51.2	kW	Turbine mechanical power = $\rho g H$-net turbine $\times Q \times \eta_r$

TABLE 6: Physical baseline bucket dimensions result for the selected site data.

Parameters, formula [6]	Calculation	Dimensions/result	Unit
Height of bucket, $h = 0.34D$	$h = 0.34 \times 500$	170	mm
Cavity length: $h1 = 5.6\%D$	$h1 = (0.056) \times 500$	28	mm
Length to impact point: $h2 = 0.114D$	$h2 = 0.114 \times 500$	57	mm
Width of bucket opening, $a = 0.14D$	$a = 0.14 \times 500$	56	mm
Bucket thickness: $t1 = 0.002D$	$t1 = 0.002 \times 500$	1	mm
Approximate number of buckets, Z	$Z = D/2d + 15$	18	
Depth of the bucket, $t = 0.121D$	$t = 0.121 \times 500$	60.5	mm
Width of the bucket, $b = 0.38D$	$b = 0.38 \times 500$	190	mm

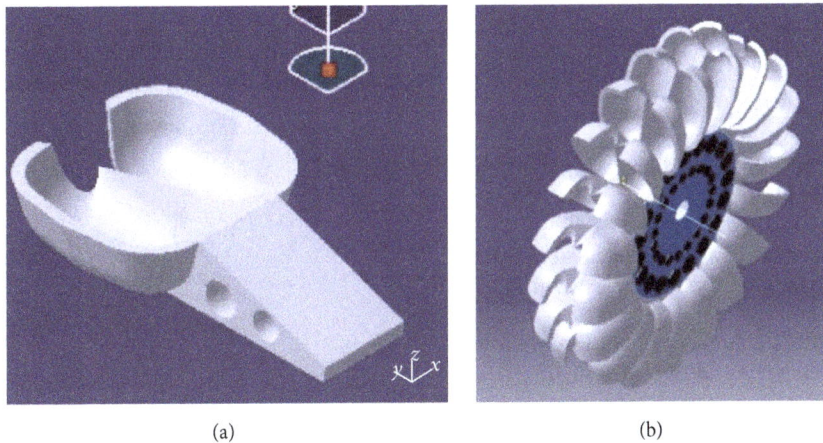

(a)   (b)

FIGURE 6: Solid model of Pelton turbine for the selected site data, (a) bucket and (b) 3D view of the runner right.

mount the buckets circularly as shown in Figure 6(b). The discs were to sandwich the buckets into place.

3. Computational Analysis

For the selected Melkey Herra's micro hydro power site, Pelton turbine was the most suitable hence chosen for our numerical performance analysis. The main dimensions of the turbine examined here correspond to this ideal plant. The numerical techniques created during this section included many numerical and physical assumptions to simplify the problem. This was necessary because accurate modeling of impulse turbines (Pelton in this case) that include complex phenomena like free surface flow, multifluid interaction, rotating frame of reference, and unsteady time dependent flow is a challenge from a computational cost point of view [10–18].

3.1. Physical Assumptions and Scaling Down. The computational domain was created removing the features that were

TABLE 7: Turbine geometry and setup values for prototype and model.

Parameters	Symbol	Unit	Prototypes' values	Selected model operating condition
Flow rate	Q	Lts/s	140	20.84278
Head	H	m	47.5	13.34275
PCD	D	mm	500	265
Bucket width	B	mm	190	100.7 > 80 mm
Power	P	KW	51.2	2.14
Number of buckets	Z		**18**	**18**

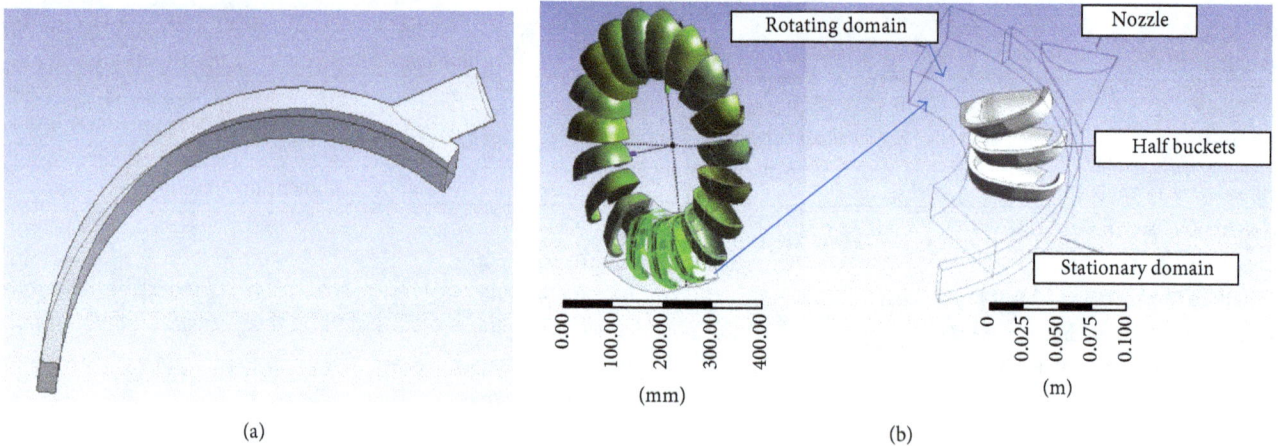

FIGURE 7: Domain geometries: stationary (a) and assembly of the rotating and stationary domain (b).

assumed to have no or minor effect when comparing the runner designs as follows. The following are some of the simplifying assumptions made in the analysis.

No Casing. Modeling the Pelton turbine without casing, similar methods can be found in the literatures [10, 11].

Symmetry. To reduce computational cost, the buckets, nozzle, and water-jet are cut in half at the symmetry axis [10, 11].

Single Jet. Modeling of only the single jet operation was found in most of the publications reviewed in Section 1 [10, 11].

No Hub. The flow will not be interacting with any other part of the runner except for the bucket. Hence, there is no need to include the hub into the CFD model, as suggested in the literatures [10, 11, 13].

Periodic Torque. Three buckets are enough to recreate the complete runner torque and are used also in this research [10, 11].

"Similitude". Similitude in a general sense is the indication of a known relationship between a model and a prototype. Equation (7) represents head coefficient, flow coefficient, and power coefficient for model studies [10]. In order to achieve similarity between model and prototype behavior, all the corresponding terms must be equated between model and prototype. So, the model is presumed to have the same values of speed ratio, flow ratio, and specific speed.

$$\left(\frac{H}{D^2 N^2}\right)_{\text{prototype}} = \left(\frac{H}{D^2 N^2}\right)_{\text{model}},$$

$$\left(\frac{Q}{ND^3}\right)_{\text{prototype}} = \left(\frac{P}{ND^3}\right)_{\text{model}}, \qquad (7)$$

$$\left(\frac{P}{D^5 N^3}\right)_{\text{prototype}} = \left(\frac{P}{D^5 N^3}\right)_{\text{model}}.$$

Scaling Down. Scaling down of the prototype is also important to reduce the time consumption and to ease the computational processing in normal computers. On the basis of the above considerations, the scaling factor of 0.53 was used to meet the minimum bucket width standard for model testing, and the Reynolds number also needs to be greater than 2×10^6. This is based on the international standard IEC 60193 of the International Electrotechnical Commission which applies to laboratory testing of model turbines [10, 23]. In this standard, the factor greater than 0.28 was found to satisfy IEC 60193 criteria for Pelton turbines.

Table 7 shows the values for prototype and model with the same speed of 520 rpm.

3.2. Computational Domain Creation. Figure 7 depicts the completed CAD drawings of the rotating and stationary

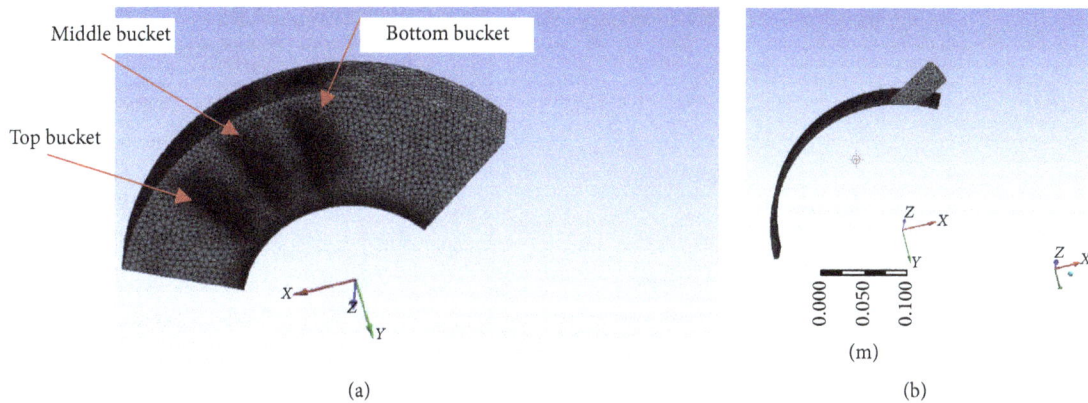

FIGURE 8: Meshed rotating (a) and stationary domain (b) sizing and inflations were applied.

domain which is modeled separately and assembled here to show their relative initial positions. Stationary domain contains half a cylinder for the inlet and a ring to accommodate an interface between the two domains.

Imagine that the rotating domain in Figure 7 is aligned with the stationary domain. When the rotating domain leaves this position, another identical domain is introduced at the top (the same geometry). Thus, one achieves a continuous simulation of a runner [11].

3.3. Meshing. Figure 8 shows both stationary and rotating domains meshed as they are imported into CFX-Pre. As suggested in the literature [10, 19], unstructured tetrahedral elements were used for the rotating domain meshing because of more complex geometry to be captured by the mesh and also to allow automatic meshing for all the upcoming geometry modifications. In order to determine the minimum grid size or mesh resolution required to resolve the boundary layer and the mean flow features, a grid independence study was conducted. This allows minimization of errors and uncertainties in the predicted results, for example, the runner power output or the efficiency. Therefore, grid convergence analysis has been carried out considering power output (in this study by monitoring torque) as a parameter of significant interest. The results of grid independence study are presented in Section 3.6.

3.4. Physical Setup with ANSYS Preprocessing. In this section, the essentials of the ANSYS Pre setup are presented.

Analysis Type. In each "Flow Analysis" in ANSYS Pre there is a tab called "Analysis Type." This is where one defines whether or not the simulation is transient or steady-state, control simulation time and time steps. In this case, the transient option was chosen. Time steps are interval for which CFX solver calculates flow parameters in transient analysis. The time step was 1/20 total time, which corresponds to 0.001714 sec to capture 20 time frames per rotation.

Multiphase Model. The flow through the penstock pipe is having only single phase for the fluid. As the water-jet comes

out of the nozzle, it is directly freed to the atmosphere and the effect of free water-jet will come into picture. This flow passes through runner buckets, and it will be a free surface flow. This free surface flow glides through internal surface of the buckets and the momentum transfer takes place. The flow leaving the runner bucket will be converted into dispersed water droplets as the fluid leaves contact from the buckets. This way the flow through the Pelton turbine is multiphase in nature in agreement with the literatures [10, 11, 13, 14]. To capture the flow accurately, homogenous multiphase analysis is performed in this research.

Volume Fraction of Water. As suggested by the literature [5], the volume fraction of water should change from 0 to 1. In each control cell, the volume fractions of the water and air sum to 1, $\alpha_w + \alpha_g = 1$. Then $\alpha_w = 0$ denotes cells filled with air, while $\alpha_w = 1$ denotes cells filled with water and $0 < \alpha_i < 1$ denotes that the cell contains an interface between the water and air. In the beginning of the simulation, both domains are full of air with 0 m/s velocity. Therefore, the given initial conditions are 0 m/s velocity and 0 Pa for relative pressure. Initial water volume fraction is 0 and initial air volume fraction is 1.

Turbulence Model. Shear Stress Turbulence (SST) model is able to capture turbulent scales in flow in high shear stress regions [10, 11]. So SST turbulence model is chosen for further simulations.

Domain Interface. The interface type between the stationary and rotating domains was Fluid-Fluid [10, 11]. Interface model for this type was General Connection and the Transient Rotor Stator option was selected for Frame Change/Mixing Model. Pitch ratio was maintained to value 1 between domain interfaces by maintaining equal area in interface region. To apply a rotation to the rotating domain, domain motion in the rotating domain was set to rotating, and angular velocity was defined by the expression "Omega." In this case, the angular velocity is negative because the domain was modeled to rotate in the negative rotation direction about the z-axis. This method is also found in the literatures [10, 11, 14].

TABLE 8: Expressions defined in ANSYS Pre for the baseline design [5, 11].

Name	Expression	Description
Gravity	9.82 [ms^{-2}]	Acceleration due to gravity
Head	13.34 [m]	Model head based on scaling
Turbine radius	132.5 [mm]	Model turbine radius based on scaling
Inlet velocity	$(2 \times gravity \times head)^{0.5}$ [m/s]	Water velocity at the nozzle inlet
Omega	Inlet velocity/(2 × turbine radius) [rad/s]	Angular velocity: rotation in the negative direction is selected
Torque middle bucket wall	*Torque_z@Middel Bucket Wall*	The entire middle bucket is selected
Total time	((120 × pi)/180)/Omega	Total simulation running time

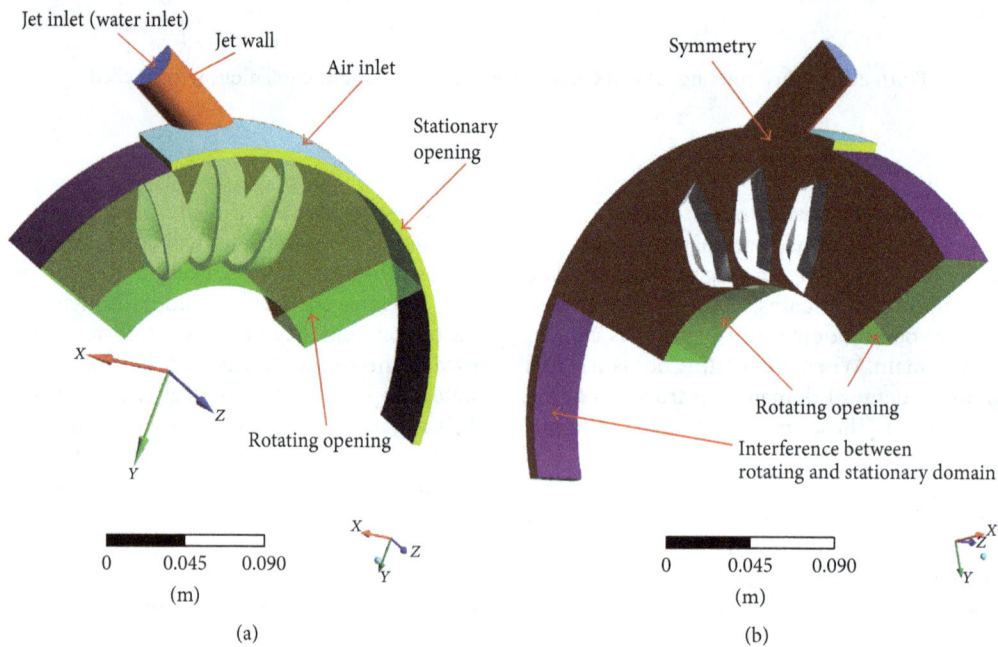

FIGURE 9: Boundaries applied on the domains.

Boundary Condition in Rotating Domain. This section contains a list of boundaries and their conditions. No boundary for outlet is defined in the rotating domain. Rather, opening type boundary condition has been defined. The bucket surface is defined by wall type boundary condition and an interface type boundary condition at interface between the rotating and stationary domain. All other remaining boundaries are defined as opening type since it is unpredictable about the actual outlet and flow pattern of fluid through the runner. An overview of the boundary conditions can be found in Figure 9.

Solver Control and Output. The chosen advection scheme was High Resolution according to the CFX Modeling Guide. This gives a good compromise between robustness and accuracy. Second-Order Backward Euler option was selected for the Transient Scheme as it is generally recommended for most transient runs in CFX [5, 10, 11, 13, 14].

Monitor Points. The main output of the simulation is the total torque on the middle bucket. A monitor for the expression *"Torque on Middle Bucket Wall"* was added, which logged the

torque and made it possible to monitor the torque during the simulation. The calculated torque data on the middle bucket was extracted from the simulation using an inbuilt torque function applied to the Named Selections created for the regions for the middle bucket. This function is then plotted as a Monitor Point, while the simulation runs, allowing the calculated torque at each time step to be exported.

Expressions. Defining expressions is a good way of streamlining a CFD case. The expressions in Table 8 are used in the setup. This way of defining expressions can be also found in the literatures [5, 11].

3.5. Result and Discussion. In general rule, a small runner is cheaper to manufacture than the larger one. It takes less material to cast it and the housing and associated components also can be smaller. For this reason, consequences of reducing PCD (=400 mm) from the baseline design (PCD = 500 mm) are evaluated and results are compared with those published by Panthee et al. [10] and Panagiotopoulos et al. [19], respectively. The geometrical characteristics of their Pelton turbine model used for their study correspond to a Pelton

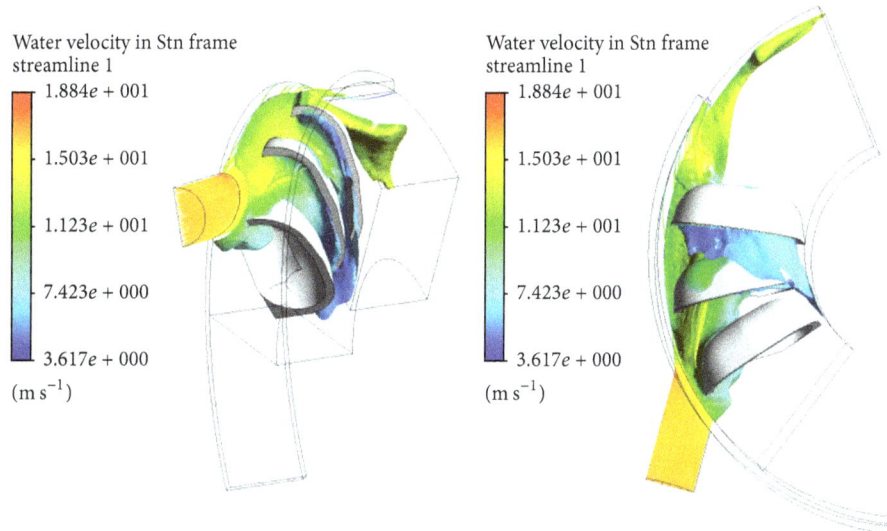

FIGURE 10: Flow visualization of the baseline design, side views, and face views (PCD = 500 mm).

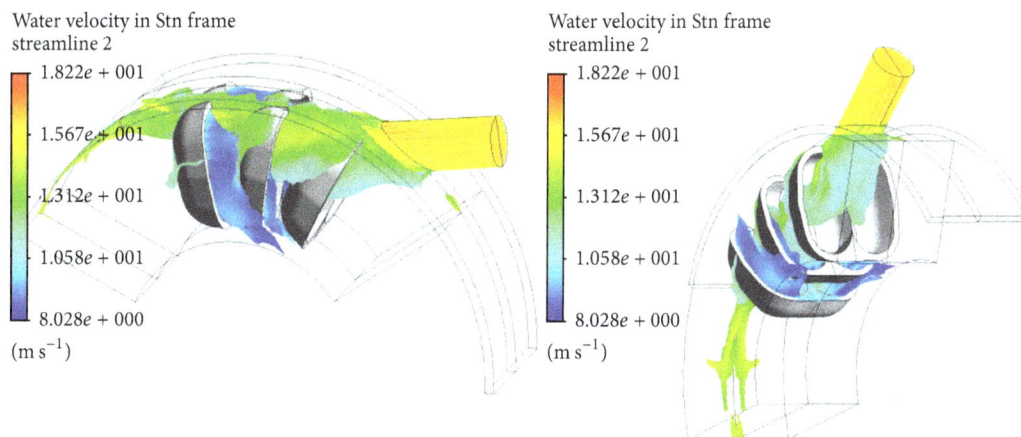

FIGURE 11: Flow distribution of the runner: water velocity in Stn frame, side views, and face views (PCD = 500 mm).

turbine installed in Khimti Hydropower in Nepal [10] and the National Technical University of Athens [19], respectively. The PCD of the model runner for both power plants was 400 mm, and the axis is horizontal with two injectors.

Next, a visualization of the flow in the turbine buckets for this study can be seen in Figure 10.

As seen in Figures 10 and 11, only very small amount of "leaked" water is present next to the bottom bucket with low jet velocity. These visualizations give good information about the flow pattern. The streamlines that came from the nozzle entered into the runner, hitting the buckets and dividing the jet into three portions (Figures 10 and 11). First, the portion from the bottom of the jet touched the outside and inside of the top bucket that is close to the nozzle. The second portion of the jet was the one from the middle, which hit the next middle bucket. Finally, the rest portion of the jet crossed into the latter bucket. On the face and side view on Figure 10, one can observe the part of the flow escaping from the cutout of the bucket as well as the lateral sheet flow.

It can be observed on Figure 12 that the maximum pressure point, in red, corresponds to the PCD of the bottom bucket and the tip of the middle bucket which is aligned with the axis of the impinging jet. Similar finding is observed in the literature [10].

As seen in Figure 13, when the PCD is reduced to 400 mm, a large amount of water does not leave the buckets, causing severe backwash on the backside of the bucket, which varies from certain sound value at the place of impingement to the reduced values as water goes outward (Figure 13(b)). However, there is a small amount of the flow that leaves through the cutout with high velocity without being utilized (cutout leakage). This phenomenon has to be reduced during the optimization stage.

The water is unevenly distributed across the buckets and there are several local accumulations of cells with a high volume fraction of water, most visible in the bottom bucket in Figure 14. This volume of fraction of water will vary with time as the buckets rotate. This occurs when the portion of

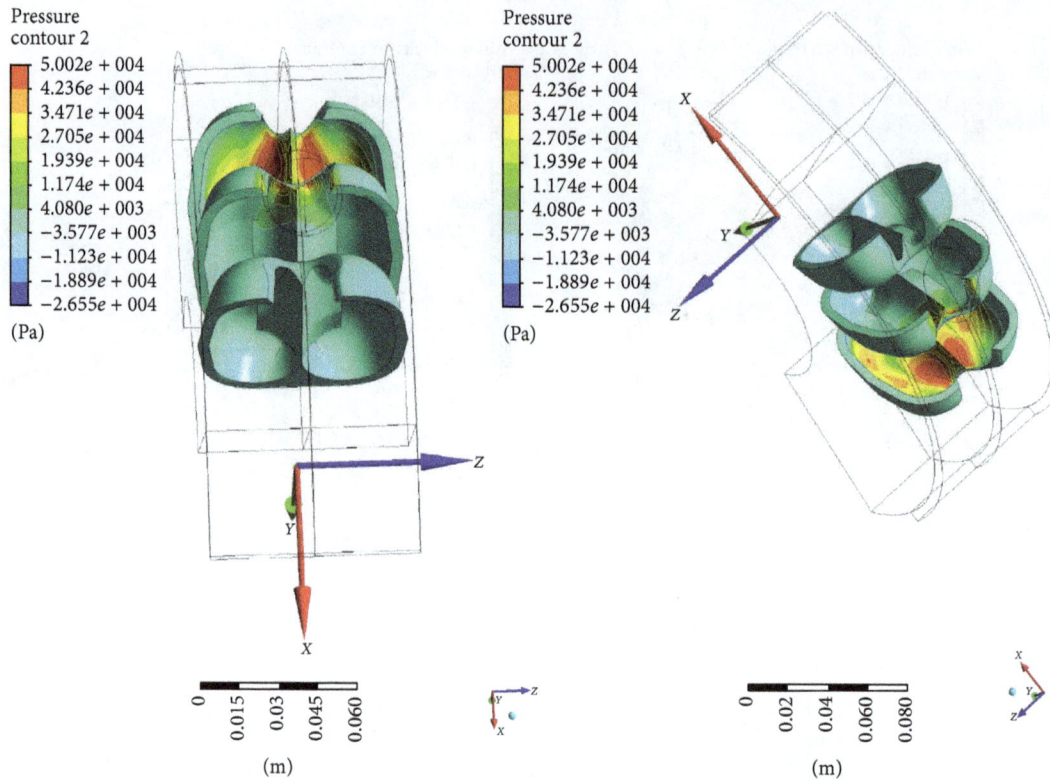

FIGURE 12: Pressure contours on the bucket (PCD = 500 mm).

the jet is going to leave out completely from the bottom bucket and completely enters the next top and consequential buckets. The red and blue regions show, respectively, water and air in the computational domain. In a more complex situation, the atmosphere also exerts pressure on the surface (water-vapor interface) [24].

Looking at Figures 13(b) and 14, it is clear that, as the water starts entering into the top bucket, the inner surface of the middle and bottom bucket will still be having some water volume fraction inside it. It means that, for a particular instance of time and runner rotation, more than one of the runner buckets are having water volume fraction. The amount of the water flow touching the runner bucket sides and back surface of the bucket (backwash) is also visible. This water fraction at the back side and side surface of the bucket will impart some force to the buckets and if this fraction is high, then the intensity of force on the bucket will be higher and it will reduce the strength of the buckets [11, 13, 14]. Another effect of this is that the water is acting as a brake on the runner, rather than helping it turn, and this gives a serious loss of power. These energy losses occur with jet entering the bucket and providing some amount of counter-torque as the outer side of the bucket hits by the surface of the jet.

The pressure distribution in the bucket was due to impact of high jet. This pressure distribution applied on the bucket again varies with the time due to the rotation of the runner. It was found that the pressure peaks are obtained at bucket tip and PCD of the runner. The pressure peak in bucket tip is due to flow disturbance when jet strikes bucket tip. It is obvious to

obtain the pressure peak at the runner PCD since the Pelton runners are designed such that it would convert most of the hydraulic energy to mechanical energy when the jet strikes the runner PCD. Result showed high pressure with the value of 3.559×10^4 Pa for the first top and the second middle bucket at some degree of rotation of the runner as shown in Figure 15. This pressure value is lower than the previous value when PCD = 500 mm (Figure 12). This might indicate that there is energy loss in the runner.

The images in Figure 15 are taken of the right bucket half; then it is mirrored here to ease the comparison through symmetry. The next step will be focused on future simulations regarding the study of the hydraulic efficiency; a better approximation of the numerical torque.

The hydrodynamic torque and the hydraulic efficiency of the runner are computed after completing the evaluation of a jet-bucket interaction flow, starting from the moment of impingement (jet cut in) until the evacuation of the bucket (jet cutout) [11]. The calculations are then continued for the particles of the oncoming frames, until all particles of a frame impinge on the next coming bucket (jet cut). As can be observed in Figure 15, the shape of the torque curve is also similar to the previous studies by different authors [10, 19], but the irregularity of the curve in this area was attributed due to the Coanda effect and the impact of the back flow which creates counter-torque on the bucket and makes it slightly different from the previous studies [5, 11, 19]. Some indicative pictures from the turbine examined here show how the energy transfer occurred in a single bucket as illustrated in Figure 16.

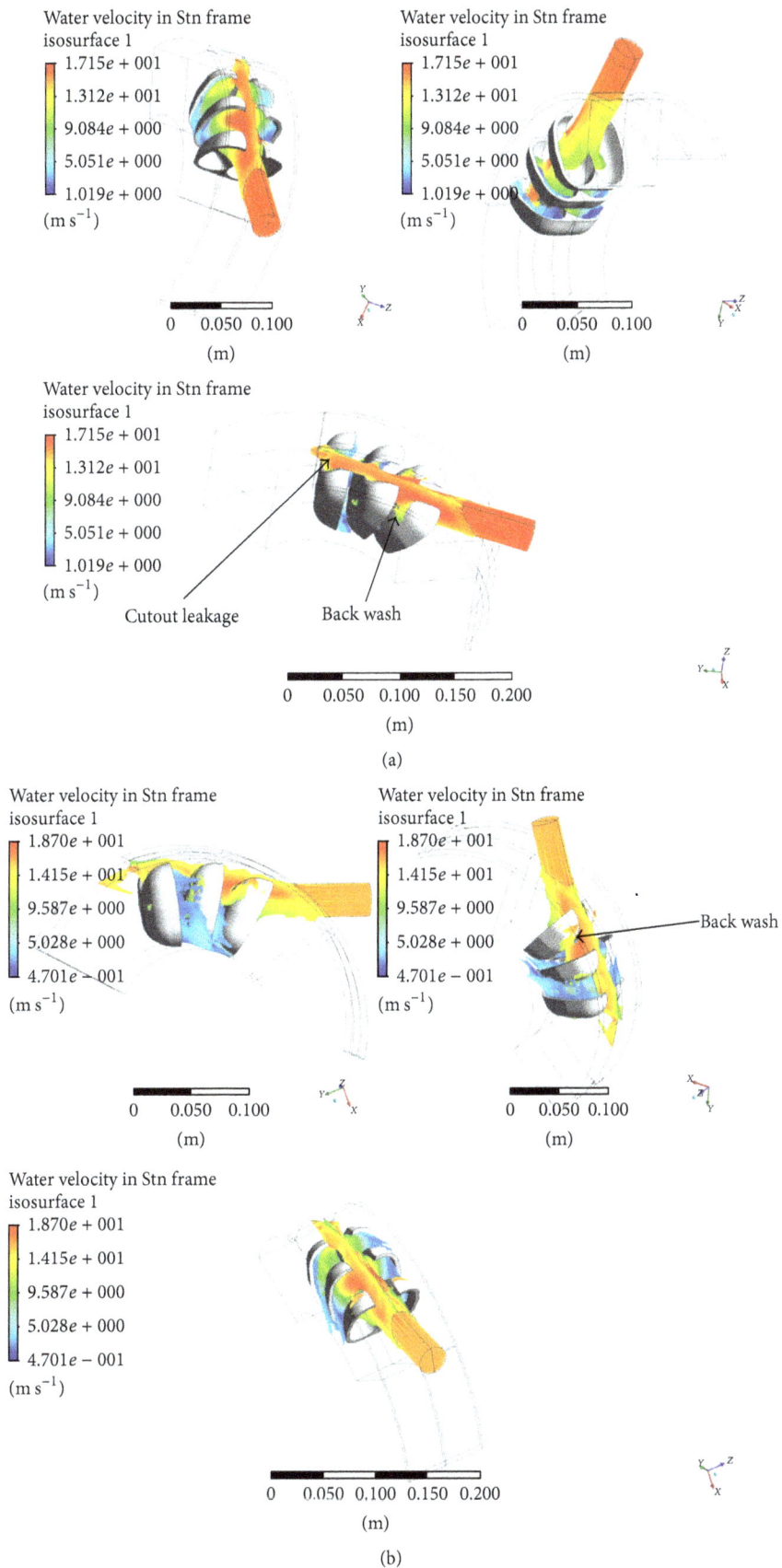

FIGURE 13: Flow distribution of the runner: water velocity in Stn frame, side views, and face views (PCD = 400 mm).

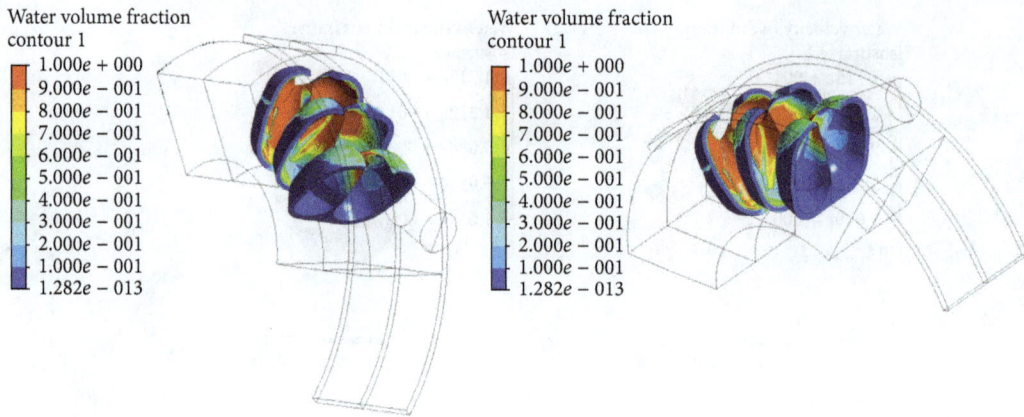

FIGURE 14: Volume fraction of water for reference bucket design at a particular time step (PCD = 400 mm).

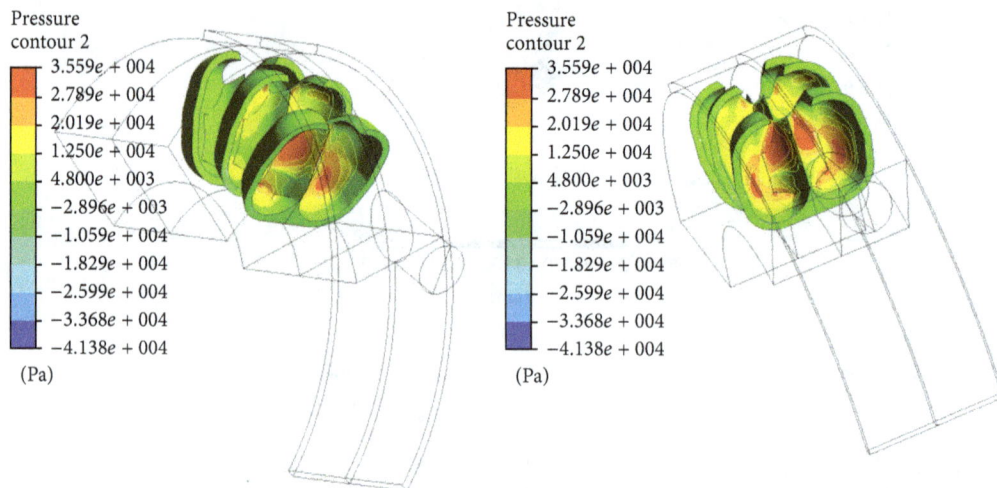

FIGURE 15: Pressure distribution at some degree of rotation of reference bucket design (PCD = 400 mm).

At the start (cut in) a counter-torque can also be observed caused by the interaction of the jet with the back surface of the bucket; this value is larger for PCD = 400 mm comparing to the baseline design (Figure 16). Just afterwards, the torque is increasing as more water interacts with the inner surface until the full jet interacts with the bucket to produce the maximum torque. The maximum torque produce by the baseline design (500 mm) is larger than that of a reduced PCD (400 mm). The result in Figure 16 shows also that smaller turbine is faster than the larger one which is true.

Next, the total curves of the complex unsteady flow in the bucket for the seven buckets analyzed can be acquired with the aid of time history views like the ones in Figures 17 and 18. The total torque curve is shifted by the single blade passage phase for the whole range of total torque values and summed to give the total torque acting on the turbine shaft. In this research paper, the torque generated by the middle bucket is replicated over time to determine total torque generated by Pelton turbine. This is done by assuming that at stable conditions every bucket is producing identical torque periodically. The replication was done till the summation graph gave steady values which occurs after

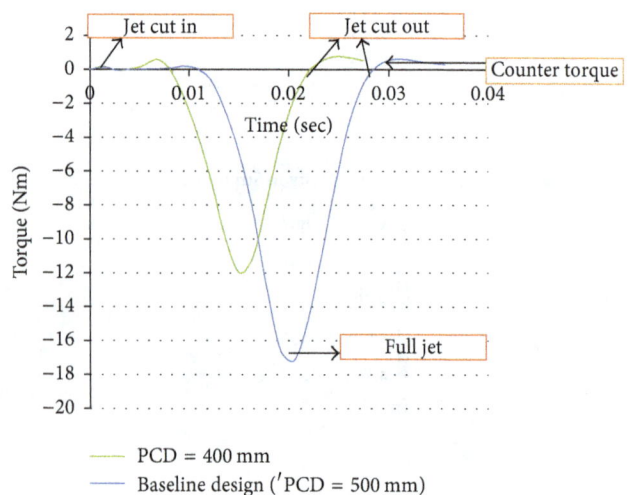

FIGURE 16: Torque generated by middle bucket versus time.

three buckets covered by the water sheets at 0.03 sec of the rotation of the runner. The plot obtained was due to half

TABLE 9: Mesh dependent test analyzed for PCD = 500 mm.

Mesh type	M1	M2	M3	M4	M5	M6	M7	M8
Total number, elements	294952	520687	2011665	2239650	2398044	2861243	3955723	5050204
Calculated total torque (Nm)	−30.93	−32.20	−34.32	−34.50	−34.69	−34.92	−35.02	−35.25
Standard torque (Nm)	−35.06	−35.06	−35.06	−35.06	−35.06	−35.06	−35.06	−35.06
Torque error percent (%)	11.8	8.2	2.1	1.6	1.1	0.4	0.1	0.5

TABLE 10: Mesh dependent test analyzed for PCD = 400 mm.

Mesh type	M1	M2	M3	M4	M5	M6	M7	M8
Total number, Elements	128251	128546	130961	2321026	3019180	4052422	4771226	6855793
Calculated total torque (Nm)	−18.13	−18.15	−18.29	−21.84	−22.15	−22.27	−22.37	−22.41
Standard torque (Nm)	−28.05	−28.05	−28.05	−28.05	−28.05	−28.05	−28.05	−28.05
Torque error percent (%)	35.4	35.3	34.8	22.1	21.0	20.6	20.2	20.1

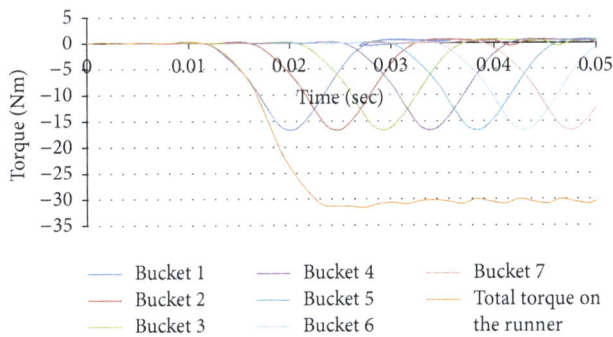

FIGURE 17: Dynamic runner and bucket torque over time for (PCD = 500 mm).

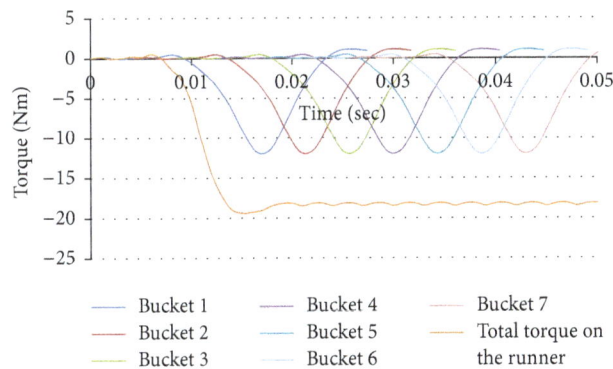

FIGURE 18: Dynamic runner and bucket torque over time for (PCD = 400 mm).

nozzle and the maximum torque on the runner at a time is given by peak value multiplied by 4 as there are two nozzles in each unit. This runner torque has been taken for power output calculations by taking the average value from 0.03 to 0.05 sec of the runner rotation. Methods of calculating the power output from a single bucket torque readings similar to those described above are quite common and can be found in the literature [10–12].

To calculate the efficiency, power input has to be calculated as well which for a complete turbine is calculated using two variables describing the flow conditions: the net pressure head and the flow rate [11].

$$P_{\text{input}} = \rho g Q H_{\text{net}},$$

$$P_{\text{out}} = \frac{2 \times \Pi \times N \times T}{60}. \tag{8}$$

Therefore, (hydraulic) runner efficiency in this model was calculated using (9) as follows:

$$\eta = \frac{P_{\text{output}}}{P_{\text{input}}}. \tag{9}$$

The next validation phase is to compare efficiency of this model to the published results by Panthee et al. [10] and Panagiotopoulos et al. [19]. This will be done after checking the accuracy of the model, in agreement with the literature [10, 19].

3.6. Mesh Independency Study and Model Validation. Mesh or grid independent study is done in order to get a solution that does not vary significantly even when we refine our mesh further. Eight different mesh sizes were tested with an effective head of 47.5 m for two different cases. The mesh size on the rotating domain is controlled by element size. The relevance was increased for finer mesh. Afterwards, the orthogonal quality of the mesh has been checked and it was in an acceptable range (0.15–1.00) for each mesh developed. Each mesh was also created with the same physical setup and boundary conditions. During the simulations, results obtained were directly dependent on the accuracy in quality of mesh. And it was performed while analyzing the torque variation by developing the SST turbulence model.

Tables 9 and 10 indicate the grid information, the calculated torque, the standard torque, and the resulting modeled

TABLE 11: Validation and performance prediction of Pelton runner.

Parameters	Unit	Baseline design test cases			Off- design test cases			Published result by Panthee et al. [10]	Published result by Panagiotopoulos et al. [19]
Head	m	47.5	53.9	100	47.5	53.9	100	53.9	100
Flow rate	m³/s	0.14	0.05	135% BEP	0.14	0.05	135% BEP	0.05	135% BEP
PCD	mm	500	500	500	400	400	400	400	400
Runner speed	Rpm	520	520	520	650	600	1000	600	1000
Number of buckets	—	18	18	18	18	22	22	22	22
Model efficiency	%	78.8	83.5	84.6	62.6	66.1	71.6	82.5	86.7

runner efficiency for two different cases, namely, the baseline design (PCD = 500 mm) and the reduced size (PCD = 400 mm), respectively.

The standard torque was calculated with the power obtained from power coefficient and the angular velocity of runner for comparison. One can observe that, from the result in Tables 9 and 10, a finer mesh will give a solution of a little higher accuracy than required but at the expense of computational power and time. About 3.9 million elements were required to obtain mesh independent result, for PCD = 500 mm (Table 9). Similarly, for the PCD = 400 mm turbine, a total number of 6.8 million mesh elements were required to obtain mesh independent results for the torque output (Table 10). A solution was considered grid independent for less than 0.3 difference in the torque output between three different consecutive mesh sizes. It appears that no significant enhancement is to be expected from a further refinement of the mesh. As seen from the results, a relative error of 20.1%, between the CFD and the analytical data, was found for turbine with PCD = 400 mm and 0.1% relative error for the baseline turbine (PCD = 500 mm) at the design-point operating condition (Q = 0.14 m³/s and H = 47.5 m). Hence, high value of relative torque error indicates that there are losses in the reduced size of the turbine, influencing the efficiency characteristics. The torque variation curves of the runner obtained using different density meshes are also presented in Figures 19 and 20.

Table 11 shows the performance prediction and comparison of the model results with two published results.

The computational analysis results presented in Table 11 showed that the performance of the baseline's turbine design was very good since the relative error is only 0.4% at the design operating conditions. Torque value from mesh (M7) resulted in 78.8% of model runner efficiency for PCD = 500 mm (baseline design). The postprocessing visualization (Figures 10 and 11) was able to qualitatively show that the reason for the better performance for the case of baseline design may be explained by less cutout leakage and a more optimal bucket design for its intended use. As presented in Table 11, the model hydraulic efficiencies for reduced size of the turbine (PCD = 400 mm) for mesh (M8) were only 62.6%, 66.1%, and 71.64%. It can be observed that, from the results of the different cases analyzed, these hydraulic efficiencies were remarkably lower than that of the baseline

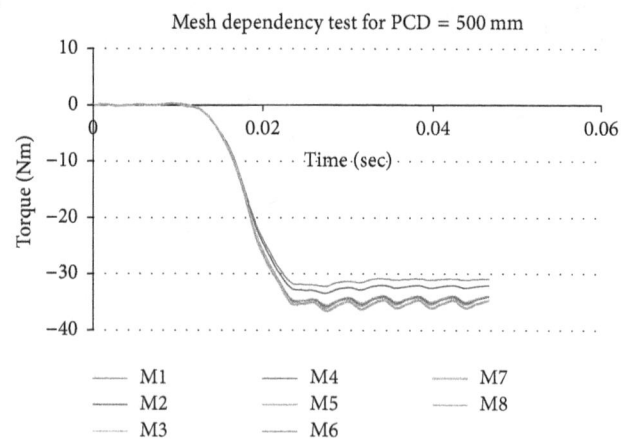

FIGURE 19: Total torque variations for different mesh sizes for PCD = 500 mm.

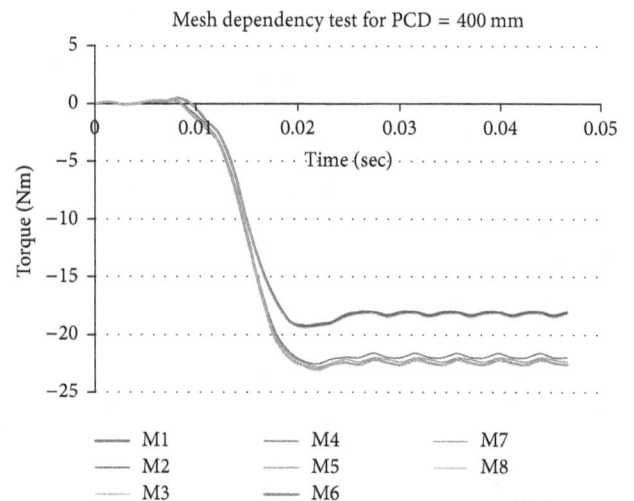

FIGURE 20: Total torque variations for different mesh sizes for PCD = 400 mm.

design. Moreover, it can be observed that for a reduced flow rate and an increase in number of buckets equal to 0.05 m³/s, 22, respectively, more water was caught by the inner surface of the buckets. This could explain why the predicted efficiency increases rapidly for volume flow less than the design flow rate. The lower efficiency of the model at off-design condition (PCD = 400 mm) is mainly due to a little backsplash (the breaking effect) on the bucket and the missing jet velocity (leakage flow rate) because of a nonoptimized bucket design. As this phenomenon highlighted before, the qualitative views of Figures 13–15 are reasonable for the lower efficiencies in the off-design cases. Thus, the goal was to compare flow visualization at these two conditions, using image postprocessing, in order to explain the difference in runner performance. An explanation for the discrepancy is that a decrease in PCD from the baseline design (PCD = 500 mm) to PCD = 400 mm would lead to high hydraulic losses, even though it will result in less value of manufacturing cost. It is important to note that the bucket shape used in the baseline design (as shown in Figure 4) in this paper is based on Jeremy Thake 2000 book [6]. In this book, it is suggested that "the design jet diameter is 11% of PCD" which will give a PCD of approximately 500 mm. Therefore, if we optimize/change the shape of this bucket for each case analyzed above, the optimum results presented in both published results above and the reduced size of the turbine (PCD = 400 mm) might be different from this study. For future work, the design optimization of this turbine bucket will be made by keeping the PCD to be 400 mm, to lower the manufacturing cost and improve efficiency of the runner. CFD results by Židonis and Aggidis [25] also show that less number of buckets were required to increase the hydraulic efficiency. Therefore, as the final step, it was decided to change the length, depth, and angular position (jet-bucket interaction) and alter the surface close to the lip and redefine the shape of the lip curve to reduce the leakage, intern to reduce the energy loss.

4. Conclusion

The next-generation turbine designs for small-scale hydropower systems seek higher efficiency and low manufacturing costs. The (size of Pelton turbine) PCD is an important parameter to lower the manufacturing cost of a Pelton turbine runner. However, no consistent guidance based on numerical research data is available in the public domain. This study provides a guideline for selecting, designing, modeling, and performance analysis of a micro hydro Pelton turbine and provides an example of how the size of turbine (PCD) can affect the efficiency of the turbine.

The paper describes also the methods used for CFD analysis of scaled model Pelton turbine which is ideally designed for Melkey Herra Village hydropower plant using ANSYS CFX software. The bucket model was designed according to the baseline design or calculated parameters. The time and cost in CFD analysis of Pelton turbine are also reduced by selecting 3 buckets to predict the flow behavior of complete turbine. One of the objectives was to understand how the turbine performance will change when the flow field is perturbed by reducing the size of turbine for a given operating condition (i.e., flow rate and head). Results obtained from the baseline design (PCD = 500 mm) have been successfully compared to that of the reduced size (PCD = 400 mm) of the turbine. The results of the two turbine designs at the design flow rate and head of turbine are as follows: one turbine (PCD =

400 mm) has a maximum efficiency of 62.6% and the baseline design turbine (PCD = 500 mm) has a maximum efficiency of 78.8%. The flow visualization study of this research provides insights into the reasons for efficiency as well as guidance for improving the efficiency. The low efficiency in the reduced size of the turbine is mainly caused by a large amount of water leaving the bucket through the lip and hence transferring close to zero of its energy to the shaft. The problem was therefore the choice of the runner PCD that could give the best advantages in terms of efficiency on the whole plant's operational field. For the purpose of validation and performance characterization, two Pelton turbines reported in the literature were considered. The first turbine has PCD = 400 mm, a maximum efficiency of 82.5%, which was studied by Panthee et al., 2014, for Khimti Hydropower in Nepal [10]. The second turbine has also the same PCD (=400 mm) but with a maximum efficiency of 86.7%, which was studied by Panagiotopoulos et al., 2015 [19], using a reference case corresponding to a Pelton turbine installed in the LHT, at the National Technical University of Athens. These were essential for the purpose of computational validation and performance characterization.

The design optimization, production, and experimental testing of a model runner based on this design may also be realized in the next part of this paper. Based on the result obtained, this study planned a modification on the baseline design and will consist of the following new designs being tested:

(i) Changing the length, depth, angular position (Jet-bucket interaction), and number of the buckets while keeping all other parameters constant

(ii) Altering the surface close to the lip and redefining the shape of the lip curve to reduce the leakage

Based on the results presented in this study, there are opportunities for improving the maximum efficiency as well as reducing manufacturing cost of Pelton turbines, if the design is prescribed using the above criteria, followed by high-fidelity computational simulations. It would then no longer be necessary to start the design and numerical simulation each time from scratch.

A natural extension of this paper would be also to validate the model for other selected sites. Achieving such a goal would be a great step towards improving the understanding of the micro hydro power development and making tools for validating CFD results, so that the benefits of this technology can be brought to rural populations.

Abbreviations

CFD: Computational fluid dynamics
CFX: CFD code by ANSYS
SST: Shear stress transport
VOF: Volume of fluid
MHP: Micro hydro power
MHT: Micro hydro turbine
MWIE: Ministry of Water, Irrigation and Energy
BEF: Best efficiency point.

Conflicts of Interest

The authors declare that they have no conflicts of interest.

Acknowledgments

This paper was funded by Ministry of Water, Irrigation and Energy. The authors would like to thank Ministry of Water, Irrigation and Energy Office for the fund and for all information and data provided to accomplish the research work.

References

[1] Energy Situation, https://energypedia.info/wiki/Ethiopia.

[2] A. Dalelo, Rural Electrification in Ethiopia, opportunities and bottlenecks, Addis Ababa University, College of Education.

[3] World Small Hydropower Development Report 2013– The Need for Further Resource Assessments, L. Esser, International Center on Small Hydro Power, Division of Multilateral Development, Hangzhou, Ethiopia.

[4] S. Melessaw, "Ethiopia's Small Hydro Energy Market-GiZ–target market analysis-hydro ethiopia," http://www.german-energy-solutions.de/enwww.gtz.de/projektentwicklungsprogram.

[5] L. F. Barstad, *CFD Analysis of a Pelton Turbine [Master, thesis]*, Norwegian University of Science and Technology, 2012.

[6] J. Thake, *The Micro-Hydro Pelton Turbine Manual: Design, Manufacture and Installation for Small-Scale Hydro-Power*, ITDG publishing, London, UK, 2000.

[7] Z. Zhang, Pelton turbines book, published by Springer Nature, the registered company is Springer International Publishing, AG, Switzerland, 2016.

[8] MHPG, Series, "Harnessing Water Power on a Small Scale" Vol. 9 (Micro Pelton Turbines), Published by SKAT, Swiss Center for Appropriate Technology, 1991.

[9] A. Harvey, Micro hydro design manual, a guide to small scale water power schemes.

[10] A. Panthee, B. Thapa, and H. P. Neopane, "CFD Analysis of pelton runner," *International Journal of Scientific and Research Publications (IJSRP)*, vol. 4, no. 8, 2014.

[11] A. Zidonis, A. Panagiotopoulos, G. A. Aggidis, J. S. Anagnostopoulos, and D. E. Papantonis, "Parametric optimisation of two Pelton turbine runner designs using CFD," *Journal of Hydrodynamics*, vol. 27, no. 3, pp. 840–847, 2015.

[12] J. D. Anderson, *Computational Fluid Dynamics: The Basics with Applications*, McGraw-Hill, New York, NY, USA, 1995.

[13] A. Perrig, *Hydrodynamics of the free surface low in Pelton turbine buckets [Ph. D. Thesis]*, École polytechnique féd, 2007.

[14] A. Perrig, F. Avellan, J.-L. Kueny, M. Farhat, and E. Parkinson, "Flow in a Pelton turbine bucket: numerical and experimental investigations," *Journal of Fluids Engineering*, vol. 128, no. 2, pp. 350–358, 2006.

[15] Y. X. Xiao, T. Cui, Z. W. Wang, and Z. G. Yan, "Numerical simulation of unsteady free surface flow and dynamic performance for a Pelton turbine," in *Proceedings of the 26th IAHR Symposium on Hydraulic Machinery and Systems*, chn, August 2012.

[16] M. Eisenring, Micro Pelton turbines. St. Gallen, Switzerland: Swiss Center for Appropriate Technology, 1991.

[17] B. Janetzky, H. Ruprecht, C. Keck, C. Schärer, and E. A. Göde, "Numerical simulation of the flow in a Pelton bucket," in *Proceedings of the in. IAHR*, pp. 276–283, 1998.

[18] J. S. Anagnostopoulos and D. E. Papantonis, "A fast Lagrangian simulation method for flow analysis and runner design in Pelton turbines," *Journal of Hydrodynamics B*, vol. 24, no. 6, pp. 930–941, 2012.

[19] A. Panagiotopoulos, A. Židonis, G. A. Aggidis, J. S. Anagnostopoulos, and D. E. Papantonis, "Flow Modeling in Pelton Turbines by an Accurate Eulerian and a Fast Lagrangian Evaluation Method," *International Journal of Rotating Machinery*, vol. 2015, Article ID 679576, 2015.

[20] B. W. Solemslie and O. G. Dahlhaug, "A reference Pelton turbine design," in *Proceedings of the IAHR, 26thIAHR Symposium on Hydraulic Machinery and Systems, IOP Conf.*, Earth and Environmental Science, Beijing, China, August 2012.

[21] S. Pudasaini, H. P. Neopane, Amod P. et al., "Computational Fluid Dynamics (CFD) analysis of Pelton runner of Khimti Hydro-power Project of Nepal," in *Rentech Symposium Compendium*, vol. 4, September 2014.

[22] N. Tilahun, W. Bogale, F. Bekele, and E. Dribssa, "Feasibility study for power generation using off- grid energy system from micro hydro-PV-diesel generator-battery for rural area of Ethiopia: The case of Melkey Hera village, Western Ethiopia," *AIMS Energy*, vol. 5, no. 4, pp. 667–690, 2017.

[23] Hydraulic Turbines, Storage pumps and pump turbines- Model acceptance tests (IEC 60193), 1999.

[24] F. M. White and I. Coreld, *Viscous Fluid Flow*, vol. 3, McGraw-Hill, New York, NY, USA, 2006.

[25] A. Židonis and G. A. Aggidis, "Pelton turbine: Identifying the optimum number of buckets using CFD," *Journal of Hydrodynamics*, vol. 28, no. 1, pp. 75–83, 2016.

6

Aerodynamic Optimization Design of a Multistage Centrifugal Steam Turbine and its Off-Design Performance Analysis

Hui Li[1,2] and Dian-Gui Huang[1,2]

[1]*School of Energy and Power Engineering, University of Shanghai for Science and Technology, Shanghai 200093, China*
[2]*Shanghai Key Laboratory of Power Energy in Multiphase Flow and Heat Transfer, Shanghai 200093, China*

Correspondence should be addressed to Dian-Gui Huang; dghuang@usst.edu.cn

Academic Editor: Ryoichi Samuel Amano

Centrifugal turbine which has less land occupation, simple structure, and high aerodynamic efficiency is suitable to be used as small to medium size steam turbines or waste heat recovery plant. In this paper, one-dimensional design of a multistage centrifugal steam turbine was performed by using in-house one-dimensional aerodynamic design program. In addition, three-dimensional numerical simulation was also performed in order to analyze design and off-design aerodynamic performance of the proposed centrifugal steam turbine. The results exhibit reasonable flow field and smooth streamline; the aerodynamic performance of the designed turbine meets our initial expectations. These results indicate that the one-dimensional aerodynamic design program is reliable and effective. The off-design aerodynamic performance of centrifugal steam turbine was analyzed, and the results show that the mass flow increases with the decrease of the pressure ratio at a constant speed, until the critical mass flow is reached. The efficiency curve with the pressure ratio has an optimum efficiency point. And the pressure ratio of the optimum efficiency agrees well with that of the one-dimensional design. The shaft power decreases as the pressure ratio increases at a constant speed. Overall, the centrifugal turbine has a wide range and good off-design aerodynamic performance.

1. Introduction

With the continuous reduction of fossil energy and the enhancement of people's environment awareness, there has been increasing attention to the high efficiency utilization of energy, where the turbine is energy conversion component; if the turbine is improved effectively, it can help to improve the efficiency utilization of energy.

Centrifugal turbine is a new type of turbine engine which has many advantages. In the centrifugal turbine, the gas flows outward the center and the channel area of the flow path increase naturally as the fluid volume flow increases, during the expansion process. That meets the principle of aerodynamic and geometric matching. In addition, centrifugal turbine can be used to achieve multistage design [1] more easily than centripetal turbine. Thus it can avoid the limitation of supersonic flow. This is more useful for design conditions, especially for off-design performance conditions.

For turbine as the key component, many scholars in this area have done some relevant researches, but there is not much research on centrifugal turbine. Domestically, mainly Jing and Peng had studied the pneumatic analysis of a rocket centrifugal turbine prototype starter and designed the modification turbine [2]; Yin-Ge et al. [3] and Xin et al. [4, 5] had researched the single-stage centrifugal turbine design and its off-design performance. Abroad, the preliminary hydrodynamic design of a small centrifugal turbine for the ORC was studied by Casati and others of the Delft University of Technology. They introduced the optimization based on the intermediate streamline method for evaluating turbine design and performance [6–10].

In this paper, one-dimensional design method of centrifugal steam turbine is proposed by drawing on the conventional turbine [11]. And numerical simulation and optimization of multistage centrifugal turbine were studied by referring to the inlet and outlet thermal parameters of a small axial steam turbine.

2. Aerodynamic Design of a Steam Centrifugal Turbine

2.1. One-Dimensional Design Method. A one-dimensional design of centrifugal turbine has been developed, which is based on the one-dimensional design method of conventional turbine. The one-dimensional thermodynamic calculation program was developed by FORTRAN. The main design principles of centrifugal turbine one-dimensional design program are as follows:

(1) The expansion flow is assumed to be adiabatic, steady, and one-dimensional in the cascade channel.

(2) The properties of working fluids are obtained by calling Refpro 9.0, which is applicable to a variety of working fluid.

(3) In order to simplify the design of the blade, the blade is designed to be of straight and constant height.

(4) Stator and rotor velocity coefficients φ and ψ are based on previous experience [12].

(5) Each stage of the rotor absolute flow angle is 90 degrees.

(6) The program is mainly composed of continuity and energy equations to achieve one-dimensional design.

Figure 1 presents the design process of the centrifugal one-dimensional aerodynamic program. The thermodynamic parameters of the centrifugal turbine, inlet stagnation temperature T_0^*, inlet stagnation pressure P_0^*, outlet pressure P_N, mass flow rate G, and rotation speed n, are based on original conventional turbine. The other parameters, impeller diameter ratio b (the influence of diameter ratio on the centrifugal turbine wheel efficiency is referred to in [3]; the optimum wheel efficiency calculated at the optimum degree of reaction and speed ratio increases as the diameter ratio decreases; $b_N = D_{N,\text{out}}/D_{N,\text{in}}$), each stage outlet flow angle of stator α_1, the radial gap δ ($\delta = R_{r,\text{in}} - R_{s,\text{out}}$ or $\delta = R_{s,\text{in}} - R_{r,\text{out}}$, because the passage area along the flow path remains constant in axial flow turbine or the flow area reduces in radial flow turbine, so the radial gap can be large in traditional turbine, while, in centrifugal turbine, the flow area increases with the working fluid expansion. If the radial gap is too large, it will make the working fluid compressive in the gap, which is bad for turbine working. So the radial gap for centrifugal turbine should be small.), and stage number N, are previously estimated. Iterative and screening methods are used to search the maximum wheel efficiency of the centrifugal turbine.

Figure 2 shows the overall stages H-S diagram. The superscript $*$ represents the stagnant state. The second numbers of subscript are as follows: 1 represents the stator and 2 represents rotor. Taking the first stage as an example, the introduction of design process is as follows:

(1) In the stator, the steam expands from state 0 to state 1. Lines 0-1 represent the actual expansion and

lines 0-1s is the ideal expansion. In this process, the pressure energies are transferred to kinetic energies. Then the thermal parameters, velocity and geometry parameters can be calculated by using

isentropic expansion: $s_0 = s_{1,1s} = f\left(P_0^*, T_0^*\right)$

$$h_{1,1s} = f\left(s_{1,1s}, P_{1,1}\right)$$

energy equation: $h_0^* = h_{1,1s} + \dfrac{c_{1,1s}^2}{2} = h_{1,1} + \dfrac{c_{1,1}^2}{2}$

$$c_{1,1} = \varphi c_{1,1s} \qquad (1)$$

$$s_{1,1}, \rho_{1,1} = f\left(h_{1,1}, P_{1,1}\right)$$

continuity equation: $G = \rho_0 c_0 A_0 = \rho_{1,1} c_{1,1} A_1 \sin \alpha_1$

$$\chi_1 = \frac{u_{1,1}}{c_{1,1s}} = \frac{\pi n D_{1,1\text{out}}}{60 c_{1,1s}}.$$

(2) Then, in the rotor, the steam continues to expand from state 1 to state 2. Lines 1-2 represent the actual expansion and lines 1-2s is the ideal expansion. The fluid kinetic energies are transferred to mechanical energies, which make the turbine outputs shaft power. The thermal parameters and the velocity triangles can be calculated by using

$$w_{1,1} = \sqrt{c_{1,1}^2 + u_{1,1}^2 - 2c_{1,1}u_{1,1}\cos \alpha_1}$$

$$\sin \beta_1 = \frac{c_{1,1} \sin \alpha_1}{w_{1,1}}$$

continuity equation: $\rho_{1,1} c_{1,1} A_1 \sin \alpha_1$

$$= \rho_{1,2} A_2 \sin \alpha_2 \sqrt{\psi^2 \left(w_1^2 + 2\Delta h_{2s} + u_{1,2}^2 - u_{1,1}^2\right) - u_{1,2}^2}$$

energy equation: $h_{1,1} + \dfrac{w_{1,1}^2}{2} - \dfrac{u_{1,1}^2}{2}$

$$= h_{1,2s} + \frac{w_{1,2s}^2}{2} - \frac{u_{1,2}^2}{2} \qquad (2)$$

$$w_{1,2} = \psi w_{1,2s}$$

isentropic expansion: $s_{1,2s} = s_{1,1} = f\left(P_{1,1}^*, T_{1,1}^*\right)$

$$P_{1,2} = f\left(s_{1,2s}, h_{1,2s}\right)$$

$$s_{1,2}, \rho_{1,2}' = f\left(P_{1,2}, h_{1,2}\right)$$

$$\alpha_2 = 90°,$$

$$c_{1,2} = \sqrt{w_{1,2}^2 - u_{1,2}^2}$$

$$tg\beta_2 = \frac{c_{1,2}}{u_{1,2}}.$$

(3) The rotor actual outlet density $\rho_{1,2}'$ is compared with estimated density $\rho_{1,2}$. If $|\rho_{1,2}' - \rho_{1,2}| < \varepsilon$ (ε is

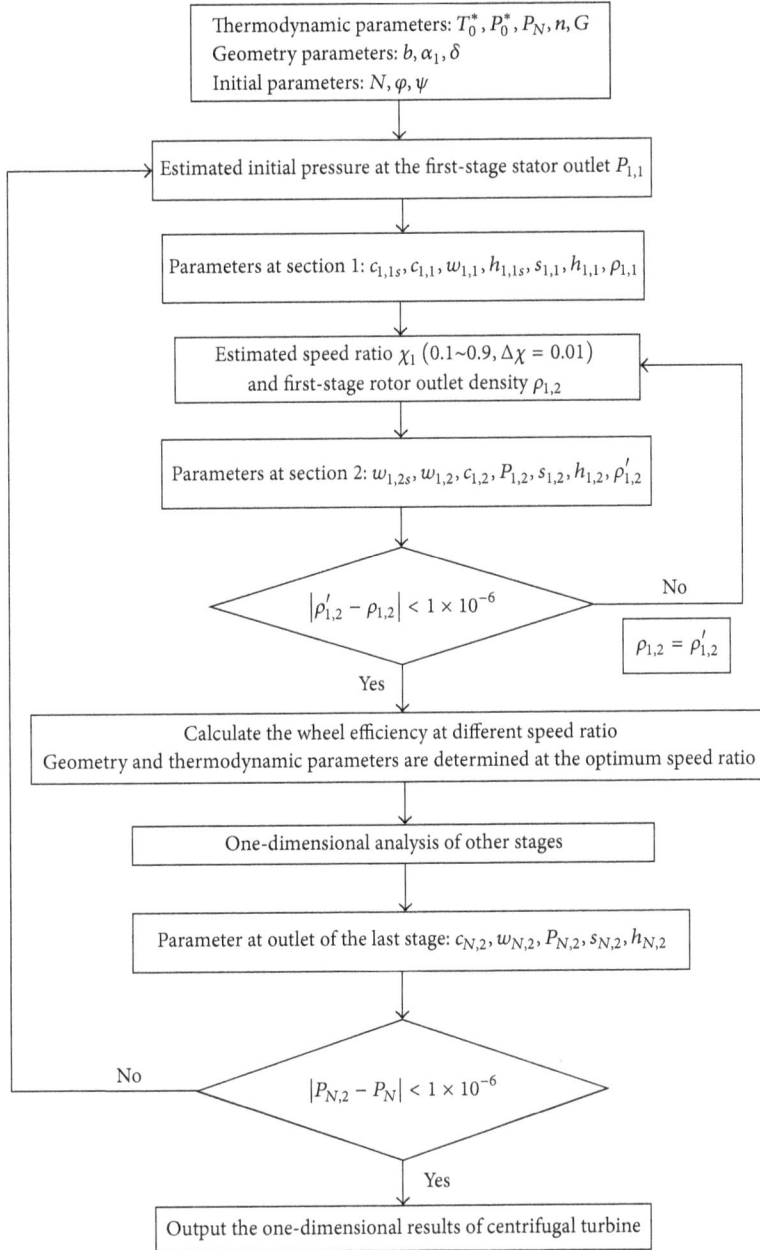

FIGURE 1: One-dimensional design process of centrifugal turbine.

a minimum), the assumption density $\rho_{1,2}$ is valid. Then the rotor outlet point is determined. If it is not satisfied, then reassume $\rho_{1,2}$ and repeat the above steps until the conditions are met.

(4) Similar to the axial flow turbine, the relevant adiabatic efficiency is given by (3), which is referred to in [13].

$$\eta = \frac{h_0 - h_{1,2}}{h_0 - h'_{1,2s}}. \tag{3}$$

The wheel efficiency is calculated at different speed ratio to select the optimum speed ratio and other corresponding parameters at maximum efficiency. Geometry and thermodynamic parameters are identified at the optimum speed ratio.

The above-mentioned steps are also adopted to design other stages.

2.2. One-Dimensional Design Results. The one-dimensional design program of the centrifugal turbine is used to design the multistage centrifugal turbine. The initial design thermal parameters are derived from a small axial turbine and summarized in Table 1.

Because there is no stage number information about the original axial turbine from the product information, the different stage number was tested to design the centrifugal turbine by using the one-dimensional design program. The results show that if the stage number is 1 and 2, there is supersonic in the centrifugal turbine. As the stage number

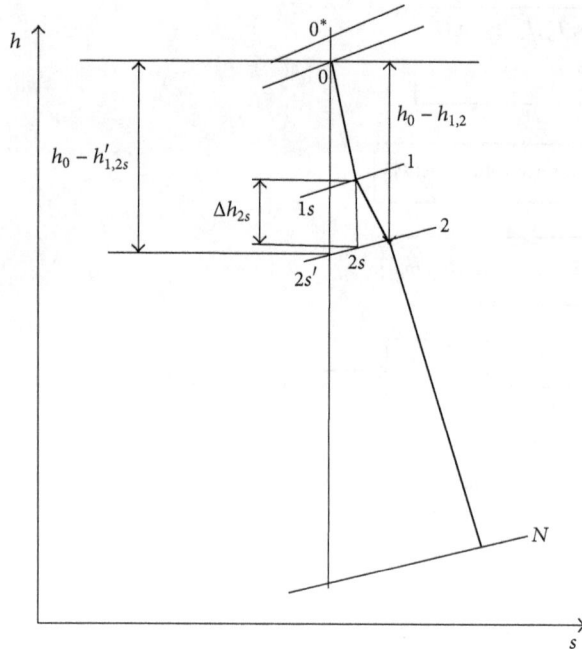

FIGURE 2: Turbine expansion process.

TABLE 1: The initial design parameters of the centrifugal turbine.

Total pressure P_0 (Pa)	1275000
Total temperature T_0 (K)	613
Back pressure P_N (Pa)	294000
Rotational speed n (rpm)	6500
Mass flow G (kg/s)	16.7
Rated power W (kW)	750

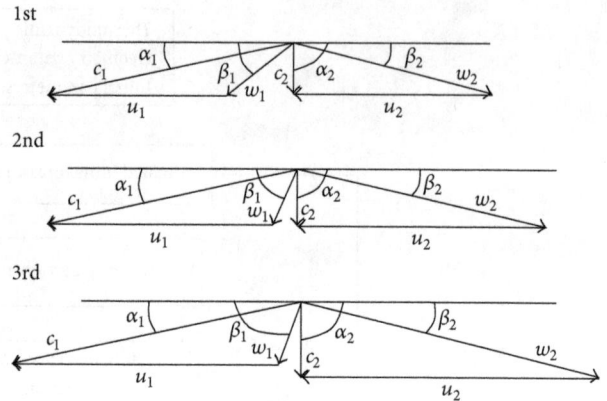

FIGURE 3: The speed triangle of the centrifugal turbine.

3. Airfoil Design and Optimization

The working fluid flows are outward the center in the centrifugal turbine. The passage area along the flow path increases naturally with the expansion of the working fluid, which makes the variety of specific volume matches the change of flow passage area. It can be seen that the centrifugal turbine is much superior to the conventional turbine from the structure. In the centrifugal turbine, the blade plane flow channel is expanding outward with the increase of the radius. It is obviously inappropriate to use the airfoil of conventional turbine at this time. So designing a suitable airfoil for centrifugal turbine is needed.

3.1. Parametric Expression of Airfoil. Angle and thickness design method is adopted to design airfoil, by using the inlet and outlet geometry angle, blade height, and leading and trailing edge diameters, which are based on one-dimensional design results in Tables 3 and 4. On the BladeGen platform, two-dimensional structure of the stator and rotor is identified. Stator and rotor of the first stage are an example, as shown in Figures 4(a) and 4(b). The blade surface includes four patches, namely, the leading edge and the trailing edge and the suction side and the pressure side. The mean camber line is controlled by cubic Bezier curve. The leading and trailing edges are both ellipse. Blade number is determined initially by referring to the relative pitch and expelling coefficient and the data is provided in [12]. The final numbers of stator and rotor for each stage are both 65.

3.2. Blade Optimization for Three-Stage Centrifugal Steam Turbine. In the optimization process of blade, the blade number, the blade inlet and outlet geometry angles, and the leading and trailing edge thicknesses are the fixed parameters, and two control points of the tangent angle and two control points of the blade thickness are taken as the optimization parameters. Workbench is as the optimization software, and NLPQL algorithm is used as optimization method.

In the stator optimization process, the minimum loss coefficient is the object and stator back pressure is the constraint condition. Stators of three-stage centrifugal turbine

increased, there is no supersonic in the centrifugal turbine, but the size of the centrifugal turbine becomes more and more large. When the stage number is 3, and the diameter ratio is 1.12, the centrifugal turbine is subsonic, and it can satisfy the enthalpy drop as well as high efficiency. So the centrifugal turbine is designed with three stages.

Then, the same blade height and the absolute airflow angle of each stage are 90 degrees as the criteria, impeller diameter ratio is estimated to be 1.12 for each stage, each stage outlet flow angle of stator is assumed to be 12 degrees, and radial clearance between stator and rotor is set to be 2 mm. The optimal efficiency and reasonable structure are the objectives. The design parameters are calculated by using the above-mentioned one-dimensional calculation program, with iterative and screening methods. In this paper, the designed three-stage centrifugal turbine just meets the required enthalpy drop and has a very high wheel efficiency. The main aerodynamic parameters of each stage are shown in Table 2, and the main aerodynamic parameters of overall centrifugal turbine are shown in Table 3. The geometry parameters are shown in Table 4, and the speed triangle data is shown in Table 5. The speed triangle schematic diagram is shown as Figure 3.

TABLE 2: The main aerodynamic parameters of each stage.

Stage number	1st	2nd	3rd
Speed ratio Xi	0.69	0.86	0.87
Reaction degree Ω	0.322	0.453	0.468
Flow coefficient φ	0.97	0.97	0.97
Flow coefficient ψ	0.95	0.95	0.95
Mach number of stator Ma	0.55	0.58	0.76
Shaft power W (kW)	1205	1551	2436
Wheel efficiency η	91.20%	90.97%	90.98%

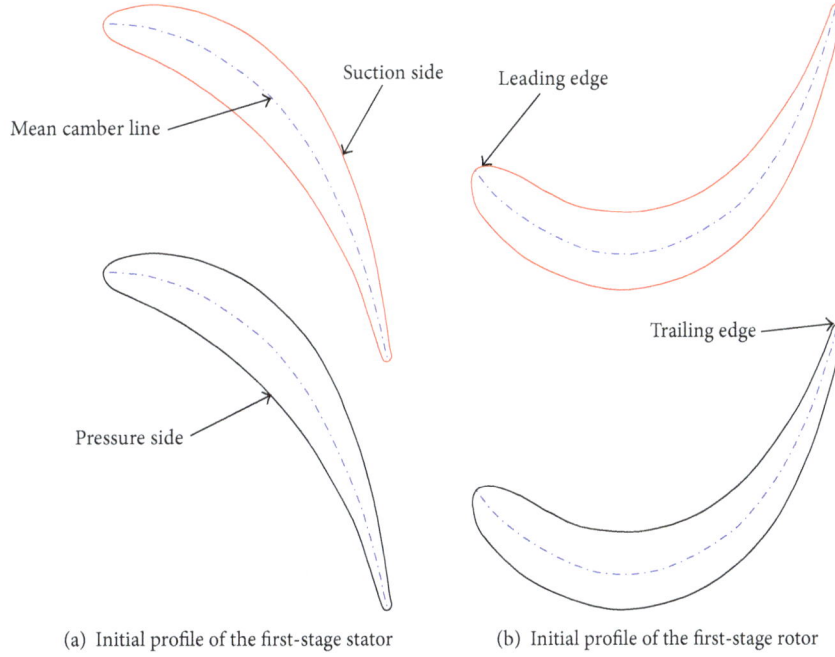

(a) Initial profile of the first-stage stator   (b) Initial profile of the first-stage rotor

FIGURE 4

TABLE 3: The main aerodynamic parameters of overall centrifugal turbine.

Overall wheel efficiency η	90.50%
Overall shaft power W (kW)	5192
Mass flow G (kg/s)	16.7
Back pressure P_N (Pa)	294000

are optimized, respectively. The optimization mathematical model of stator is expressed as

$$\text{Object:} \quad \text{Min (lossCoefficient)}$$

$$\text{Constraint:} \quad P_{te,1} \leq P_{N,1} \tag{4}$$

$$= f\left(X_{\theta[1,2]}, Y_{\theta[1,2]}, X_{t[1,2]}, Y_{t[1,2]}\right).$$

The design variables $X_{\theta[1,2]}$, $Y_{\theta[1,2]}$, respectively, are the horizontal and vertical coordinates of control points of the tangent angle. $X_{t[1,2]}$ and $Y_{t[1,2]}$ are the horizontal and vertical coordinates of control points of the blade thickness. A total of eight variables are involved in optimization (the stator

of the first stage is an example, as shown in Figure 5). $P_{te,1}$ represents the pressure at the trailing edge of the stator, and $P_{N,1}$ represents the stator back pressure of one-dimensional design (N represents stage number).

Then the rotor is added behind the optimized stator. The maximum shaft power is the object and rotor back pressure is the constraint condition. The optimization mathematical model of rotor is expressed as

$$\text{Object:} \quad \text{Max (ShaftPower)}$$

$$\text{Constraint:} \quad P_{te,2} \leq P_{N,2} \tag{5}$$

$$= f\left(X_{\theta[1,2]}, Y_{\theta[1,2]}, X_{t[1,2]}, Y_{t[1,2]}\right).$$

The same as the stator, the design variables $X_{\theta[1,2]}$ and $Y_{\theta[1,2]}$ are the horizontal and vertical coordinate points for controlling the tangent angle of the rotor mean camber line, respectively. $X_{t[1,2]}$ and $Y_{t[1,2]}$ are the horizontal and vertical coordinate points for controlling the thickness of the blade. A total of eight variables are involved in optimization (the rotor of the first stage is an example, as shown in Figure 6). $P_{te,2}$ represents the pressure at the trailing edge of the rotor, and

TABLE 4: The main geometrical parameters of the centrifugal turbine.

Stage number	1st	2nd	3rd
Stator inlet diameter $D_{s,in}$ (m)	0.597	0.758	0.959
Rotor outlet diameter $D_{r,in}$ (m)	0.754	0.955	1.208
Radial clearance δ_N (m)	0.002	0.002	0.002
Blade height H (m)	0.03	0.03	0.03
Symmetrical cone angle γ_N (degree)	0	0	0
Diameter ratio of stator $D_{s,out}/D_{s,in}$	1.12	1.12	1.12
Diameter ratio of rotor $D_{r,out}/D_{r,in}$	1.12	1.12	1.12

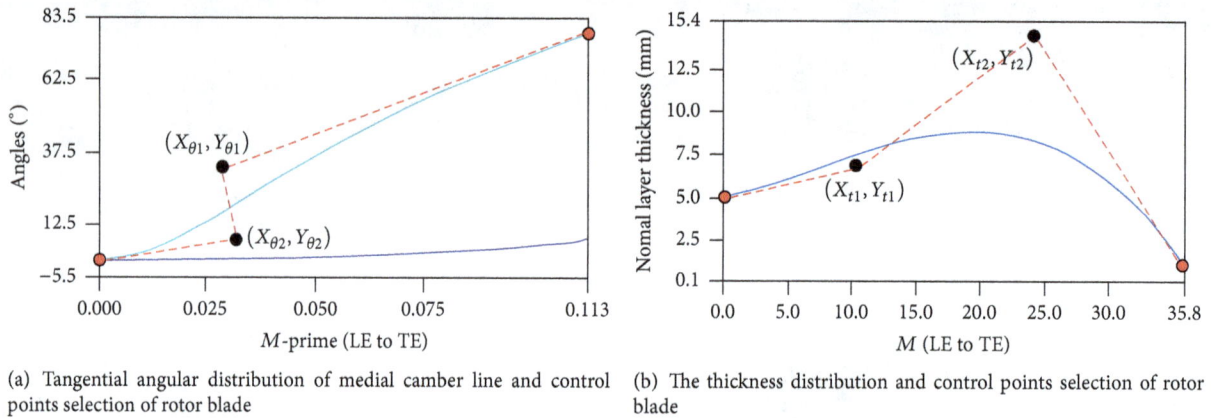

(a) Tangential angular distribution of medial camber line and control points selection of rotor blade

(b) The thickness distribution and control points selection of rotor blade

FIGURE 5

$P_{N,2}$ represents the rotor back pressure of one-dimensional design.

The optimization strategy and process are shown in Figures 7 and 8, respectively.

Figure 8 shows the blade optimization process. All of those were simulated automatically on the workbench. Firstly, the value range of optimization parameters ($X_{\theta[1,2]}$, $Y_{\theta[1,2]}$, $X_{t[1,2]}$, and $Y_{t[1,2]}$), constraint condition, and object are set artificially. Then, Design Exploration selects a set of data in the value range of optimization parameters, automatically. The blade geometry and parameterization are obtained by geometry software. After that, the blade mesh is generated by TurboGrid, and centrifugal turbine is simulated by CFX. Finally, if the results satisfy the constraint and object, the parameter values of blade control points are the optimization values. If not, the Design Exploration will select another set of data in the value range of optimization parameters, automatically. Then repeat the above steps until the conditions are met. Since the optimized stages deviate from the design condition, when they are calculated together, the leading and trailing edge thicknesses of stator and rotor are needed to be changed slightly for local adjustment, then the centrifugal turbine can obtain a better performance.

Since the SST model is used in turbomachinery, in most cases, it requires Y^+ to be very small ($Y^+ < 2$). So the layer mesh quality requirement of SST model is higher than k-epsilon model. If the SST uses the same mesh generated for k-epsilon, the layer mesh quality for SST is not good and cannot satisfy the requirement. The global size factors were changed to increase the mesh number, and the mesh quality was

improved. So the grid number of SST is more than k-epsilon. It is about 8627,000, while the grid number of k-epsilon model is 4010,000. If the multistage centrifugal turbine simulation turbulence model was SST, a lot of computing memory will be needed; meanwhile a lot of calculating time will be cost. We found that the calculation results of SST and k-epsilon models are almost the same, as shown in Table 6, so there is no difference between SST model and k-epsilon model to simulate the multistage centrifugal turbine. In order to relax the demand on computer memory and to raise the efficiency when the computational grid number is very large, the k-epsilon model is used as the turbulence model to simulate the centrifugal turbine.

The optimized centrifugal turbine is simulated by CFX, based on Navier-Stokes equations. Stators and rotors use structured grid generated in TurboGrid, and the total grid number is about 4010,000; the blade global size factors of mesh are 1.2, 1.2, 1.25, 1.25, 1.25, and 1.2 for each stator and rotor, respectively. The computational model mesh used for simulation is shown in Figure 9. The boundary conditions are inlet total pressure, total temperature, back pressure, and adiabatic wall. The turbine rotation speed is 6500 rpm. A no-slip boundary condition is applied at all the solid walls. The stator-rotor interface is Frozen Rotor. Separate periodic conditions are used for rotor and stator regions. K-epsilon is used as the turbulence model, and the fluid is Water Ideal Gas.

The optimal parameters of stator and rotor are almost consistent with one-dimensional aerodynamic design. The optimized stator and rotor of whole stages are both straight blades and the blade diagram is referred to in Figure 10.

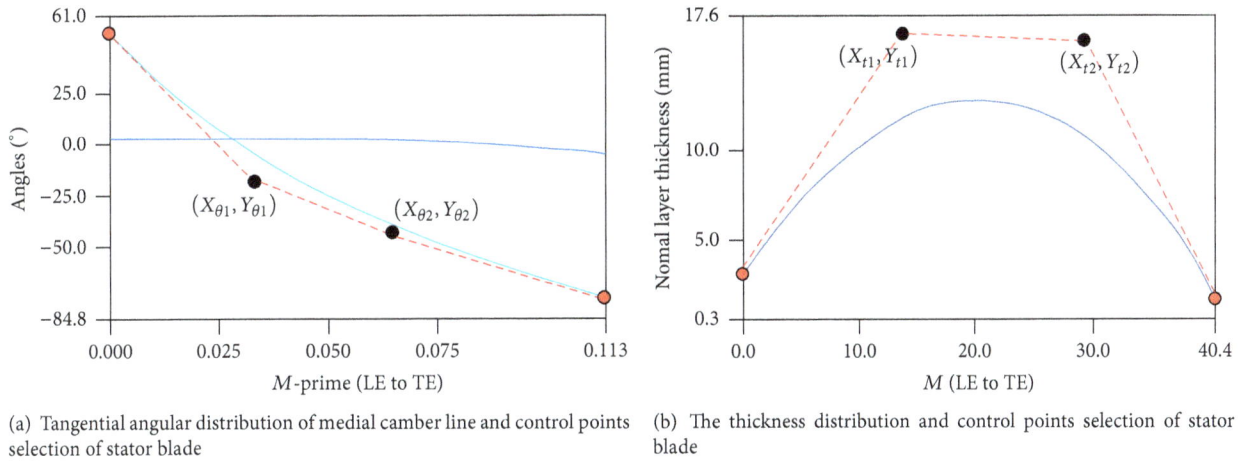

(a) Tangential angular distribution of medial camber line and control points selection of stator blade

(b) The thickness distribution and control points selection of stator blade

FIGURE 6

TABLE 5: The speed triangle data of the centrifugal turbine.

Stage number	1st	2nd	3rd
α_1 (°)	12	12	12
β_1 (°)	37.93	64.47	68.44
C_1 (m/s)	322.1	328.7	408.1
W_1 (m/s)	108.9	75.74	91.22
U_1 (m/s)	229.1	290.3	367.0
α_2 (°)	90	90	90
β_2 (°)	14.18	12.32	13.84
C_2 (m/s)	64.81	71.02	101.3
W_2 (m/s)	264.6	332.8	423.3
U_2 (m/s)	256.6	325.1	411.0

TABLE 6: The contrast of SST and k-epsilon models.

Variable	K-epsilon	SST	Δ
Wheel efficiency η	93.29%	93.01%	0.29%
Shaft power W (kW)	5387	5256.8	2.42%
Mass flow G (kg/s)	16.708	16.443	1.59%
Back pressure P_N (Pa)	291170	293999	0.97%

4. Results and Discussion

4.1. Design Condition Aerodynamic Performance. The numerical simulation results of the optimized centrifugal turbine are basically consistent with the one-dimensional aerodynamic design, which shows that the one-dimensional design program is reliable and effective. The overall performance data of centrifugal turbine is shown in Table 7. Comparison of performance data at each stage is shown in Table 8. Seen from Table 8, the performance data has some deviations at each stage. This is because the optimized stages influence each other during the matching process. So it results in some deviations in the performance data. And these deviations are gradually accumulated in the process of fluid flow. As the stage increases, the deviations are more. But, from the overall performance data seen in Table 7, the overall performance

data is within 4% deviation. This is due to the fact that there are positive and negative deviations at each stage, which can offset each other. So the overall performance data of one-dimensional design and simulation results is not quite different. From the total shaft power and wheel efficiency, the simulation results of overall centrifugal turbine are better than one-dimensional design values. The three-stage centrifugal turbine meets the design requirements and has good performance.

Figures 11–13 show the distribution of pressure, Mach number, and velocity streamline at the middle plane of the three-stage centrifugal turbine, respectively. From the diagrams, the pressure distribution is uniform; the main flow in the impeller is the pressure flow. The flow field in the cascade channel is reasonable, and the streamline is smooth and the aerodynamic performance is in line with the expectation. As

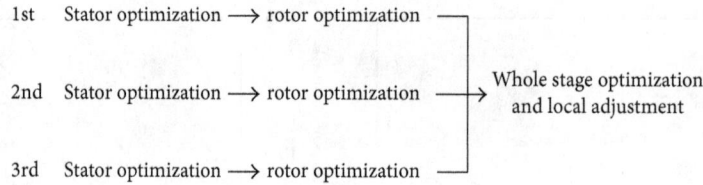

FIGURE 7: Optimization strategy.

TABLE 7: The contrast of design value and CFD value of the overall performance of the centrifugal turbine.

Variable	Design value	Simulation value	Δ
Wheel efficiency η	90.50%	93.29%	+3.08%
Shaft power W (kW)	5192	5387	+3.75%
Mass flow G (kg/s)	16.7	16.708	+0.05%
Back pressure P_N (Pa)	294000	291170	−0.96%

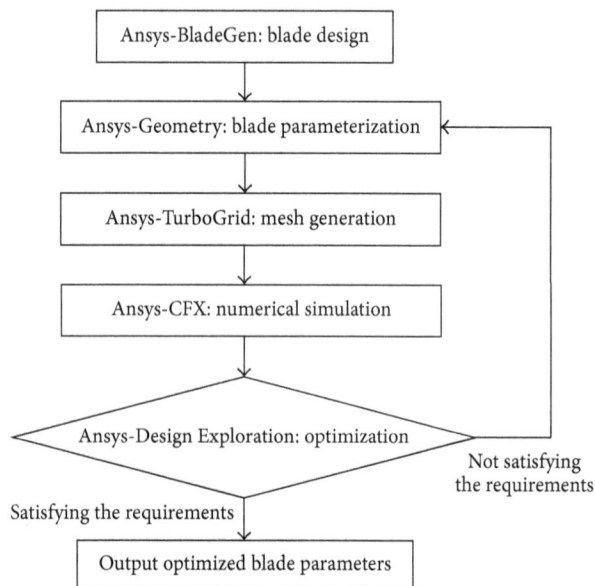

FIGURE 8: Optimization process.

can be seen from Table 8, the stator and rotor outflow angles of the first and second stage have deviation from the one-dimensional design value. From the analysis of the flow field diagram, there is a flow separation in the small area near the trailing edge; the fluid expands at the chamfered part of the stator outlet and the gap between the rotor and stator, which results in deflection of the flow angle. The flow loss caused by the separation and the deviation of the inflow angle caused by the flow deflection are not yet considered in the one-dimensional design procedure. This is also the reason why the simulation results deviate from the one-dimensional design values.

4.2. Off-Design Condition Aerodynamic Performance. Figures 14, 15, and 16 show the relationship between mass flow G, efficiency η, and shaft power W at different rotation speeds with variable pressure ratio $\pi = P_N/P_0^*$. The results indicate that as π decreases, the mass flow increases until the critical flow is reached and then the mass flow remains constant, when the speed is constant. This is because the flow process appeared supersonic, and the flow reached its maximum. And the flow passage had the blocking phenomenon, but the turbine was still able to work. There is an optimal efficiency point on the efficiency with pressure ratio curve. At both sides of the optimal efficiency point, the efficiency decreases with the increase or decrease of π. This is due to the change of the back pressure, and the enthalpy drop is changed, the speed ratio is deviated from the optimum value, the flow angel is deflected, and the blade surface occurs the flow separation. Those lead to the flow loss increase and efficiency decrease. For shaft power, the power decreases as π increases. Because of increase in back pressure, the enthalpy drop of the whole stage is reduced. Then the work capacity of working fluid is reduced. So the shaft power decreases with the increasing π. When the rotation speed changes, the optimum efficiency point moves in the direction of pressure ratio decreasing and the optimum efficiency value decreases with the increase of the rotation speed. But the efficiency trend with the pressure ratio is consistent at different speeds. For the mass flow, the flow characteristics at different rotation speeds are basically the same. It can be seen that the mass flow variation at different rotation speeds is the same and the change has little effect on mass flow. At different rotation speeds, the trends of shaft power curves are basically the same. But the curve of shaft power with the pressure ratio is steeper with the increase of rotation speed.

5. Conclusions

The three-stage centrifugal turbine is simulated by CFX with Water Ideal Gas as working fluid, drawing on the conventional turbine aerodynamic design method and off-design condition performance research method. Conclusions are as follows.

The numerical simulation results of the whole stages are basically consistent with the one-dimensional design results, and the aerodynamic performance meets the expected requirements, which indicates the reliability and effectiveness

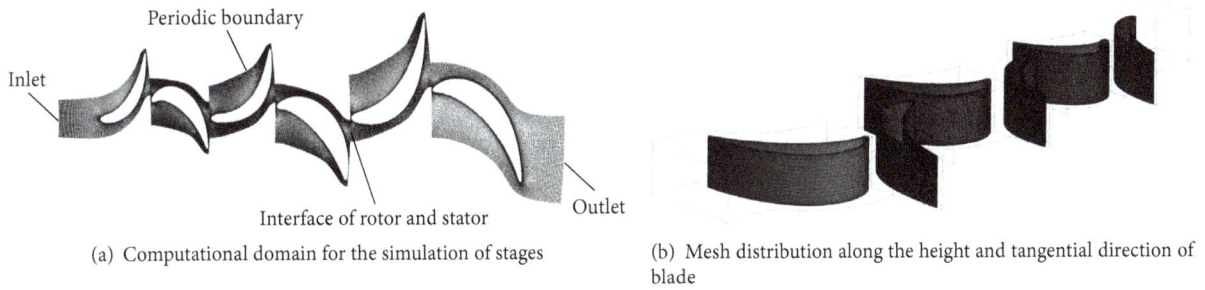

(a) Computational domain for the simulation of stages

(b) Mesh distribution along the height and tangential direction of blade

FIGURE 9

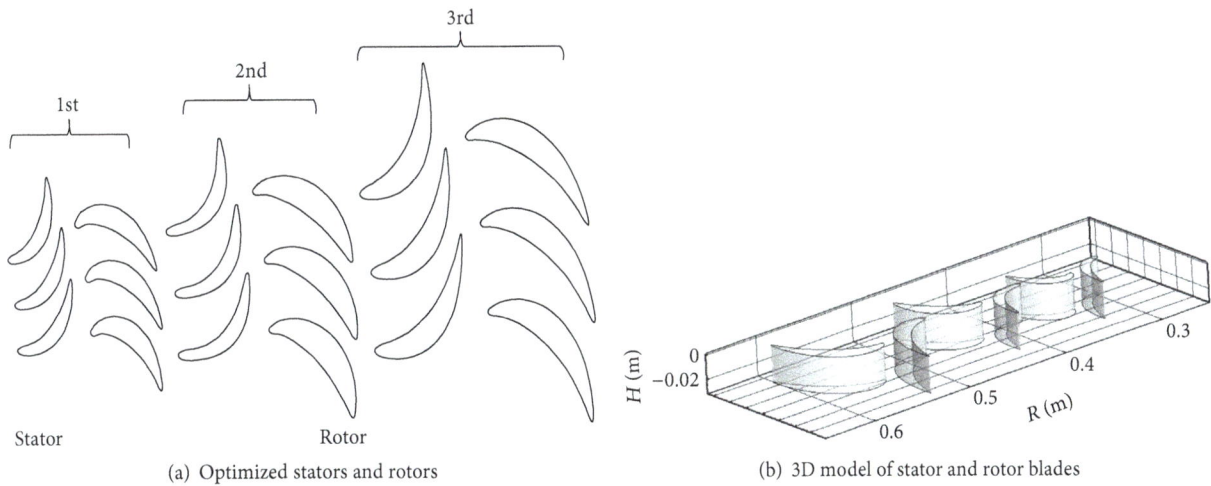

(a) Optimized stators and rotors

(b) 3D model of stator and rotor blades

FIGURE 10

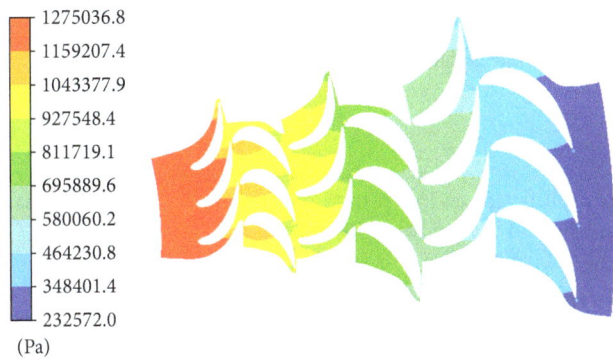

FIGURE 11: Static pressure distribution.

FIGURE 12: Mach number distribution.

of the centrifugal turbine design. Because each stage of centrifugal turbine is optimized, the simulated overall efficiency is 3.08% higher than the one-dimensional design efficiency and the shaft power is 3.75% more than the one-dimensional design shaft power. At the design condition, the streamline of three-stage centrifugal steam turbine is smooth at cascade flow channel, the pressure distribution is uniform, and the flow field is reasonable.

At the off-design condition, when the speed is constant, the pressure ratio π reaches the critical ratio and the mass flow reaches the maximum. If π continues to decrease, the maximum flow value remains the same, and the flow passage

has the blocking phenomenon. But the turbine is still able to work. At the same π, the speed decreases; then the mass flow increases; but the impact is quite small.

At a constant speed, the efficiency is highest at optimum pressure ratio π. The pressure ratio of the optimum efficiency agrees well with that of the one-dimensional design. When the speed decreases, the efficiency curve moves to the place where π increases, and the corresponding maximum efficiency increases.

The shaft power curve trend with pressure ratio is similar at different rotation speeds. But as the rotation speed

TABLE 8: The contrast of design value and CFD value of each stage performance of the centrifugal turbine.

| Variable | Stage number | | | | | | | | |
| | 1st | | | 2nd | | | 3rd | | |
	Design value	Simulation value	Δ (%)	Design value	Simulation value	Δ (%)	Design value	Simulation value	Δ (%)
Ma	0.55	0.55	0	0.58	0.52	−10.34	0.76	0.82	+7.89
Ω	0.32	0.30	−6.25	0.45	0.51	+13.33	0.47	0.42	−10.64
χ	0.69	0.63	−8.69	0.86	0.73	−15.12	0.87	0.67	−22.99
W (kW)	1205	1231	+2.16	1551	1384	−10.77	2436	2771	+13.75
η (%)	91.20	93.34	+2.35	90.97	87.36	−3.97	90.98	89.59	−1.53
α_1 (°)	12.0	12.3	+2.5	12.0	14.2	+18.3	12.0	12.0	0
α_2 (°)	90.0	78.9	−12.3	90.0	80.3	−10.8	90.0	90.0	0

FIGURE 13: Streamline distribution.

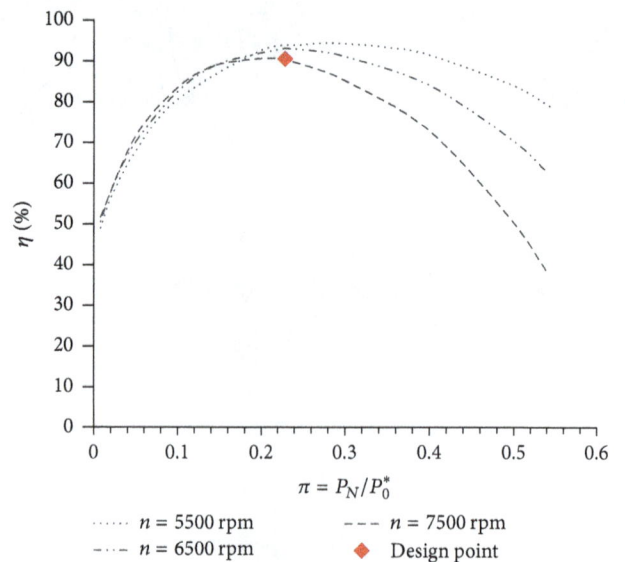

- ······ n = 5500 rpm
- ─ ─ ─ n = 7500 rpm
- ─·─· n = 6500 rpm
- ◆ Design point

FIGURE 15: The variation of efficiency with pressure ratio at different speeds.

numerical simulation results of the three-stage centrifugal turbine and the analysis of the off-design conditions.

Nomenclature

T: Temperature, K
P: Pressure, Pa
h: Enthalpy, kJ/kg
s: Entropy, kJ/(K·kg)
n: Rotational speed, r/min
G: Mass flow, kg/s
b: Diameter ratio
D: Diameter, m
R: Radius, m
c: Absolute velocity, m/s
w: Relative velocity, m/s
u: Circumferential velocity, m/s
H: Height, m
N: Stage number
W: Shaft power, kW
Ma: Mach number.

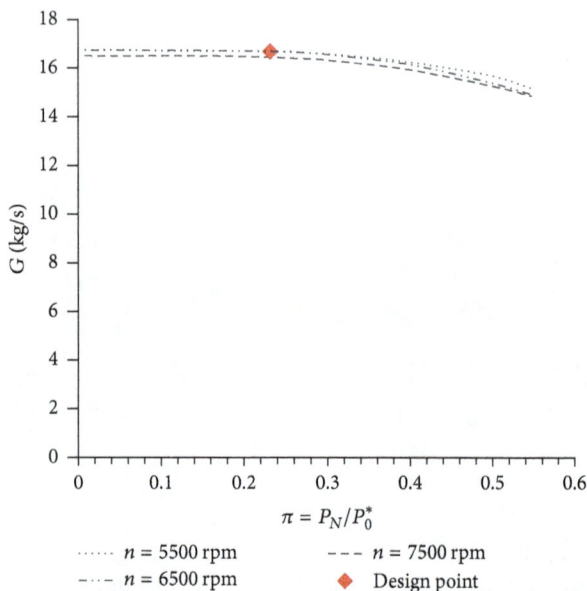

- ······ n = 5500 rpm
- ─ ─ ─ n = 7500 rpm
- ─·─· n = 6500 rpm
- ◆ Design point

FIGURE 14: The variation of mass flow with pressure ratio at different speeds.

increases, the power curve drops faster when the pressure ratio increases.

Overall, it can be seen that the centrifugal turbine has a wide range and good off-design performance, from the

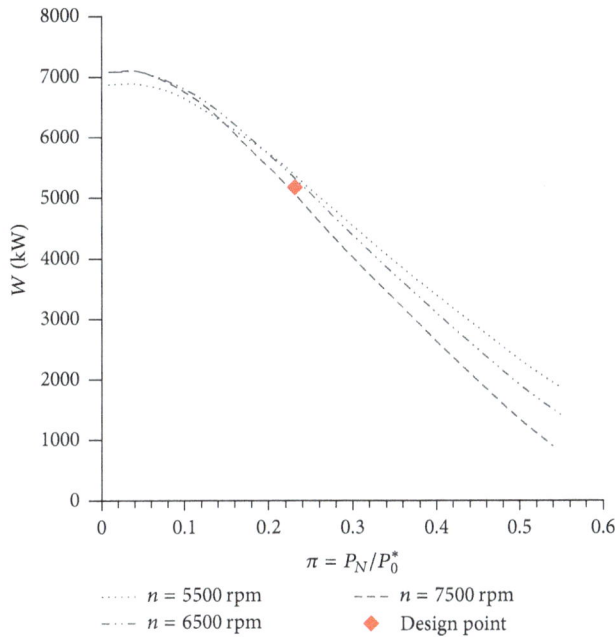

FIGURE 16: The variation of shaft power with pressure ratio at different speeds.

Greek Letters

ρ: Density, kg/m^3
γ: Radial symmetry cone angle, °
δ: Radial clearance between stator and rotor, m
α: Absolute angle, °
β: Relative angle, °
φ: Stator flow coefficient
ψ: Rotor flow coefficient
χ: Speed ratio
η: Wheel efficiency
Ω: Reaction degree
π: Pressure ratio.

Subscripts

0: First stage
in: Inlet
out: Outlet
s: Stator
k: Stage number.

Superscripts

∗: Stagnation condition.

Conflicts of Interest

The authors declare that they have no conflicts of interest.

Acknowledgments

This work was supported by National Natural science Foundation of China no. 51536006 and Shanghai Science and Technology Committee with Grant no. 17060502300 and funded by Opening Project of Shanghai Key Laboratory of Multiphase Flow and Heat Transfer in Power Engineering no. 13DZ2260900.

References

[1] E. Coomes, R. Dodge, and D. Wilson, "Design of a high power-density ljüngstrom turbine using potassium as a working fluid," in *Proceedings of the 21st Intersociety Energy Conversion Engineering Conference*, vol. 3, San Diego, CA, USA, 1986.

[2] L. Jing and S. Peng, "Modification design and flowfield analysis of a centrifugal turbine stage with discrete nozzles," *Journal of Aerospace Power*, vol. 23, no. 6, pp. 1047–1053, 2008.

[3] L. Yin-Ge, T. Xin, and L. Xian-Qiao, "The thermal design and analysis of centrifugal turbine," *Journal of Engineering Thermophysics*, vol. 37, no. 10, pp. 2103-2010, 2016.

[4] T. Xin, L. Yin-Ge, L. Xian-Qiao et al., "Research on off-design characteristics of centrifugal turbine," *Journal of Engineering Thermophysics*, vol. 37, no. 6, pp. 1201–1208, 2016.

[5] T. Xin, L. Hui, S. Yan-Ping et al., "Design of the auxiliary components of the centrifugal turbine," *Journal of Engineering Thermophysics*, vol. 0253-231X, pp. 04–0000, 2017.

[6] M. Pini, G. Persico, E. Casati et al., "Preliminary design of a centrifugal turbine for organic rankine cycle applications," *Engineering for Gas Turbines & Power*, vol. 135, no. 4, 2013.

[7] E. Casati, S. Vitale, M. Pini, G. Persico et al., "Centrifugal turbines for mini-organic rankine cycle power systems," *Journal of Engineering for Gas Turbines & Power*, vol. 136, no. 136, Article ID 122607, 11 pages, 2014.

[8] G. Persico, M. Pini, V. Dossena, and P. Gaetani, "Aerodynamic design and analysis of centrifugal turbine cascades," in *Proceedings of the ASME Turbo Expo: Turbine Technical Conference and Exposition, GT '13*, vol. 6C, 12 pages, San Antonio, Texas, Tex, USA, 2013.

[9] G. Persico, M. Pini, V. Dossena et al., "Aerodynamics of centrifugal turbine cascades," *Journal of Engineering for Gas Turbines & Power*, vol. 137, no. 11, Article ID 112602, 2015.

[10] M. Pini, A. Spinelli, G. Persico, and S. Rebay, "Consistent look-up table interpolation method for real-gas flow simulations," *Computers & Fluids*, vol. 107, pp. 178–188, 2015.

[11] L. Yan-Sheng and L. Gui-Lin, *Radial Turbine and Centrifugal Compressor*, Machinery Industry Press, Bei Jing, China, 1987.

[12] W. Xin-Jun and L. Liang, *Principle of Steam Turbine*, XI'an Jiaotong University Press, Xi'an, Shaanxi, China, 2014.

[13] S. L. Dixon and C. A. Hall, *Fluid Mechanics and Thermodynamics of Turbomachinery*, Elsevier Inc Press, 2010.

7

Enhancing LVRT Capability of DFIG-Based Wind Turbine Systems with SMES Series in the Rotor Side

Xiao Zhou, Yuejin Tang, and Jing Shi

State Key Laboratory of Advanced Electromagnetic Engineering and Technology, R&D Center of Applied Superconductivity, Huazhong University of Science and Technology, Wuhan 430074, China

Correspondence should be addressed to Jing Shi; shijing@mail.hust.edu.cn

Academic Editor: Lei Chen

The necessary Low Voltage Ride Through (LVRT) capability is very important to wind turbines. This paper presents a method to enhance LVRT capability of doubly fed induction generators- (DFIGs-) based wind turbine systems with series superconducting magnetic energy storage (SMES) in the rotor side. When grid fault occurs, series SMES in the rotor side is utilized to produce a desired output voltage and absorbs energy. Compared with other methods which enhance LVRT capability with Superconducting Fault-Current Limiter-Magnetic Energy Storage System (SFCL-MESS), this strategy can control the output voltage of SMES to suppress the transient AC voltage component in the rotor directly, which is more effective and rapid. Theoretical study of the DFIG under low voltage fault is developed; the simulation results are operated by MATLAB/Simulink.

1. Introduction

Wind energy generation has experienced a fast development in the last two decades [1]. DFIG is the most widely used wind turbine, but it is vulnerable to low voltage fault. Many wind turbines trip off from the grid due to the lack of LVRT capability [2].

There are many enhanced ways to solve this problem. Crowbar is a commonly used protection method to implement the LVRT capability [3–5]; it can effectively suppress overvoltage and overcurrent but it needs to absorb a large amount of reactive power from the grid, which will do harm to the grid. Improving control strategy of grid converter or rotor converter has been studied by many researchers [6–8], but it cannot work effectively when deep voltage drop occurs. Energy storage device is suitable for dynamic matching of intermittent wind power. The use of supercapacitors or batteries for WTGs has been studied by some researchers [9–11]. SMES used in DFIGs to reduce power fluctuation and alleviate the influence on power quality has been studied in some papers [12]. The application of SFCL-SMES in the rotor side of the DFIG has been carried out. But it is connected with the DFIG through a three-phase diode rectifier; thus the SFCL-MES is inserted into rotor circuit as an impedance, which will not provide help in enhancing the controllability of the RSC.

This paper presents a new method to enhance LVRT capability of DFIG with SMES. SMES, with its high efficiency and quick response ability to power compensation, could be a good choice for ESD in DFIG. Most studies applied ESD to solve the problem which are usually in parallel with the DC side. The proposed method in this paper is to generate a desired output voltage utilizing series SMES in the rotor side. Through the analysis of the rotor side voltage under low voltage fault, counter-electromotive force generated by SMES can counteract overvoltage and then overcurrent quickly and effectively. The mathematical model of DFIG will be illustrated first. Then, behavior of the DFIG in normal situation and under low voltage fault will be analyzed and the control strategy of VSC-SMES will be introduced.

2. Math Model of DFIG

The stator windings of the DFIG wind turbine are directly connected to the grid. The rotor winding is connected to the grid via back-to-back transformers, where the converter near the grid side is called grid sided converter (GSC) while near the rotor side converter it is called rotor sided converter (RSC). The stator voltage is provided by the grid. The rotor voltage is provided by the back-to-back transformers, which can be adjusted in frequency, phase, and amplitude.

In this paper, the control system is implemented considering a stator voltage oriented (SVO) control philosophy in the d/q reference frame [13]. In this kind of systems, the d/q-axes are aligned with the stator voltage. Thus, the park model of the DFIG can be expressed as follows:

$$
\begin{aligned}
u_{dr} &= R_r i_{dr} + \frac{d\psi_{dr}}{dt} - (\omega_1 - \omega_r)\psi_{qr} \\
u_{qr} &= R_r i_{qr} + \frac{d\psi_{qr}}{dt} + (\omega_1 - \omega_r)\psi_{dr} \\
u_{ds} &= R_s i_{ds} + \frac{d\psi_{ds}}{dt} - \omega_1 \psi_{qs} \\
u_{qs} &= R_s i_{qs} + \frac{d\psi_{qs}}{dt} + \omega_1 \psi_{ds} \\
\psi_{ds} &= L_s i_{ds} + L_m i_{dr} \\
\psi_{qs} &= L_s i_{qs} + L_m i_{qr} \\
\psi_{dr} &= L_m i_{ds} + L_r i_{dr} \\
\psi_{qr} &= L_m i_{qs} + L_r i_{qr},
\end{aligned}
\tag{1}
$$

where ψ is the magnetic flux, L is the inductances, u and i are voltage and current, respectively, R is the resistance, subscripts s, r, m represent the stator, rotor, and mutual, respectively, and ω_1 and ω_r are the synchronous and rotating angular frequencies ($\omega_s = \omega_1 - \omega_r$).

Once there is a grid fault, the stator voltage will change immediately because it is connected to the grid directly. But, according to the principle of flux conservation, the amplitude of stator flux will not transition. At the moment when the voltage drops, the stator flux linkage contains an AC component and a DC component, which will induce overvoltage in rotor side with different frequencies.

As analyzed in [14], the rotor back EMF voltage can be expressed as follows:

$$
\begin{aligned}
u_{rdq} = {}& j\frac{L_m}{L_s}\omega_s\psi_{s2dq} \\
& - j\frac{L_m}{L_s}\omega_r\left(\psi_{s0dq} - \psi_{s2dq}\right)e^{-\sigma t}e^{-j\omega_s t},
\end{aligned}
\tag{2}
$$

where ψ_{s0dq} and ψ_{s2dq} are normal-state stator flux and fault-state stator flux. The rotor back EMF voltage is mainly decided by part two of (2) because the slip frequency is much smaller than rotor frequency. And the second part in (2) is decided by the stator flux which can be several times the default value.

3. Methodology of DFIG and SMES Model

Unlike SFCL-MES, this paper controls the SMES output voltage through converter switch control, which could suppress the overvoltage in rotor side.

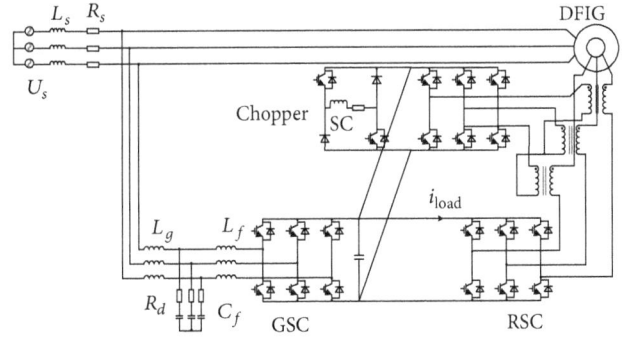

FIGURE 1: Topology of the whole system.

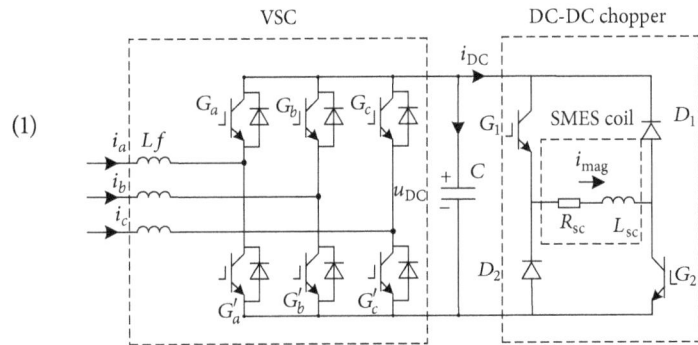

FIGURE 2: Topology of VSC-SMES.

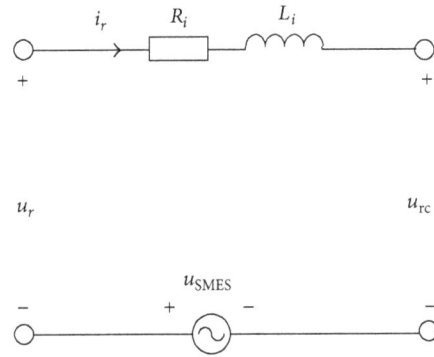

FIGURE 3: Equivalent circuit of the series SMES.

The topology of the whole system is shown in Figure 1. SMES is connected to the rotor side by a Voltage Source Converter (VSC). Topology of VSC-SMES is given in Figure 2. SMES can exchange active and reactive power with the system quickly and independently [13].

In the case of normal power supply, SMES is in standby mode. In this case, the series SMES will introduce resistance and reactance to DFIG. Thus the rotor voltage will not equate to rotor converter voltage. As shown in Figure 3, R_i and L_i are the introduced resistance and reactance, respectively. In order to eliminate the influence of the rotor voltage by series SMES, the control of series SMES is to realize

$$
u_r = u_{rc}.
\tag{3}
$$

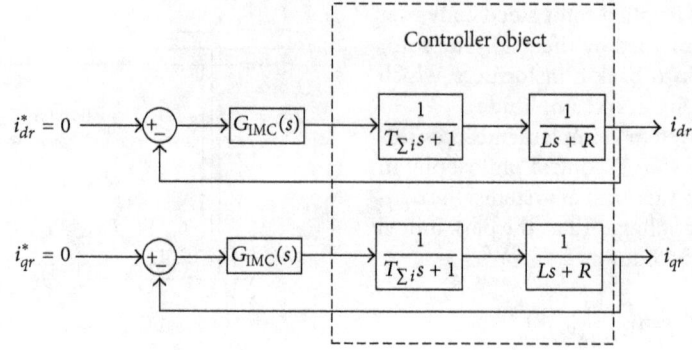

FIGURE 4: The current inner loop control stratagem.

When low voltage fault occurs, the SMES outputs the voltage to suppress the rotor back EMF voltage induced by stator; thus the rotor current is supposed to be zero when fault occurs and the rotor back EMF voltage induced by stator $u^*{}_r$ can be shown as follows:

$$u^*{}_{dr} = L_m \frac{di_{ds}}{dt}$$

$$u^*{}_{qr} = L_m \frac{di_{qs}}{dt}. \tag{4}$$

The stator current under grid fault after doing Laplace transform can be expressed as follows [15]:

$$i_{ds} = \frac{(L_s + R_s) u_{ds} + \omega_1 L_s u_{qs}}{\left(L_s{}^2 s^2 + 2L_s R_s s + R_s{}^2 + \omega_1{}^2 L_s{}^2\right)}$$

$$- \frac{\left(L_s s^2 + R_s s + \omega_1{}^2 L_s\right) L_m i_{dr} - R_s \omega_1 L_m i_{qr}}{\left(L_s{}^2 s^2 + 2L_s R_s s + R_s{}^2 + \omega_1{}^2 L_s{}^2\right)}$$

$$i_{qs} = \frac{(L_s + R_s) u_{qs} - \omega_s L_s u_{ds}}{\left(L_s{}^2 s^2 + 2L_s R_s s + R_s{}^2 + \omega_1{}^2 L_s{}^2\right)}$$

$$- \frac{\left(L_s s^2 + R_s s + \omega_1{}^2 L_s\right) L_m i_{qr} + R_s \omega_1 L_m i_{dr}}{\left(L_s{}^2 s^2 + 2L_s R_s s + R_s{}^2 + \omega_1{}^2 L_s{}^2\right)}. \tag{5}$$

Considering that $R^2{}_s, R_s \omega_s L_m$ are approximately 0, the equation can be simplified as:

$$i_{ds} = \frac{1}{L_s{}^2} \frac{L_s s + R_s}{s^2 + 2\left(R_s/L_s\right) + \omega_1{}^2} \cdot u_{ds} - \frac{L_m}{L_s} \cdot i_{dr}$$

$$i_{qs} = \frac{1}{L_s{}^2} \frac{(-L_s \omega_1)}{s^2 + 2\left(R_s/L_s\right) + \omega_1{}^2} \cdot u_{ds} - \frac{L_m}{L_s} \cdot i_{qr}. \tag{6}$$

According to the above analysis, if the rotor side converter could provide relative excitation voltage to $u^*{}_r$, the low LVRT capability of DFIG will be enhanced. Thus set $u^*{}_r$ as the reference voltage of VSC-SMES under grid fault. The VSC-SMES reference voltage calculation process is presented in Figure 5.

3.1. SMES Control

3.1.1. VSC Control. As shown in Figure 2, through the park transformation, the mathematical model of VSC in the dq reference frame is expressed as follows:

$$L \frac{di_d}{dt} = -Ri_d + \omega Li_q + u_{sd} - u_{DC} s_d$$

$$L \frac{di_q}{dt} = -Ri_q - \omega Li_d + u_{sq} - u_{DC} s_q \tag{7}$$

$$C \frac{du_{DC}}{dt} = \frac{3}{2} \left(i_q s_q + i_d s_d\right) - i_{chopper}.$$

The current inner loop control of VSC adopts the internal model control strategy [14]; the control block diagram is shown in Figure 4. The closed-loop transfer function of the current inner loop is

$$W_{ci}(s) = \frac{1}{T_f^2 s^2 + 2\xi T_f s + 1}. \tag{8}$$

For achieving the purpose of rapid response of the current inner loop, let T_f be equal to 0.0002 s.

3.1.2. DC-DC Chopper Control. Chopper has two basic modes of operation: (1) magnetizing mode; (2) releasing magnetic mode. The mathematical model of chopper is

$$L_{sc} \frac{di_{mag}}{dt} = -R_{sc} i_{mag} + (D_1 + D_2 - 1) u_{DC}$$

$$C \frac{du_{DC}}{dt} = -(D_1 + D_2 - 1) i_{mag} + i_{DC}, \tag{9}$$

where i_{mag} is the current of magnet and D is the switch state of each switch; if switch 1 is on, D_1 is equal to 1.

In normal operation, the control of chopper is to limit the magnet current in a setting scale. When low voltage fault happens, the control of chopper is to maintain the stability of the DC voltage. The control strategy of SMES is shown in Figure 5.

3.2. RSC Control.

In normal operation, the control of d/q component of rotor current is to control the active and

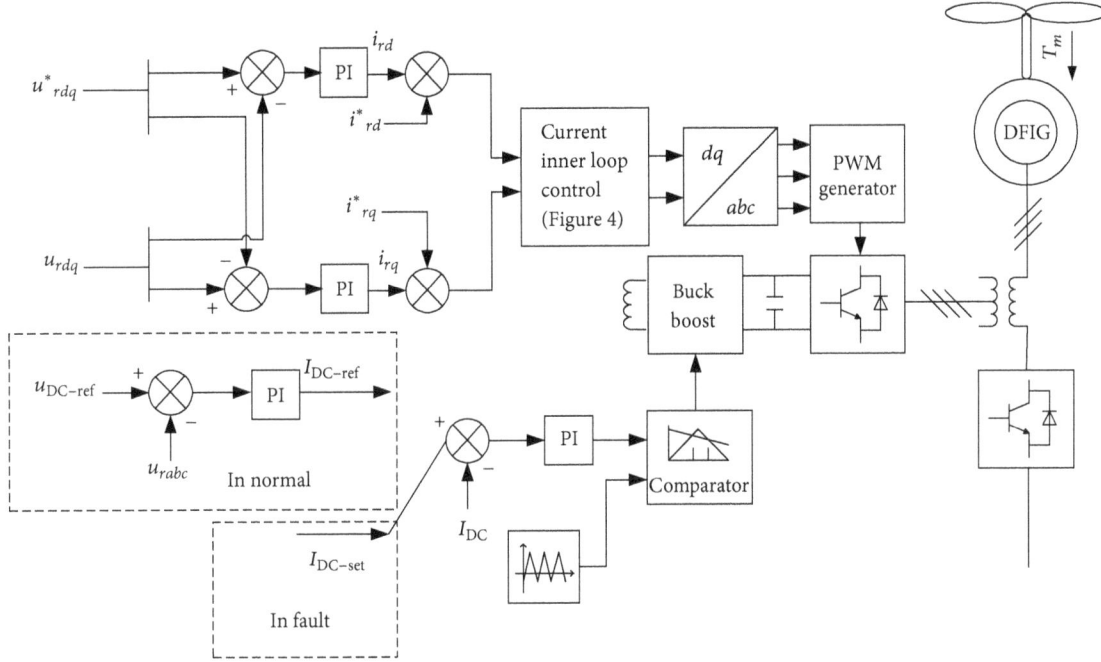

FIGURE 5: The control stratagem for SMES.

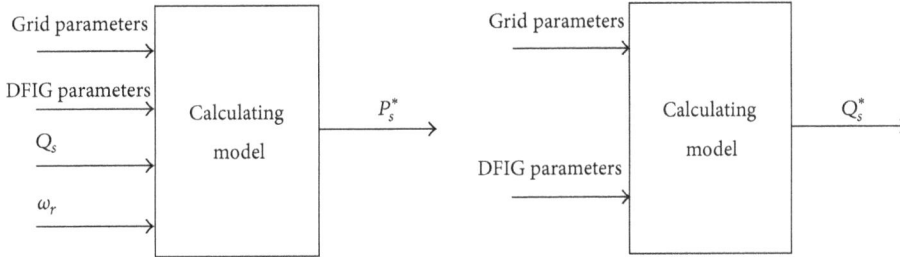

FIGURE 6: Reference power calculation model of DFIG.

reactive power of DFIG. Generally, the purpose of control system is to obtain the max wind power.

The essence of the maximum wind energy tracking is to control the DFIG output active power and control electromagnetic torque to achieve the best speed control. When the mechanical output of the wind turbine achieves maximum output power, the reference active power of DFIG P_s^* is equal to stator power P_s, as shown as follows:

$$P_s^* = \frac{1}{1-s}\left(P_{\max} - P_{ms}\right) - P_{cus}, \tag{10}$$

where P_{ms} is the mechanical loss of wind turbine and P_{cus} is the copper loss of stator, which is related to I_s and R_s. After doing identity transformation, (9) can be expressed as follows:

$$A\left(P_s^*\right)^2 + BP_s^* + C = 0$$

$$A = \frac{R_s}{3U_s^2}, \ B = 1, \ C = \frac{1}{s-1}\left(k_w\omega_w^3 - P_{ms}\right) + \frac{R_s}{3U_s^2}Q_s. \tag{11}$$

The reactive reference power value Q_s^* can be calculated by achieving the lowest loss of DFIG or improving the system's ability to regulate power. In this paper, Q_s^* is calculated to achieve the lowest loss of DFIG. Thus the expression of Q_s^* is shown as follows:

$$Q_s^* = -\frac{3X_s R_r U_s^2}{R_s X_m^2 + R_r R_s^2 + R_r X_s^2}. \tag{12}$$

The reference power calculation model of DFIG is shown in Figure 6.

When a fault occurs, the control strategy of rotor converter is irrelevant. Considering SMES will absorb the wind energy, i_{dr}^* and i_{qr}^* can be set to zero to prevent the energy through the converter. Thus, the capacitor in DC side will not absorb a lot of energy and its voltage will not increase sharply. The control stratagem of RSC, in normal situation and when voltage drop happens, is shown in Figure 7.

3.3. GSC Control. The purpose of GSC control is to maintain DC capacitor voltage stability and control the factor of input

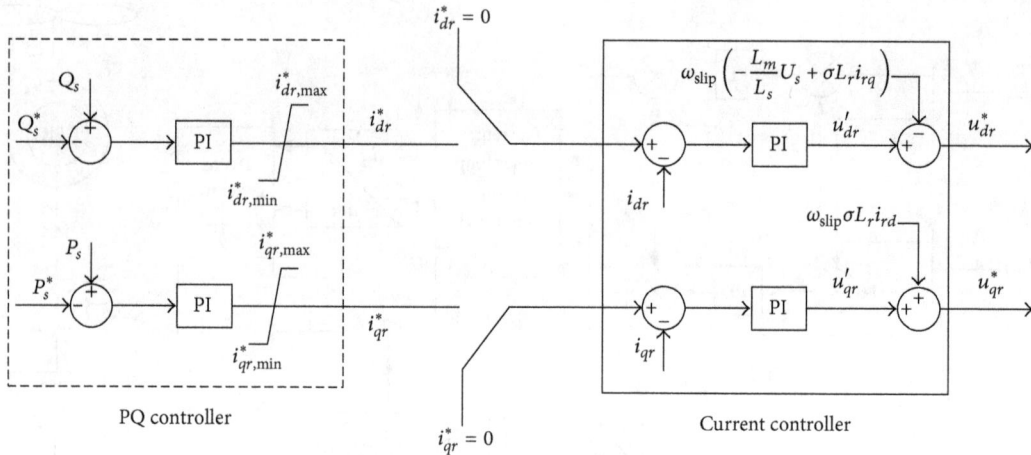

FIGURE 7: Control block diagram of the RSC.

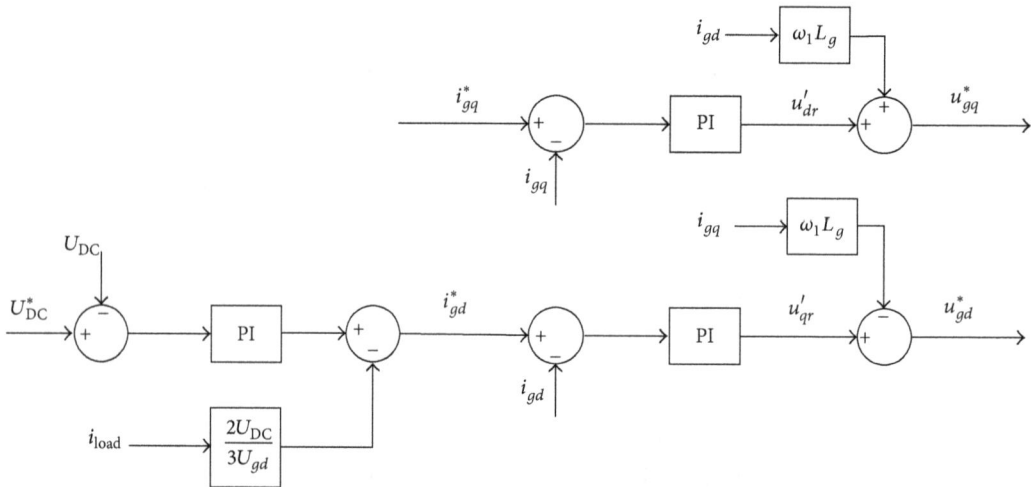

FIGURE 8: Control block diagram of the GSC.

power. The GSC control system is divided into two parts: the outer voltage loop and the inner current loop, which is shown in Figure 8. Compensation of load disturbances is achieved by feedforward of the load current i_{load}. System is decoupled by introduced current state feedback $w_1 L_g i_{gd}$, $w_1 L_g i_{gq}$.

4. Economic Analysis

Compared to other energy storage devices, SEMS has the advantage of high response speed, high efficiency, high-power density, and high cycle life characteristics. But the application of SMES in DFIG will increase the cost of the system. Therefore, the capital cost of SMES applied in DFIG is an indispensable part of DFIG's technical performance. In terms of energy storage devices, the capital costs contain energy cost $/kWh, power cost $/kW, and cycle cost. As the energy storage device is used in DFIG, high-power characteristic is necessary. The power costs of SMES are less

than their energy costs, and this is an indication that they are suitable for high-power applications. Table 1 shows the cost comparison for the storage technologies [16]. As can be seen from the table, SMES has a lower power cost compared to pumped-hydro storage, lithium-ion batteries and other energy storage devices. In addition, the long life of SMES makes it possible to participate in the protection of the entire life of the DFIG. In general, considering the economic and other technical performances, SMES is a good choice for ESD in DFIG.

5. Simulation Analysis and Conclusion

The simulations are carried out in MATLAB/Simulink. The simulation parameters are shown in Table 2. The simulation parameters of SMES are shown in Table 3. The purpose of the simulation is to observe the performance of the back-to-back converter when the grid fault occurs.

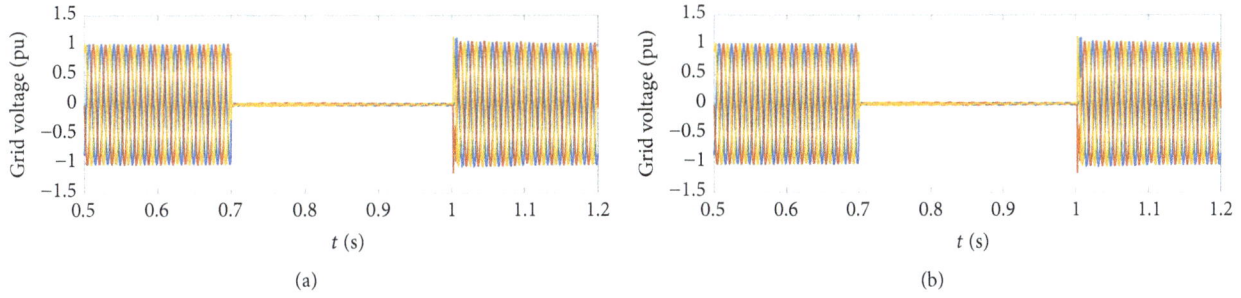

FIGURE 9: The grid voltage.

TABLE 1: Cost comparison for the storage technologies.

ESD	Capacity	Life	Power cost ($/kW)	Energy cost ($/kWh)
Pumped-hydro storage	100–2000 MW	>20 years	2000–4300	10–100
Compressed-air energy storage	100–300 MW	>20 years	400–1000	2–50
Flywheel energy storage	5 kW–1.5 MW	>15 years	250–350	1000–10000
Lithium-ion batteries	1–1000 kW	3000 cycles	1000–3000	1000–3800
Supercapacitors	1–100 kW	>50000 cycles	100–300	300–2000
Superconducting magnetic energy storage	10–500 kW	>20 years	200–400	500–1000
Vanadium redox battery	<10 MW	10–30 years	600–1500	150–1000

TABLE 2: DFIG parameters.

Symbol	Name	Quantity
S_N	Capacity	1.5 MW
f_N	System frequency	60 Hz
R_s	Stator resistance	0.016 p.u.
L_{ls}	Stator leakage inductance	0.16 p.u.
R_r	Rotor resistance	0.016 p.u.
L_{lr}	Rotor leakage inductance	0.16 p.u.
L_m	Magnetizing inductance	2.9 p.u
S_b	Base capacity	1.5/0.9 MVA
f_b	Base frequency	60 Hz
V_{s_nom}	Base stator voltage (V_{rms})	575 V
V_{r_nom}	Base rotor voltage (V_{rms})	1975 V

TABLE 3: SMES parameters.

Parameter	Value
HTS material	Bi 2223
Capacity	40 kJ/600 kW
Magnet inductance (L)	0.5 H
Magnet resistance	$2 \times 10^{-5}\ \Omega$
Filter capacitance (C)	0.02 F
Series transformer ratio	1
Transformer resistance	$1 \times 10^{-4}\ \Omega$

The transient grid voltage fault occurs at 0.7 s and lasts for 300 ms. The grid voltage during fault drops to 0. The grid voltage is shown in Figure 9. Figures 10 and 11 compare the currents of stator converter and rotor converter with and without SMES. When there is no SMES, in the moment of low voltage faults, the overcurrent of the stator and rotor side converters is almost five times larger than the normal operating range. With the proposed method, the currents of back-to-back converters are always in the affordable range. Moreover, as shown in Figure 13, the DC capacitor voltage is well suppressed during the fault. The output voltage of SMES in d/q frame is shown in Figure 12. The base voltage is the same as base rotor voltage. Figure 14 shows the transient SMES current response.

Compared to traditional control stratagems, the new control strategy with SMES can suppress the overcurrent in stator and rotor within the affordable range even when grid voltage drops to 0. The DC bus voltage can also be limited effectively.

This paper presents a new control stratagem to enhance the LVRT capability of DFIG with series SMES. By controlling the SMES output voltage with the proposed controller,

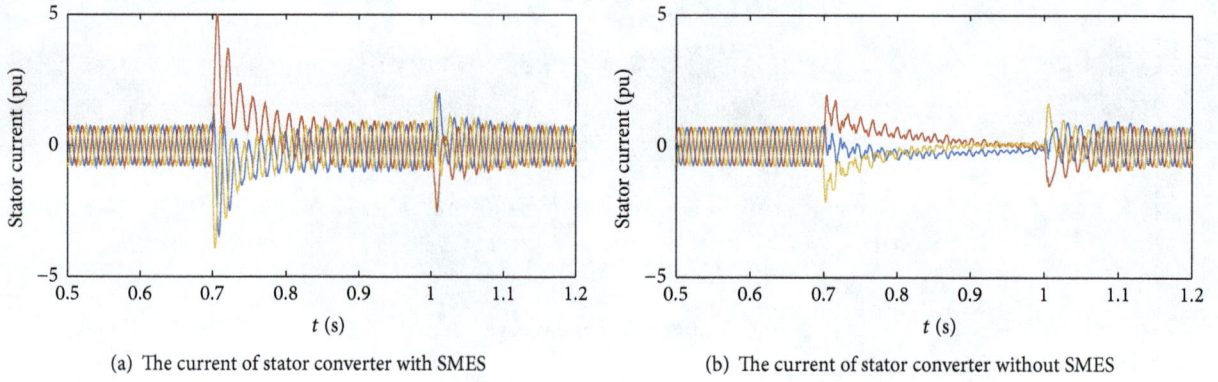

(a) The current of stator converter with SMES

(b) The current of stator converter without SMES

FIGURE 10

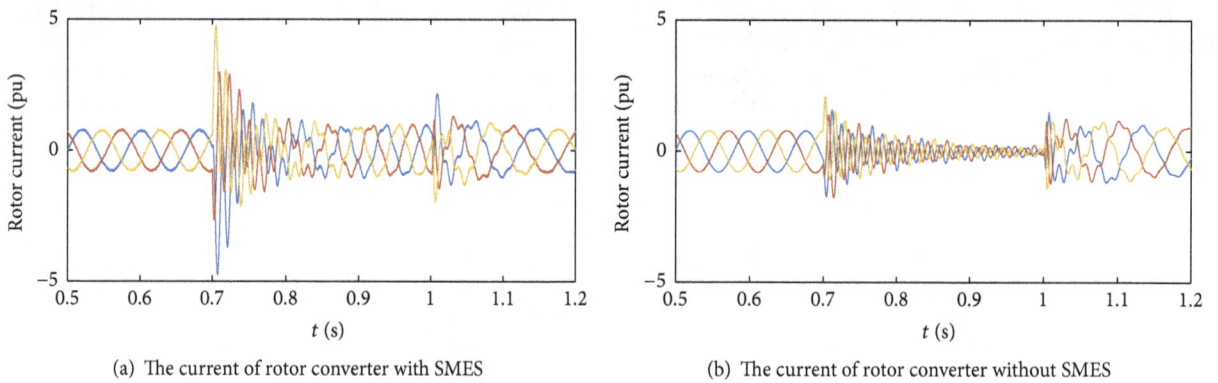

(a) The current of rotor converter with SMES

(b) The current of rotor converter without SMES

FIGURE 11

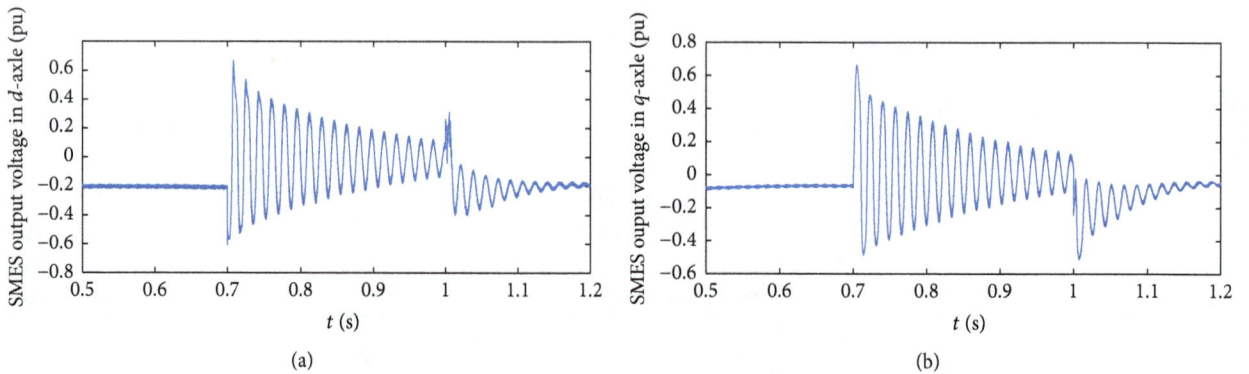

(a)

(b)

FIGURE 12: The SMES output voltage.

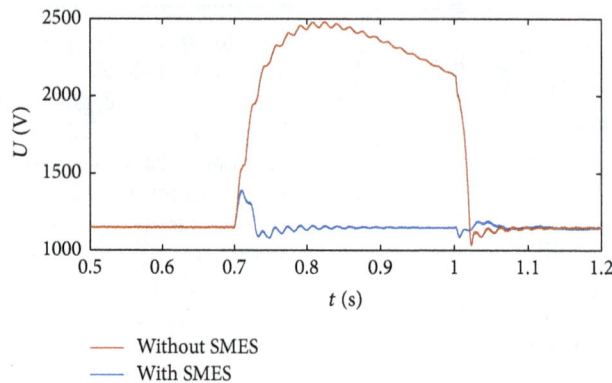

Without SMES
With SMES

FIGURE 13: The DC voltage.

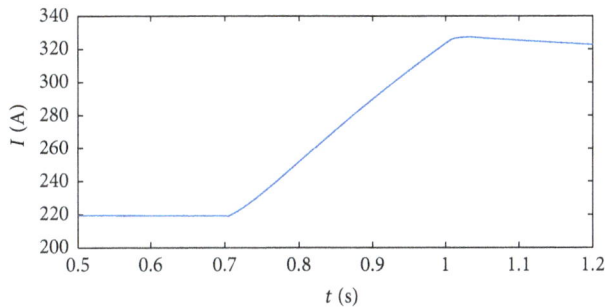

FIGURE 14: The magnet current of SMES.

the converter currents and capacitor voltage in fault are suppressed effectively. The method has been validated by simulations in MATLAB/Simulink.

Conflicts of Interest

There are no conflicts of interest regarding the publication of this paper.

References

[1] Y.-W. Shen, D.-P. Ke, W. Qiao, Y.-Z. Sun, D. S. Kirschen, and C. Wei, "Transient reconfiguration and coordinated control for power converters to enhance the LVRT of a DFIG wind turbine with an energy storage device," *IEEE Transactions on Energy Conversion*, vol. 30, no. 4, pp. 1679–1690, 2015.

[2] J. Liu, H. Zhang, B. Yang, and Y. Min, "Fault ride through under unbalanced voltage sag of wind energy conversion system using superconducting magnetic energy storage," in *Proceedings of the 2nd IEEE International Future Energy Electronics Conference*, Taipei, Taiwan, November 2015.

[3] J. Niiranen, "Voltage dip ride through of a doubly-fed generator equipped with an active crowbar," in *Proceedings of the Nordic Wind Power Conference Nwpc*, 2004.

[4] L.-L. Sun, P. Yang, and Y. Wang, "Simulation research for LVRT of DFIG based on rotor active crowbar protection," in *Proceedings of the International Conference on Sustainable Power Generation and Supply, SUPERGEN 2012*, pp. 1–7, Hangzhou, China, September 2012.

[5] P. Su and K.-S. Zhang, "Simulation research for LVRT of DFIG with active IGBT Crowbar," *Power System Protection and Control*, vol. 38, no. 23, pp. 164–171, 2010.

[6] J. Hu, H. Wang, Y. He, and L. Xu, "Improved rotor current control of wind turbine driven doubly fed induction generators during network unbalance," in *Proceedings of the 1st International Conference on Sustainable Power Generation and Supply, SUPERGEN '09*, pp. 1–7, Nanjing, China, April 2009.

[7] T. K. A. Brekken and N. Mohan, "Control of a doubly fed induction wind generator under unbalanced grid voltage conditions," *IEEE Transactions on Energy Conversion*, vol. 22, no. 1, pp. 129–135, 2007.

[8] A. G. Abo-Khalil, D.-C. Lee, and J.-I. Jang, "Control of back-to-back PWM converters for DFIG wind turbine systems under unbalanced grid voltage," in *Proceedings of the 2007 IEEE International Symposium on Industrial Electronics, ISIE 2007*, pp. 2637–2642, Vigo, Spain, June 2007.

[9] C. Abbey and G. Joos, "Supercapacitor energy storage for wind energy applications," *IEEE Transactions on Industry Applications*, vol. 43, no. 3, pp. 763–776, 2007.

[10] B. S. Borowy and Z. M. Salameh, "Dynamic response of a stand-alone wind energy conversion system with battery energy storage to a wind gust," *IEEE Transactions on Energy Conversion*, vol. 12, no. 1, pp. 73–78, 1997.

[11] A. Yazdani, "Islanded operation of a doubly-fed induction generator (DFIG) wind-power system with integrated energy storage," in *Proceedings of the 2007 IEEE Canada Electrical Power Conference, EPC 2007*, pp. 153–159, Montreal, Canada, October 2007.

[12] J. Shi, Y. Tang, Y. Xia, L. Ren, and J. Li, "SMES based excitation system for doubly-fed induction generator in wind power application," *IEEE Transactions on Applied Superconductivity*, vol. 21, no. 3, pp. 1105–1108, 2011.

[13] Y. Zhang, Y. Tang, J. Li, J. Shi, and L. Ren, "Superconducting magnet based VSC suitable for interface of renewable power sources," *IEEE Transactions on Applied Superconductivity*, vol. 20, no. 3, pp. 880–883, 2010.

[14] W. Guo, L. Xiao, and S. Dai, "Enhancing low-voltage ride-through capability and smoothing output power of DFIG with a superconducting fault-current limiter-magnetic energy storage system," *IEEE Transactions on Energy Conversion*, vol. 27, no. 2, pp. 277–295, 2012.

[15] F. K. A. Lima, A. Luna, P. Rodriguez, E. H. Watanabe, and F. Blaabjerg, "Rotor voltage dynamics in the doubly fed induction generator during grid faults," *IEEE Transactions on Power Electronics*, vol. 25, no. 1, pp. 118–130, 2010.

[16] D. O. Akinyele and R. K. Rayudu, "Review of energy storage technologies for sustainable power networks," *Sustainable Energy Technologies and Assessments*, vol. 8, pp. 74–91, 2014.

8

Detection of Emerging Faults on Industrial Gas Turbines using Extended Gaussian Mixture Models

Yu Zhang,[1] **Chris Bingham,**[1] **Miguel Martínez-García,**[1] **and Darren Cox**[2]

[1]*School of Engineering, University of Lincoln, Lincoln LN6 7TS, UK*
[2]*Siemens Industrial Turbomachinery Ltd., Lincoln LN5 7FD, UK*

Correspondence should be addressed to Yu Zhang; yzhang@lincoln.ac.uk

Academic Editor: P. Stephan Heyns

This paper extends traditional Gaussian mixture model (GMM) techniques to provide recognition of operational states and detection of emerging faults for industrial systems. A variational Bayesian method allows a GMM to cluster with its mixture components to facilitate the extraction of steady-state operational behaviour; this is recognised as being a primary factor in reducing the susceptibility of alternative prognostic/diagnostic techniques, which would initiate false-alarms resulting from control set-point and load changes. Furthermore, a GMM with an outlier component is discussed and applied for direct novelty/fault detection. An advantage of the variational Bayesian method over traditional predefined thresholds is the extraction of steady-state data during both full- and part-load cases, and a primary advantage of the GMM with an outlier component is its applicability for novelty detection when there is a lack of prior knowledge of fault patterns. Results obtained from the real-time measurements on the operational industrial gas turbines have shown that the proposed technique provides integrated preprocessing, benchmarking, and novelty/fault detection methodology.

1. Introduction

Industrial gas turbines (IGTs) are utilised globally with units generally acting as prime-movers for either pumps, generators, or compressors, for both on- and off-shore systems. Various root-cause factors responsible for failures on such systems include vibration, shock, noise, heat, cold, dust, corrosion, humidity, rain, oil debris, flow, pressure, and excessive operating speed [1]. A key challenge of condition monitoring in order to provide an "early warning" of faults is to distinguish between sensor-based failure, component failure, and the normal operational transient behaviour of the system, for example, due to control or load changes. With advances in instrumentation, communications hardware, and computational capability [2], there has been an increased ability to realize such prognostic and diagnostic methods and provide remedial action and informed flexible maintenance scheduling prior to encountering unplanned downtime. This is especially pertinent as a result of the increasing operational speeds of IGTs [3].

Two popular categories of monitoring techniques have emerged over recent years, namely, model-based and signal processing-based approaches [4]. Model-based approaches construct models (or virtual sensors) to estimate physical variables from which residuals are calculated and used as indicators of emerging failure modes [5]. However, to build an accurate dynamic model that can accommodate the full operating envelope of IGTs is, in general, computationally demanding. In such circumstances, direct signal processing and data fusion methods often provide for more practical and effective monitoring solutions [6]. It is this latter category of techniques that is considered here.

Traditional signal processing-based methodologies use techniques such as principal component analysis (PCA) [7, 8], artificial neural networks (ANNs) [9], and data filter methods [10]. However, monitoring systems based on these algorithms are generally only applicable during steady-state operating conditions, since measurement transients caused by changes of loading or control action can generate "false alarms." Many techniques are therefore only effective under very constrained operating regimes. For instance, [11] employed ANNs for fault detection on gas turbines during the engine start-up phase, whilst [12] only considers solutions during steady-state operation. Typically, such techniques do not attempt to address the issue by incorporating implicit methods that discriminate between steady-state and transient behaviour as part of the fault detection system, and it is this aspect that is initially considered here [13].

By recognizing that steady-state measurement data is often superimposed by noise having a characteristic Gaussian distribution [14], that is, containing practical "measurement transients" that are not due to operational variations, the signals can be modelled through use of Gaussian Mixture Models (GMMs) [15]. This characterization of signals has previously been reported [16, 17] and used in a related family of clustering methods based on GMMs, regarded as "soft clustering" techniques, with the benefits for condition monitoring and fault detection explored in [16–18].

Here then, it is initially shown that GMMs provide a convenient mechanism to effectively discriminate between data relating to steady-state operation and that relating operation with transients and are specifically regarded in this instance as a preprocessing tool for subsequent operational pattern discrimination [13]. An essential part of developing the GMM methodology is parameter fitting, which is often carried out using an expectation-maximization (EM) method [19]. However, this requires the number of the mixture components (MCs) in the model to be fixed a priori [18]. Consequently, Bayesian-based frameworks are commonly used to provide a probabilistic inference on the data [20], and Variational Bayesian GMM techniques (VBGMMs) have been proposed [21] to provide improved performance [22, 23] by automatically selecting the number of MCs in the GMMs. VBGMMs are able to classify steady-state operation that can occur under full- or part-load conditions (e.g., 50% load). Additionally therefore, as well as identifying steady-state operation, the remaining data, including that associated with start-ups, shutdowns, and load changing conditions, is also naturally separated and it can therefore be used as a preanalysis tool for alternative dynamic scenarios, as reported in [24, 25], for instance.

An additional property of GMMs that facilitates novelty/fault detection is the ability to filter novel class samples when the machine learning system has no a priori knowledge [26]. Moreover, GMMs have been previously reported for the detection of "novel" vibration signatures with experimental results showing good fault detection properties with known classifications [27]. However, the technique was sensitive to vibration characteristics that are not associated with the detection of wear or damage and so remained susceptible to initiating false alarms. As a consequence of these features, the outlier components are now included as background "noise" for the GMM [28, 29], and the resulting technique is considered as a GMM with an outlier component (GMMOC). When using GMMOC, measurements remain characterized by the GMM, but novel characteristics, that is, measurements that have a very low probability of being clustered into existing distributions, are considered as outliers. It is the identification of such outliers that provides a robust mechanism for IGT fault detection considered here to provide an "early warning" fault detection tool.

Key contributions of this paper are summarized as follows:

(1) An extension to the use of GMMs, including VBGMM and GMMOC, is proposed for novelty/fault detection on industrial systems.

(2) Steady-state operation is discriminated from transient operation using VBGMM.

(3) GMMOC is used to indicate the presence of outliers and hence the emergence of faults.

(4) The efficacy of the proposed techniques is demonstrated from case studies (CSs) on IGTs, where CS1 is considered as a feasibility study of VBGMM and GMMOC for novelty detection and CS2 demonstrates a real bearing fault case study.

2. Methodology

The stages in the proposed methodology are depicted in Table 1. To provide an application focus for the development of the algorithms, measurements taken from bearing vibration probes on IGTs during commissioning are used as an illustrative example. Operational pattern separation is achieved through the use of VBGMM, where the steady-state data are distinguished for further analysis in this paper. Datasets from the identified transient operation can also be used for fault detection through start-up analysis and shutdown analysis and during load changes [6, 24, 25], which is not included in the current paper. The most relevant features are then extracted from the steady-state data and a statistical "fingerprint" for the extracted features is obtained through the application of GMMOC.

2.1. Underlying Principles of GMM. The empirical probability distribution of sampled data can be estimated by a GMM using a linear combination of Gaussian distributions $N(x \mid \mu, \sigma)$ [15], for example, as a sum of K Gaussian distributions with mean μ_k and standard deviation σ_k. The GMM with $K \times$ MCs is expressed as

$$p(x \mid \theta, \mu, \sigma) = \sum_{k=1}^{K} \theta_k N(x \mid \mu_k, \sigma_k), \qquad (1)$$

where x is a multidimensional variable and θ_k are the mixing coefficients that need to be chosen.

Let $X \equiv \{x_m\}$ represent M data samples, $m = 1, \ldots, M$, where each sample consists of a multidimensional variable x_m. Provided that x_m are statistically independent, the probability function of X can be expressed as

$$p(X \mid \theta, \mu, \sigma) = \prod_{m=1}^{M} \left(\sum_{k=1}^{K} \theta_k N(x_m \mid \mu_k, \sigma_k) \right), \qquad (2)$$

TABLE 1: Stages of the proposed methodology.

Step	Process	Purpose	Output
1	Sensor measurement collection (sampled or batch)	—	Collate data for analysis
2	VBGMM	Preprocess data to remove transient measurements	Measurements only of steady state operation
3	Statistical analysis	Extraction of important features within the data	Features for identifying emerging faults
4	GMMOC	Novelty detection of emerging faults	Early warning of emerging or imminent fault conditions

which constitutes the likelihood function and by taking the natural logarithm gives

$$L(X \mid \theta) = \ln p(X \mid \theta, \mu, \sigma)$$

$$= \sum_{m=1}^{M} \ln \left(\sum_{k=1}^{K} \theta_k N(x_m \mid \mu_k, \sigma_k) \right). \quad (3)$$

The mixture density is then

$$\rho_{mk} = \theta_k N(x_m \mid \mu_k, \sigma_k), \quad (4)$$

where k indicates the MC index and m indicates the data sample index. The conditional probability is calculated as

$$P_{mk} = \frac{\rho_{mk}}{\sum_{j=1}^{K} \rho_{mj}}, \quad (5)$$

for a selected component k on the given data sample x_m.

Evaluation of (3) typically necessitates an EM optimization procedure to maximize the log-likelihood function [18] from the maximization step (termed the M-step):

$$\frac{\partial L}{\partial \mu_k} = 0,$$

$$\frac{\partial L}{\partial \sigma_k} = 0, \quad (6)$$

$$\frac{\partial L}{\partial \theta_k} = 0.$$

The unknowns in (3) are solved in the expectation step (E-step), from (6), as follows:

(1) Choose an initialization of $\theta_k^{(0)}$, $\mu_k^{(0)}$, and $\sigma_k^{(0)}$.

(2) Iteratively update p_{mk}, θ_k, μ_k, and σ_k until convergence to a desired tolerance, using

$$P_{mk}^{(i)} = \frac{\theta_k^{(i)} N\left(x_m \mid \mu_k^{(i)}, \sigma_k^{(i)}\right)}{\sum_{k=1}^{K} \theta_k^{(i)} N\left(x_m \mid \mu_k^{(i)}, \sigma_k^{(i)}\right)},$$

$$\mu_k^{(i+1)} = \frac{\sum_{m=1}^{M} P_{mk}^{(i)} x_m}{\sum_{m=1}^{M} P_{mk}^{(i)}},$$

$$\sigma_k^{(i+1)} = \sqrt{\frac{1}{M} \frac{\sum_{m=1}^{M} P_{mk}^{(i)} \left\| x_m - \mu_k^{(i+1)} \right\|^2}{\sum_{m=1}^{M} P_{mk}^{(i)}}}, \quad (7)$$

$$\theta_k^{(i+1)} = \frac{1}{M} \sum_{m=1}^{M} P_{mk}^{(i)},$$

where M is the dimensionality of the data. For the special case $k = 1$, the whole dataset belongs to only 1 cluster, and the problem reduces to that of a Gaussian distribution fit.

2.2. Extension to VBGMM. A Variational Bayesian (VB) method can be used to determine the required number of MCs. Specifically, $M \times K$ binary latent variables $z_{mk} \in \{0, 1\}$ are used to indicate which MCs the data sample clusters into. When forming a GMM using classical methods, K is selected a priori. However, when using a VB method, K is resolved from the solutions of $\{z_{mk}\}$, termed Z. The joint probability distribution function of all variables is therefore

$$p(X, Z, \theta, \mu, \sigma)$$

$$= p(X \mid Z, \mu, \sigma) p(Z \mid \theta) p(\theta) p(\mu \mid \sigma) p(\sigma). \quad (8)$$

A lower bound on $p(X \mid \theta) = p(X \mid Z, \mu, \sigma) p(Z \mid \theta)$ can be determined using the VB method reported in [21]. Let $\Psi \equiv \{Z, \mu, \sigma\}$, and the marginal likelihood is expressed as

$$p(X \mid \theta) = \int p(X, \Psi \mid \theta) d\Psi. \quad (9)$$

Next, a variational distribution is introduced, that is, $q(\Psi) \approx p(\Psi \mid \theta)$. Through the use of Kullback-Leibler (KL) divergence [22], that is, $D_{KL}(q \parallel p)$, (9) becomes

$$\ln p(X \mid \theta) = D_{KL}(q \parallel p) + L(q), \qquad (10)$$

where $L(q)$ indicates the lower bound. Minimization of the KL divergence can be achieved by maximizing $L(q)$ by selecting appropriate q distributions. $q(\Psi)$ can be rewritten as the product of $q_i(\Psi_i)$, in terms of the subsets $\{\Psi_i\} = \{Z, \mu, \sigma\}$, giving

$$q(Z, \mu, \sigma) = q_Z(Z) q_\mu(\mu) q_\sigma(\sigma). \qquad (11)$$

The best distribution for each term, $q_i^*(\Psi_i)$, can be solved using

$$q_i^*(\Psi_i) = \frac{\exp\left(E_{j \neq i}\left[\ln p(X, \Psi)\right]\right)}{\int \exp\left(E_{j \neq i}\left[\ln p(X, \Psi)\right]\right) d\Psi_i}, \qquad (12)$$

where $E_{j \neq i}[\cdot]$ denotes the expectation of $q_j(\Psi_j)$ for all $j \neq i$.

Since $\{Z, \mu, \sigma\}$ are mutually coupled, these can be calculated using the E-step:

(1) Initialize the parameters, which are normally set to be small, real, and positive [30].

(2) Calculate $q(\Psi)$ in (11) through use of (12).

(3) Iteratively update until the predefined tolerance is met.

After calculating $q(\Psi)$, $L(q)$ can be obtained. Maximization of $L(q)$, which minimizes $\ln p(X \mid \theta)$, is again achieved by using the EM procedure. The M-step for calculating θ can be derived from (7), and the optimized distributions q_i can be obtained through the E-step as mentioned above. For brevity, the reader is directed to [30] for a more in-depth discussion of the VB framework on GMMs.

2.3. Principles of GMMOC.

GMMOC extends the original GMM [28, 29] approach by adding an outlier component that is modelled by a uniform distribution. The hybrid GMMOC model is then written as

$$p(x) = \sum_{k=1}^{K} \theta_k N(x \mid \mu_k, \sigma_k) + \left(1 - \sum_{k=1}^{K} \theta_k\right) U(x). \qquad (13)$$

Since $U(x)$ is uniformly distributed, the EM procedure described previously can still be employed to solve for the required parameters. The outlier component is normally assumed to be small initially (e.g., 0.01, and therefore the initialization of the mixing coefficients in GMM satisfies $\sum_{k=1}^{K} \theta_k = 0.99$). In this case, there are $(K+1)$ clusters, including K Gaussian distributions and one uniform distribution, the outlier component.

Parameters μ_k, σ_k, and θ_k can be estimated from (7), and if the probability of a data sample, belonging to any of the K Gaussian MCs, is smaller than a predefined threshold, it is clustered to the outlier component and therefore indicates a warning of an emerging fault, or facilitate novelty detection.

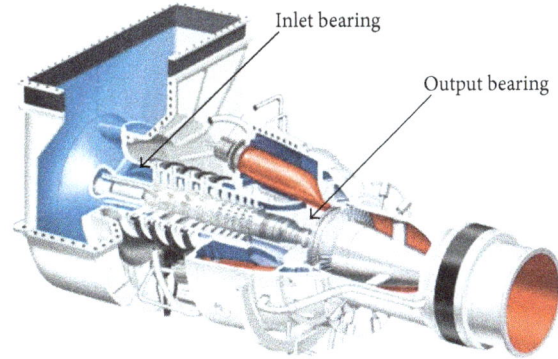

FIGURE 1: Locations of IGT bearing vibration sensors.

2.4. Feature Extraction.

Feature extraction provides an essential tool for reducing the dimensionality of raw data whilst keeping informative features [25]. Many feature extraction techniques have been reported and successfully applied, including the use of the Fast Fourier Transform (FFT) and Discrete Wavelet Transform (DWT), all involving elaborate time-frequency transforms [31, 32]. However, the data used in the following studies are taken from IGT units in the field with sampling rates in the order of minutes, which excludes the use of frequency domain based methods as they do not satisfy traditional sampling rate criteria for the measured variables. In this case, statistical features of the data in the time domain are used, for example, the peak value, root mean square, crest factor, kurtosis, clearance factor, impulse factor, shape factor, and skewness, and the most informative features can be identified through optimization methods [33]. For this study, in order not to divert the focus from the use of extended GMMs, only the most basic statistical attributes are employed; the mean (which carries information about measurement equilibrium) and standard deviation (which carries information about signal power) are used here as a proof of concept and also for practical reasons since it has been observed empirically that these features are sufficient in most cases.

3. Application Case Studies

The proposed methods are applied to monitor the vibration characteristics of fluid-film inlet- and output-bearings which typically support the compressor rotors of sub-15MW IGTs (Figure 1). The thrust bearings and journal bearings have operating speeds in excess of ~10,000 rpm. Radial and axial positions are monitored using noncontact probes. Two experimental case studies are now considered, both of which adopt the procedure in Table 1. CS1 uses measurements taken over a relatively short time period (1-month) and demonstrates the effectiveness of VBGMM for operational state discrimination (steady-state/transient behaviour), feature extraction, and the initial setup of a benchmarking ellipse using GMMOC, whilst CS2 considers the analysis of longer periods of measurement data (12-month) and aims to show the efficacy of GMMs for identifying longer-term emerging faults.

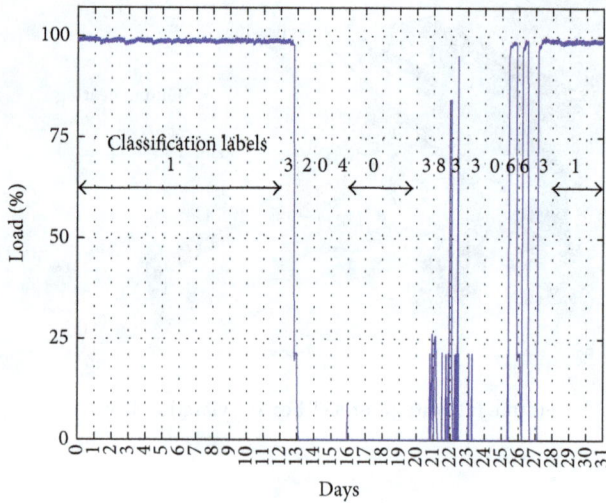

FIGURE 2: [CS1] Plot of load (%) versus time (days): identification of clusters in power/load measurements (month 1—normal operation).

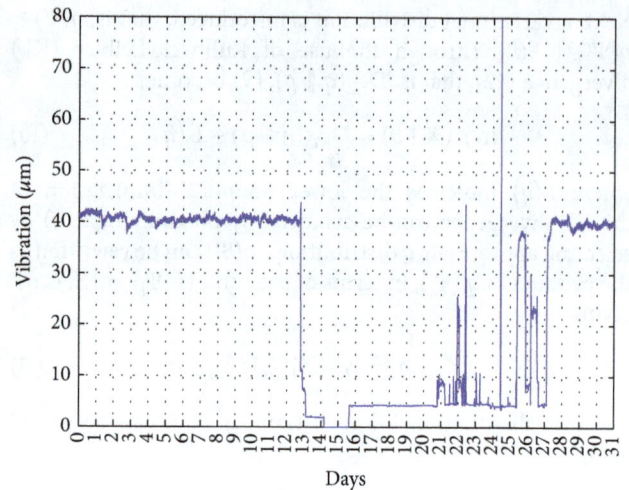

FIGURE 4: [CS1] Plot of vibration measurements versus time (days) (month 1—normal operation).

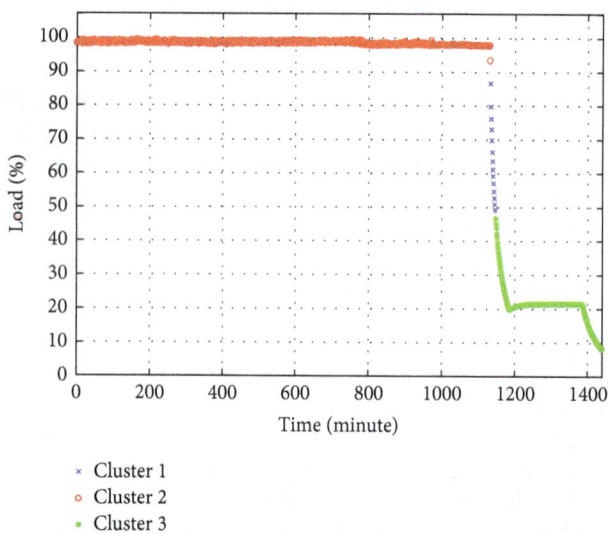

× Cluster 1
○ Cluster 2
• Cluster 3

FIGURE 3: [CS1] VBGMM clustering result: day 13, 3 clusters identified, that is, non-steady-state data incorporating load changes.

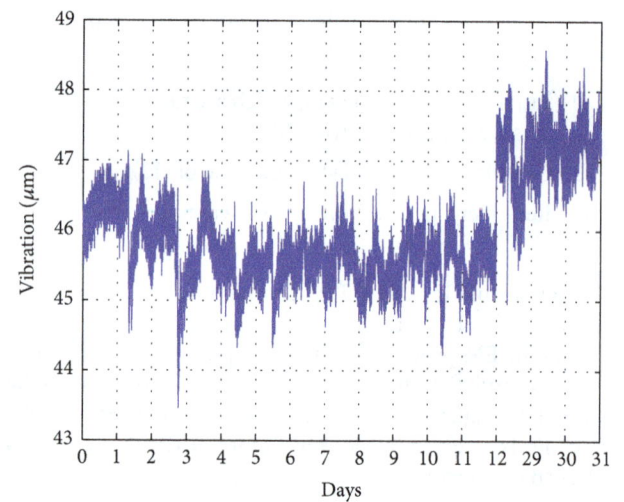

FIGURE 5: [CS1] Vibration data identified as steady-state (month 1—normal operation).

3.1. Case Study 1: Operation State Discrimination, Feature Extraction, and Novelty Detection. VBGMM is used to cluster measurements of output power (in terms of loading percentage) from a sub-15MW IGT to classify the unit's operational behaviour (Figure 2). One month (31 days) of daily data, each containing 1440 sample measurements, is used (i.e., sampling period = 1 minute). The resulting clusters from the power/load measurements are also shown in Figure 2 after applying VBGMM to each individual set of daily data. In line with the algorithm description, it can be seen that when the unit is considered to be operating normally, in steady-state, a classification label of 1 is assigned and that classification labels > 1 are assigned for all other detected cases (cluster result given in Figure 3). A classification label of 0 indicates constant null readings and is precluded from further pattern analysis as the unit is considered to be shut down.

Corresponding inlet bearing vibration measurements taken for the same 31-day period are shown in Figure 4. Having discriminated between transient and steady-state operation using VBGMM, the steady-state data (days 1–12 and days 29–31) shown in Figure 5 can be used for subsequent novelty detection.

Having identified appropriate datasets, feature extraction is used to capture important characteristics present in the data. The mean and the standard deviation of each day's data for months 1–3 are calculated and given in Figure 6, to present a benchmarking envelope representing normal operation. Having effectively obtained an operational fingerprint of behaviour, measurements taken over subsequent months are then used and compared to the fingerprint.

Considering only the magnitude of vibration amplitudes, levels up to ~50 μm are typically considered normal, with warnings at ~70 μm and unit shutdown occurring at ~90 μm.

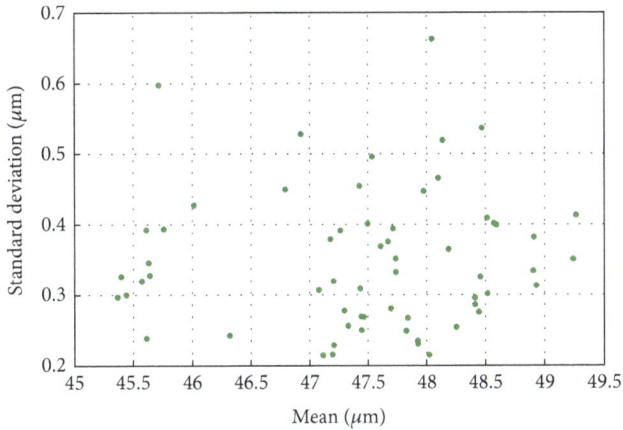

FIGURE 6: [CS1] Extracted features for 3 months' steady-state data (months 1–3—normal operation).

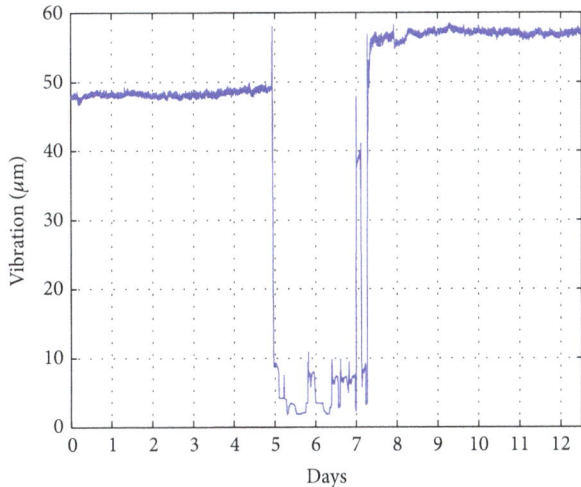

FIGURE 8: [CS1] GMMOC novelty detection results compared to those of the original GMM.

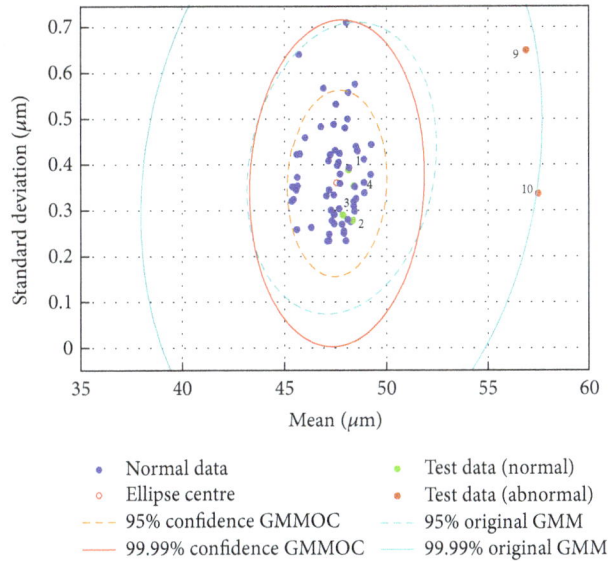

FIGURE 7: [CS1] Test data for novelty detection (in month 4).

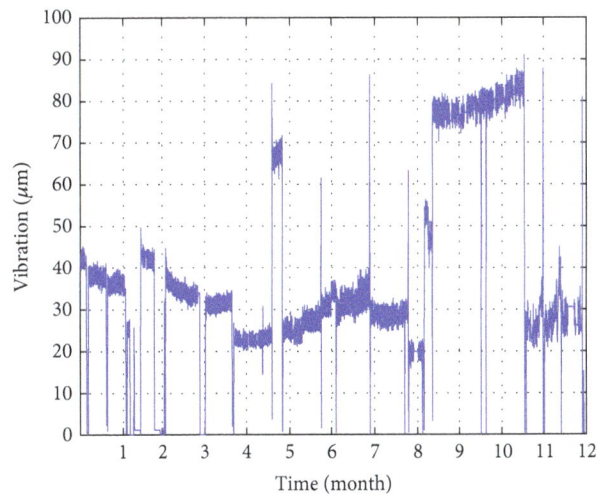

FIGURE 9: [CS2] IGT inlet bearing vibration information (months 1–12—all operations including transients).

Having obtained a fingerprint representing normal operation, GMMOC is applied to subsequent periods of data on a daily basis (in this case in month 4, Figure 7). It is well known that gradual bearing wear leading to failure is often preceded by gradual changes in vibration characteristics. Figure 8 provides an ellipse boundary drawn according to the 1-cluster GMM model (see (13)). In general, the confidence level to identify outliers will be set according to application. For instance, in this case, a 99.99% confidence level does not discriminate the outliers (here, it indicates outliers in normal operation which may be caused by sensor malfunctioning); however, a 95% confidence level is clearly seen to be appropriate in this instance. The results from GMMOC are compared with the envelopes drawn from the original GMM, as shown in Figure 8, and the advantage of GMMOC over the original GMM is evident, since the outliers, days 9 and 10, are clearly identified even with a 99.99% confidence level of GMMOC in this case.

By considering the following month of measurement data, it is notable that the measurements have correctly been identified as outliers (day 9 and day 10 in this case), whereas measurements from days 1–4 are correctly considered to correspond to normal behaviour. Although it is not known at that stage if the increase in vibration in days 9 and 10 is related to a component fault, the measurements are considered as anomalies. Although CS1 is used as a proof of concept application of VBGMM and GMMOC for novelty detection, the bearing considered in the study did fail around 3 months after initially being identified using the extended GMM.

3.2. Case Study 2. CS2 uses measurements taken over a longer period of 12 months using a lower sampling rate; specifically 1 data sample taken every 9 minutes, as shown in Figure 9. Again the procedure depicted in Table 1 is applied.

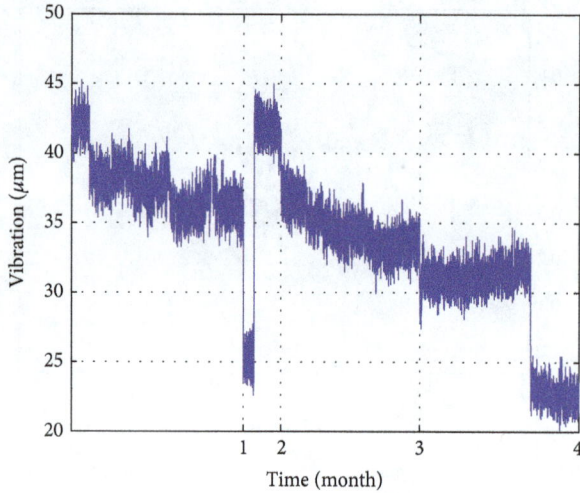

FIGURE 10: [CS2] Extracted steady-state vibration data for normal operation (months 1–4).

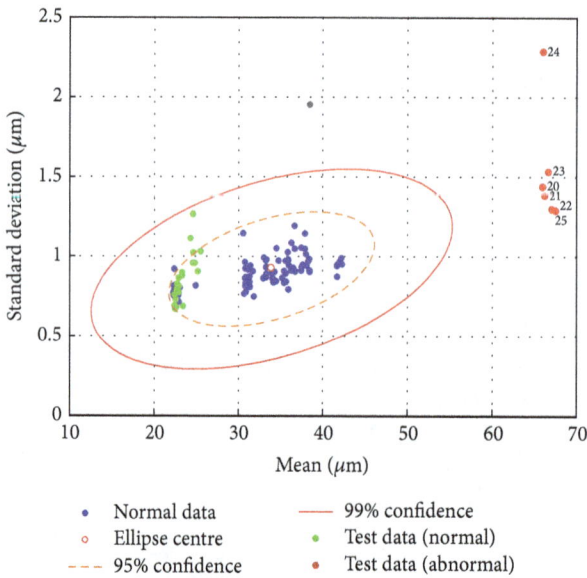

FIGURE 12: [CS2] Novelty detection and fault pattern location based on GMMOC (months 1–12).

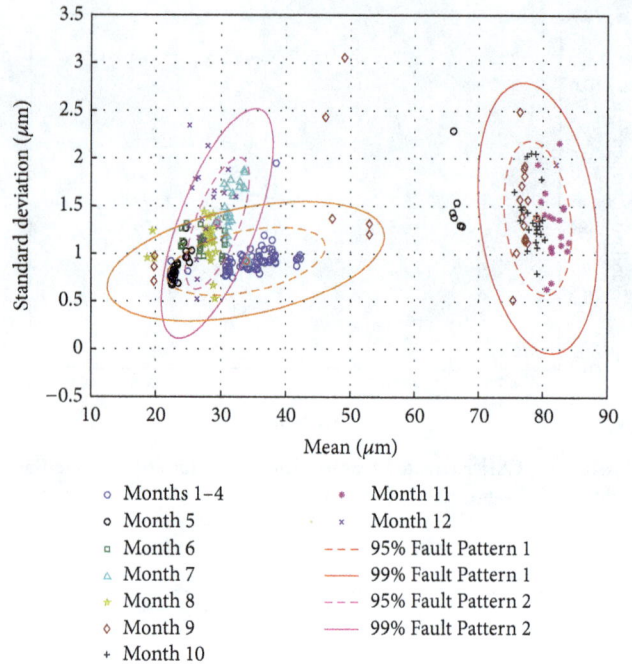

FIGURE 11: [CS2] Fingerprint envelope and novelty detection through GMMOC (months 1–4 and month 5).

Measurements from months 1–4 are deemed to describe normal operational characteristics. VBGMM is then applied to identify what is considered to be steady-state operational data for further analysis (Figure 10). Important characteristics are then determined; once again it is known from empirical studies that for this application case the mean and standard deviation are effective measures of underlying behaviour. Through the application of GMMOC and by clustering the extracted features into a single cluster, a fingerprint of normal operation is obtained, as shown in Figure 11, where the 95% confidence envelope is considered as an early warning boundary in this instance and the 99% confidence envelope as a fault detection boundary. Using measurements from month 5 as testing data, it is shown in Figure 11 that, between days 20

and 25 of the 5th month, operation falls outside the "normal fingerprint" ellipsoid and is therefore identified as an outlier, thereby providing an early warning of expected failure, which was evidenced during the field service.

Referring again to measurements shown in Figure 9, on day 140 of operation (day 20 of month 5) a transient in the vibration level is evident, and although the mean level ultimately returns to normal after month 5, the variance continued to indicate evidence of emerging failure during and after month 6 (Figure 12). Considering months 6–12 as a period of emerging fault, the number of fault patterns can be discriminated using VBGMM (2 clusters for faults in this case) and the fault pattern locations identified using GMMOC, as shown in Figure 12. It can be seen that Fault Pattern 2 overlaps with the original fingerprint from months 1–4, indicating in this case that ideally other feature extraction indices (e.g., those involved with loading conditions) could be used to further isolate this type of fault characteristic from that of normal operation. Between the 2 sets of fault patterns (Figure 12), some anomalies are apparent which are due to the increasing levels of vibration as the bearing deteriorates, such as the 20th to 25th days of the 5th month's operation.

In this instance the unit was shut down for maintenance in order to prevent a catastrophic failure. The bearing was known to be undamaged in month 1 as it was assessed in the previous service check (see Figure 13). Through subsequent decommissioning of the unit and investigation, vibration damage is evident on the inlet bearing, with excessive wear on the tilt pad shown in Figure 14. It can be clearly seen that there has been abnormal wear from the markings shown on both the shaft (Figure 15) and bearing pads (Figure 14). Through root-cause analysis, the damage was attributed to

FIGURE 13: Normal IGT bearing with undamaged tilt pads.

FIGURE 14: IGT inlet bearing showing vibration damage.

FIGURE 15: Compressor rotor inlet bearing shaft showing vibration damage.

an incorrectly specified lubricant oil cooler causing high temperatures in the lubricant oil.

4. Conclusion

The paper has developed and demonstrated extensions of GMMs to provide a highly practical preprocessing and novelty/fault detection tool. The main contributions of the paper are (1) an automatic clustering method for VBGMM which identifies steady-state operational behaviour from transient operation, allowing the extraction of steady-state measurement segments for subsequent condition monitoring and (2) a GMMOC method that has been proposed and shown to provide a valuable tool for use as an early warning system of emerging failure through novelty detection. The presented techniques are currently being utilised in an industrial environment to monitor the operational status of a global fleet of IGTs. Although the experimental trials have focused on IGTs and bearing vibration measurements in this instance, the proposed methods are much more widely applicable to other industrial components and systems for pattern analysis, benchmarking, and novelty/fault detection.

Nomenclature

IGT: Industrial gas turbine
GMM: Gaussian Mixture Model
VBGMM: Variational Bayesian Gaussian Mixture Model
GMMOC: Gaussian Mixture Model with an outlier component
PCA: Principal component analysis
ANN: Artificial neural network
MC: Mixture component
EM: Expectation-maximization
M-step: Maximization step
E-step: Expectation step
KL: Kullback-Leibler
CS: Case study.

Conflicts of Interest

The authors declare that they have no conflicts of interest.

Acknowledgments

The authors would like to thank Siemens Industrial Turbomachinery, Lincoln, UK, for providing research support and access to field data to support the research outcomes.

References

[1] R. Kurz, K. Brun, and M. Wollie, "Degradation effects on industrial gas turbines," *Journal of Engineering for Gas Turbines and Power*, vol. 131, no. 6, Article ID 062401, 2009.

[2] M. A. Zaidan, A. R. Mills, R. F. Harrison, and P. J. Fleming, "Gas turbine engine prognostics using Bayesian hierarchical models: A variational approach," *Mechanical Systems and Signal Processing*, vol. 70-71, pp. 120–140, 2016.

[3] M. Maalouf, "Gas turbine vibration monitoring," *Orbit Article*, vol. 25, no. 1, pp. 1–7, 2005.

[4] F. Lu, J. Huang, and Y. Xing, "Fault diagnostics for turbo-shaft engine sensors based on a simplified on-board model," *Sensors (Switzerland)*, vol. 12, no. 8, pp. 11061–11076, 2012.

[5] H. Hanachi, J. Liu, A. Banerjee, and Y. Chen, "Sequential state estimation of nonlinear/non-Gaussian systems with stochastic input for turbine degradation estimation," *Mechanical Systems and Signal Processing*, vol. 72-73, pp. 32–45, 2016.

[6] Y. Zhang, C. Bingham, Z. Yang, B. W.-K. Ling, and M. Gallimore, "Machine fault detection by signal denoising - With application to industrial gas turbines," *Measurement: Journal of the International Measurement Confederation*, vol. 58, pp. 230–240, 2014.

[7] S. Yin, S. X. Ding, A. Haghani, H. Hao, and P. Zhang, "A comparison study of basic data-driven fault diagnosis and process monitoring methods on the benchmark Tennessee Eastman process," *Journal of Process Control*, vol. 22, no. 9, pp. 1567–1581, 2012.

[8] Y. Zhang, C. M. Bingham, and M. Gallimore, "Fault detection and diagnosis based on extensions of PCA," *Advances in Military Technology*, vol. 8, no. 2, pp. 27–41, 2013.

[9] G. Romesis and K. Mathioudakis, "Setting up of a probabilistic neural network for sensor fault detection including operation with component faults," *Journal of Engineering for Gas Turbines and Power*, vol. 125, no. 3, pp. 634–641, 2003.

[10] D. L. Simon and J. S. Litt, "A data filter for identifying steady-state operating points in engine flight data for condition monitoring applications," *Journal of Engineering for Gas Turbines and Power*, vol. 133, no. 7, Article ID 071603, 2011.

[11] O. Uluyol, Kyusung K., and E. O. Nwadiogbu, "Synergistic use of soft computing technologies for fault detection in gas turbine engines," *IEEE Transactions on Systems, Man and Cybernetics, Part C (Applications and Reviews)*, vol. 36, no. 4, pp. 476–484, 2006.

[12] Y. Zhang, C. M. Bingham, and M. Gallimore, "Applied sensor fault detection,identification and data reconstruction, identification and data reconstruction," *Advances in Military Technology*, vol. 8, no. 2, pp. 13–26, 2013.

[13] Y. Zhang, C. Bingham, M. Gallimore, and J. Chen, "Steady-state and transient operation discrimination by Variational Bayesian Gaussian Mixture Models," in *2013 16th IEEE International Workshop on Machine Learning for Signal Processing, MLSP 2013*, gbr, September 2013.

[14] A. Jablonski, T. Barszcz, M. Bielecka, and P. Breuhaus, "Modeling of probability distribution functions for automatic threshold calculation in condition monitoring systems," *Measurement*, vol. 46, no. 1, pp. 727–738, 2013.

[15] D. A. Reynolds, *Encyclopedia of Biometrics Recognition*, Springer, Heidelberg, Germany, 2008.

[16] G. Yu, C. Li, and J. Sun, "Machine fault diagnosis based on Gaussian mixture model and its application," *International Journal of Advanced Manufacturing Technology*, vol. 48, no. 1-4, pp. 205–212, 2010.

[17] J. Yu, "Bearing performance degradation assessment using locality preserving projections and Gaussian mixture models," *Mechanical Systems and Signal Processing*, vol. 25, no. 7, pp. 2573–2588, 2011.

[18] C. M. Bishop, *Pattern Recognition and Machine Learning*, Information Science and Statistics, Springer, Heidelberg, Germany, 2006.

[19] G. J. McLachlan and T. Krishnan, *The EM Algorithm and Extensions*, Wiley-Interscience, New York, NY, UK, Second edition, 2008.

[20] A. R. Groves, *Bayesian Learning Methods for Modelling Functional MRI*, Department of Clinical Neurology, University of Oxford, Oxford, UK, 2010.

[21] A. Corduneanu and C. M. Bishop, "Variational Bayesian model selection for mixture distributions," *Artificial Intelligent Statistics*, pp. 27–34, 2001.

[22] K. Watanabe and S. Watanabe, "Stochastic complexities of Gaussian mixtures in variational Bayesian approximation," *Journal of Machine Learning Research*, vol. 7, pp. 625–644, 2006.

[23] Y. Zhang, C. Bingham, M. Gallimore, and D. Cox, "Novelty detection based on extensions of GMMs for industrial gas turbines," in *2015 IEEE International Conference on Computational Intelligence and Virtual Environments for Measurement Systems and Applications, CIVEMSA 2015*, chn, June 2015.

[24] Y. Zhang, S. Cruz-Manzo, and A. Latimer, "Start-up vibration analysis for novelty detection on industrial gas turbines," in *Proceedings of 2016 International Symposium on Industrial Electronics*, Banja Luka, Bosnia and Herzegovina, November 2016.

[25] Y. Zhang, M. Gallimore, C. Bingham, J. Chen, and Y. Xu, "Hybrid hierarchical clustering — piecewise aggregate approximation, with applications," *International Journal of Computational Intelligence and Applications*, vol. 15, no. 4, Article ID 1650019, 26 pages, 2016.

[26] M. Markou and S. Singh, "A neural network-based novelty detector for image sequence analysis," *IEEE Transactions on Pattern Analysis and Machine Intelligence*, vol. 28, no. 10, pp. 1664–1677, 2006.

[27] T. Heyns, P. S. Heyns, and J. P. de Villiers, "Combining synchronous averaging with a Gaussian mixture model novelty detection scheme for vibration-based condition monitoring of a gearbox," *Mechanical Systems and Signal Processing*, vol. 32, pp. 200–215, 2012.

[28] M. K. Titsias, *Unsupervised Learning of Multiple Objects in Images*, School of Informatics, University of Edinburgh, UK, 2005.

[29] P. Coretto, *The Noise Component in Model-Based Clustering*, Department of Statistical Science, University of London, London, UK, 2008.

[30] N. Nasios and A. G. Bors, "Variational learning for Gaussian mixture models," *IEEE Transactions on Systems, Man, and Cybernetics, Part B: Cybernetics*, vol. 36, no. 4, pp. 849–862, 2006.

[31] P. Jayaswal, A. K. Wadhwani, and K. B. Mulchandani, "Machine fault signature analysis," *International Journal of Rotating Machinery*, vol. 2008, Article ID 583982, 10 pages, 2008.

[32] S. F. Ding, H. Zhu, W. K. Jia, and C. Y. Su, "A survey on feature extraction for pattern recognition," *Artificial Intelligence Review*, vol. 37, no. 3, pp. 169–180, 2012.

[33] Z. Huo, Y. Zhang, P. Francq, L. Shu, and J. Huang, "Incipient fault diagnosis of roller bearing using optimized wavelet transform based multi-speed vibration signatures," *IEEE Access*, 2017.

9

Overvoltage and Insulation Coordination of Overhead Lines in Multiple-Terminal MMC-HVDC Link for Wind Power Delivery

Huiwen He, Lei Wang, Peihong Zhou, and Fei Yan

State Key Laboratory of Power Grid Environmental Protection, China Electric Power Research Institute, Wuhan 430074, China

Correspondence should be addressed to Huiwen He; husthhw@126.com

Academic Editor: Tao Wang

The voltage-sourced converter-based HVDC link, including the modular multilevel converter (MMC) configuration, is suitable for wind power, photovoltaic energy, and other kinds of new energy delivery and grid-connection. Current studies are focused on the MMC principles and controls and few studies have been done on the overvoltage of transmission line for the MMC-HVDC link. The main reason is that environmental factors have little effect on DC cables and the single-phase/pole fault rate is low. But if the cables were replaced by the overhead lines, although the construction cost of the project would be greatly reduced, the single-pole ground fault rate would be much higher. This paper analyzed the main overvoltage types in multiple-terminal MMC-HVDC network which transmit electric power by overhead lines. Based on ±500 kV multiple-terminal MMC-HVDC for wind power delivery project, the transient simulation model was built and the overvoltage types mentioned above were studied. The results showed that the most serious overvoltage was on the healthy adjacent line of the faulty line caused by the fault clearing of DC breaker. Then the insulation coordination for overhead lines was conducted according to the overvoltage level. The recommended clearance values were given.

1. Introduction

The voltage-sourced converter- (VSC-) based high-voltage DC (HVDC) transmission system is considered for a wide range of applications for wind power, photovoltaic energy, or other kinds of new energy delivery and grid-connection due to flexibly control of active power and reactive power output capability [1–4]. The Nanhui MMC-HVDC demonstration project and the Nanao multiple-terminal MMC-HVDC link have been put into operation already and achieved the stable and reliable wind power transmission [5, 6].

At present, the study focuses on the MMC principles and controls and less research has been done on the overvoltage of transmission line for the MMC-HVDC project. And there are few studies on the overvoltage characteristics of multi-terminal MMC-HVDC [7, 8]. State Grid Beijing Economic and Technical Research Institute has done research on the Dalian MMC-HVDC project considering the short circuit fault at AC bus, grid side and valve side of connection transformer, smoothing reactor, and DC pole bus. According to the simulation results of overvoltage, the reference voltage and switching impulse withstanding voltage of MOAs in the

converter station were determined [9]. Based on the Trans Bay MMC-HVDC project, Zhejiang University studied the 14 kinds of fault and found the most serious overvoltages were caused by the single-phase grounding fault at valve side of converter transformer, the short circuit of transformer, the ground short circuit fault at valve head, DC bus grounding fault, and DC line grounding fault. The MOA configuration and insulation coordination scheme were proposed. The withstanding voltages of equipment were determined by deterministic method and the margin coefficient between withstand voltage and impulse overvoltage could be selected as 15%, 20%, and 25% for switching impulse, lighting impulse, and steep impulse, respectively [10].

The researches mentioned above were focused on the converter station. Most of the MMC-HVDC projects adopt cables for power transmission all around the world. The cables buried in the ground are not affected by the surrounding environment, so the failure rate is lower [11, 12]. However, cables lines are expensive and the overhead lines can greatly reduce the cost of construction. The overhead lines are susceptible to the surrounding environment, such as lightning and pollution, and the single-pole grounding fault rate is much higher than using cable lines. Therefore, it is necessary to study the overvoltage generated by overhead line faults to support the design of line insulation.

This paper focuses on the overvoltage level of overhead lines in multiple-terminal MMC-HVDC project. Firstly it pointed out the main overvoltage types of overhead lines in multiple-terminal MMC-HVDC link. Based on an ±500 kV multiple-terminal MMC-HVDC for wind power delivery project, the transient simulation model was built and various overvoltage types mentioned above were studied. Then the insulation coordination for overhead lines was conducted according to the overvoltage level. In the end the recommended clearance values were given.

This paper is organized as follows. Section 2 is devoted to analyzing the main overvoltage types of overhead lines in multiple-terminal MMC-HVDC link and the differences between it and conventional point-to-point LCC-HVDC project while Section 3 presents the simulation overvoltage results based on a multiple-terminal MMC-HVDC project according to Section 2. Then insulation coordination for overhead lines is conducted and the recommended clearances are given in Section 4. Finally, conclusions are drawn in Section 5.

2. Overvoltage Types of Overhead Line in Multiple-Terminal MMC-HVDC Project

LCC-HVDC systems are usually 2 terminals adopting overhead transmission lines and the occurrence probability of single-pole grounding fault is the highest. Statistical data show that the single-pole lightning flashover rate of ±500 kV DC transmission lines is 0.28 times per 100 km every year. In a bipolar operation, a grounding fault occurring at one pole will induce the slow front overvoltage at the sound pole through the coupling between two poles and reflection and refraction of traveling wave at both ends of line. The magnitude of the overvoltage can be estimated by the following formula:

$$U_p = U_{dc}\left(1 + \frac{Z_{m0} - Z_{m1}}{Z_{m0} + Z_{m1} + 2R_e}\right), \tag{1}$$

where Z_{m0} and Z_{m1} are zero sequence and positive sequence wave impedance of transmission line, respectively; R_e is the grounding resistance.

The induced overvoltage at sound pole caused by single-pole grounding fault should also be considered for overhead lines in multiple-terminal MMC-HVDC project and the magnitude of the overvoltage can be estimated by formula (1) too. In order to improve the availability of MMC-HVDC project, the DC circuit breaker would be installed for fault clearing at DC side. So the line fault clearing overvoltage and reclosing overvoltage should be considered.

(1) The line fault clearing overvoltage mainly refers to the overvoltage on the sound pole line and the adjacent sound line generated by the fault tripping. It is equivalent to superimposing a current source with a reverse fault current on the circuit breaker when the circuit breaker interrupts the fault current. The current wave propagation and reflection on the adjacent sound line

form transient overvoltage. The fault current is related to the main circuit parameters of converter station, the operation characteristics of DC circuit breaker, and the fault grounding resistance.

(2) When the DC line is switched on or reclosing, the overvoltage is generated due to the difference between the initial voltage of the line to the ground and the forced voltage at the end of the transition process. The magnitude of the overvoltage can be estimated by the following formula:

$$U_{pc} = 2U_w - U_0, \tag{2}$$

where U_w is the forced voltage at the end of the transition process and U_0 is the initial voltage of the line to the ground.

Therefore 3 kinds of overvoltage are as follows, which should be studied for overhead lines of multiterminal MMC-HVDC systems:

(1) Induced overvoltage on healthy pole line caused by single-pole grounding fault

(2) Fault clearing overvoltage

(3) Closing and reclosing overvoltage.

3. Analysis of Overvoltage of Overhead Transmission Lines in Multiple-Terminal MMC-HVDC Project

3.1. Simulation Parameters

3.1.1. Main Circuit Parameter. A 4-terminal ±500 kV MMC-HVDC for wind power delivery project, using half bridge and real bipolar connection, has 245 levels and DC rated voltage is 535 kV. The rated voltage of AC power network is 230 kV at delivery end A, which can send wind power out with 17 kA three-phase short circuit current, and the rated capacity of A is 3000 MW under the normal operation in island. The rated voltage of AC power network is 230 kV at delivery end B, which can send wind power out with 17 kA three-phase short circuit current, and the rated capacity of B is 1500 MW under the normal operation in island. The rated voltage of AC power network is 525 kV at delivery end C, which can send wind power out with 18 kA three-phase short circuit current, and the rated capacity of C is 1500 MW. The rated voltage of AC power network is 525 kV at delivery end D, which can send wind power out with 63 kA three-phase short circuit current, and the rated capacity of D is 3000 MW.

The lengths of DC transmission lines of 4-terminal system are shown in Figure 1. The type of conductors is 4 × JL/G2A-720/50 with 12.8 m pole distance, and the type of ground wire is OPGW-150 with 12 m horizontal distance, together with the model JNRLH60/G1A-400/35 of metallic return line which has the horizontal distance of 9 m. Typical tower model of tangent tower is shown in Figure 2, and the transmission line is simulated by Frequency Dependent (Phase) Model in PSCAD. In the converter station, the

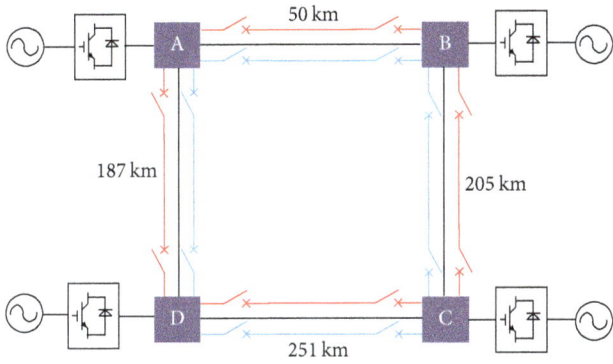

FIGURE 1: Schematic diagram of 4-terminal transmission lines.

current limiting reactors are arranged on the pole lines and the metallic return lines; the values of reactance are 150 mH on the pole line and 300 mH on the neutral line. The reference voltage of pole line arrester is 629 kV and residual voltage is 904 kV under switch operation with 2 kA current.

According to empirical equation (3), the arc resistance R_G in the air can be calculated:

$$R_G \approx 1050 \frac{L_G}{I_G}, \tag{3}$$

where L_G is the length of arc and I_G is the rms. of current with the unit ampere. In the simulation, the sum resistance of arc and grounding tower is 4 Ω.

3.1.2. Control and Protection Model of Converter Station. A double closed-loop controller is established in PSCAD to achieve converter station-level control. Double closed-loop control can be divided into an outer loop controller and an inner loop controller. The outer loop controller can calculate the current reference value of the current-mode inner loop controller according to the active and the reactive power command. The inner loop controller keeps the dq axis current tracking its reference value by adjusting the output voltage of the inverter. The control system used in the simulation is shown in Figure 3 [13–16].

According to Figure 3, the dynamic differential equation of AC side in three-phase stationary coordinate system is shown as

$$\begin{bmatrix} u_{sa} \\ u_{sb} \\ u_{sc} \end{bmatrix} = L \frac{d}{dt} \begin{bmatrix} i_{sa} \\ i_{sb} \\ i_{sc} \end{bmatrix} + R \begin{bmatrix} i_{sa} \\ i_{sb} \\ i_{sc} \end{bmatrix} + \begin{bmatrix} u_{ca} \\ u_{cb} \\ u_{cc} \end{bmatrix}, \tag{4}$$

where u_{sa}, u_{sb}, and u_{sc} are the measured phase A, phase B, and phase C voltages at grid side of converter transformer, respectively. i_{sa}, i_{sb}, and i_{sc} are the measured phase A, phase B, and phase C currents at grid side of converter transformer, respectively. u_{ca}, u_{cb}, and u_{cc} are the measured phase A, phase B, and phase C voltages at valve side of converter transformer, respectively. L is the equivalent reactance. R is the equivalent resistance.

After the park transformation, the active power P injected into the converter station from the AC system is as (5). The

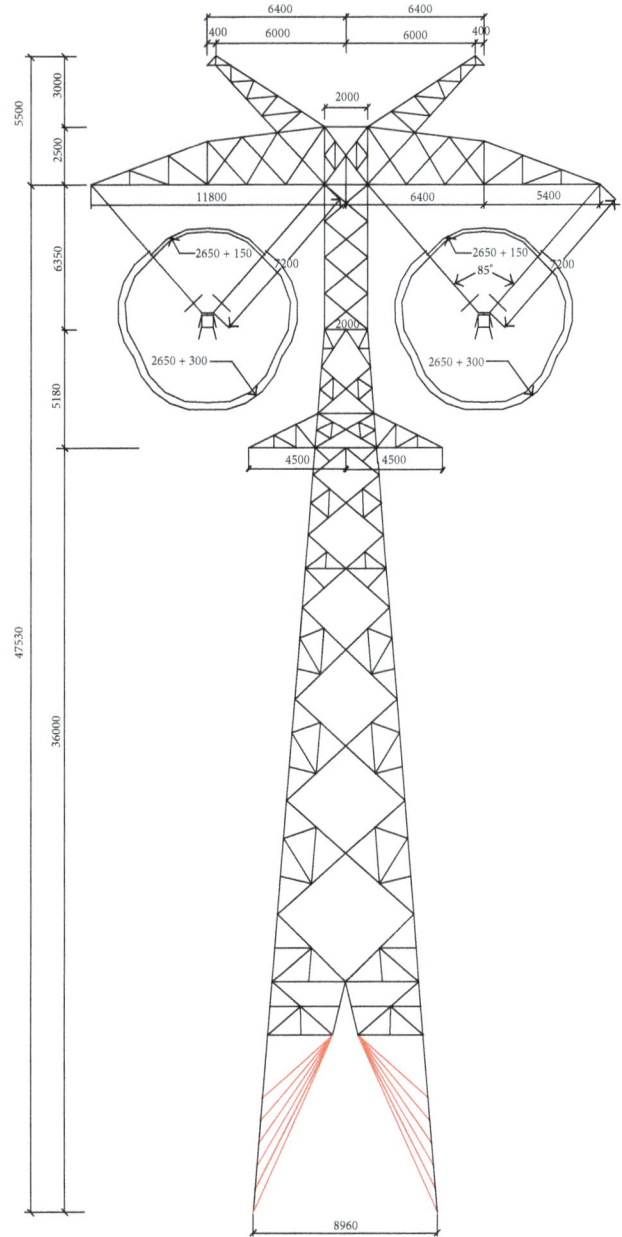

FIGURE 2: Typical tower diagram of single circuit transmission line.

reactive power Q injected into the converter station from the AC system is as (6). So the decoupled control of active power and reactive power can be realized:

$$P = \frac{3}{2} \left(u_{sd} i_{sd} + u_{sq} i_{sq} \right) = \frac{3}{2} u_{sd} i_{sd}, \tag{5}$$

$$Q = \frac{3}{2} \left(u_{sd} i_{sq} - u_{sq} i_{sd} \right) = \frac{3}{2} u_{sd} i_{sq}, \tag{6}$$

where u_{sd} and u_{sq} are the d axis and q axis voltages derived from park transformation of u_{sa}, u_{sb}, and u_{sc}, respectively. i_{sd} and i_{sq} are the d axis and q axis currents derived from park transformation of i_{sa}, i_{sb}, and i_{sc}, respectively.

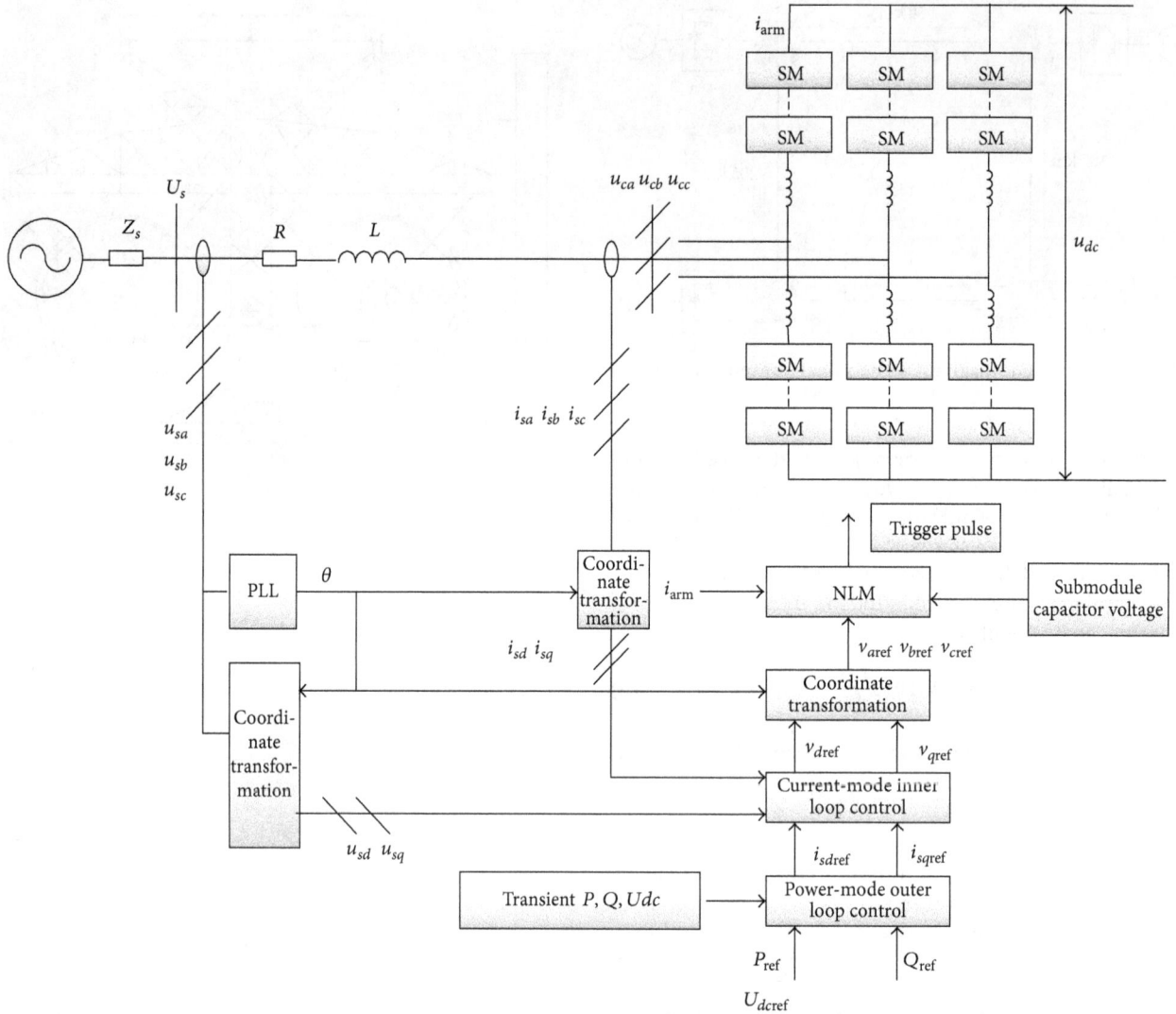

FIGURE 3: The schematic diagram of MMC control system.

3.1.3. Wind Power Model. Two wind farms provide the power sources for 2 substations of the 4-terminal ±500 kV MMC-HVDC project. In order to calculate the accurate overvoltage level, the influence of the wind farms on overvoltage was considered.

In the simulation, the electromagnetic transient model based on the $dq0$ transform of doubly fed induction generator was established. The dq coordinate system based on rotation speed of rotor was adopted. The voltage and current on the armature side and the exciting side satisfy [17]

$$V_P = -R_P i_P - \frac{d\lambda_P}{dt} + \omega N \lambda_P, \qquad (7)$$

$$V_E = -R_E i_E - \frac{d\lambda_E}{dt}, \qquad (8)$$

$$N = \begin{bmatrix} 0 & 0 & 0 \\ 0 & 0 & -1 \\ 0 & 1 & 0 \end{bmatrix}, \qquad (9)$$

where V_P and V_E are voltages on the armature side and the exciting side, respectively. R_P and R_E are resistance on the armature side and resistance on the exciting side, respectively. i_P and i_E are currents on the armature side and the exciting side, respectively. λ_P and λ_E are flux linkage matrixes on the armature side and the exciting side, respectively. ω is the angular velocity of rotor rotation.

3.2. Overvoltage Analysis. Two modes of operation are considered during the simulation:

(1) Normal mode:

All the stations access the network.

(2) Abnormal mode:

Some of the stations are unconnected to the network, which leads to a longer transmission line than normal mode.

The specific calculation process is as follows. The DC operation voltage is 535 kV before fault. The fault line is

TABLE 1: Overvoltages of transmission lines under normal mode.

Lines	Overvoltages of pole lines	
	kV	p.u.
A-D	950	1.78
C-D	974	1.82
B-C	991	1.85
A-B	888	1.66

TABLE 2: Overvoltages of transmission lines under abnormal mode (C station is unconnected).

Lines	Overvoltages of pole lines	
	kV	p.u.
A-D	998	1.87
C-D	1034	1.93
B-C	1010	1.89
A-B	896	1.68

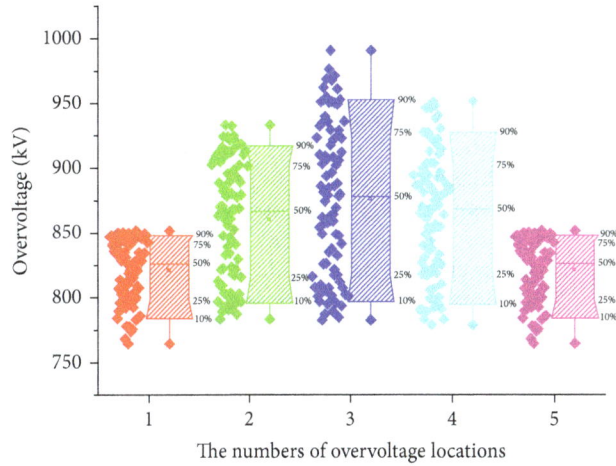

FIGURE 4: The overvoltage distribution and probability distribution of B-C line caused by A-D line fault under normal mode.

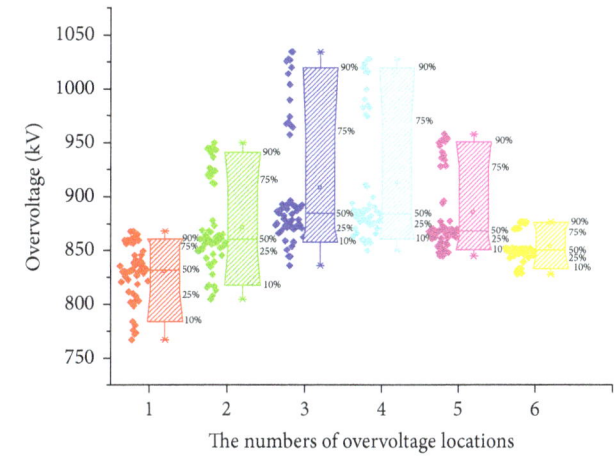

FIGURE 5: The overvoltage distribution and probability distribution of C-D line caused by A-B line fault with unconnected C station.

divided into 10 sections. Along the line each section is grounded. Then the circuit breakers at the ends of fault line trip remove the temporary ground fault, and after the delay time the DC circuit breakers reclose. Both the overvoltages caused during the grounding and reclose processes are taken into consideration. Besides, the overvoltages caused by faults on other lines should be calculated too.

3.2.1. Normal Mode. The overvoltage levels of all pole lines are shown in Table 1. The maximum overvoltage of 990.77 kV (1.85 p.u., 1 p.u. = 535 kV) appeared in the B-C line, caused by the A-D line fault. The overvoltage distribution and probability distribution of B-C line are shown in Figure 4. The statistics overvoltage along the line is umbrella type distribution. In the middle of line the overvoltage is high and to both ends of the line diminishing. This is due to the installation of the arrester on the pole line bus, which can suppress the overvoltage.

3.2.2. Abnormal Mode. The overvoltage levels of all pole lines were simulated under abnormal mode, which means some of the stations were unconnected to the DC grid during the simulations. The maximum overvoltage, out of hundreds of simulation results of different unconnected stations, occurred in case C station is unconnected. In this section, overvoltages of transmission lines when C station is unconnected are shown in Table 2. The maximum overvoltage has the peak of 1034 kV (1.93 pu), appearing in the CD line, caused by

the A-B positive line ground fault. The overvoltage distribution and probability distribution of C-D line are shown in Figure 5, under situation of the unconnected C station. The maximum overvoltage waveform and the corresponding pole bus waveform are shown in Figure 6. It can be seen that the maximum overvoltage occurs during the DC circuit breaker opening process. Depending on the time difference between peaks, the peaks of the waveform in Figure 6 were caused by the different wave velocities of reflected positive sequence and zero sequence waves.

4. Insulation Coordination of Pole Lines

According to IEC 60071-3, the switching impulse 50% flashover voltage $U_{50\%s}$ of air gap between wire and tower can be calculated in (10), which can be used for insulation coordination of pole lines:

$$U_{50\%s} = \frac{K_a K_3'}{(1 - 2\sigma_s)} U_m, \tag{10}$$

where U_m is maximum voltage of system (535 kV in this paper), K_a is discharge voltage correction coefficient of air density and humidity of the switch impulse voltage, K_3' is the per unit value of overvoltage, and σ_s is the standard deviation.

Altitude correction can be conducted by IEC 60071-2 with safety margin. The calculation formula is showed by

$$K_a = e^{m(H/8150)}, \tag{11}$$

TABLE 3: Required values of air gaps under switch impulse overvoltage for a 4-terminal MMC-HVDC project.

Lines	Switching overvoltage level (p.u.)	Required 50% switching impulse discharge voltage	Altitude (m)		
			0	1000	2000
A-D	1.9 p.u.	1130	2.73	3.13	3.63
C-D	1.93 p.u.	1147	2.79	3.21	3.73
B-C	1.9 p.u.	1130	2.73	3.13	3.63
A-B	1.7 p.u.	1011	2.4	2.75	3.2

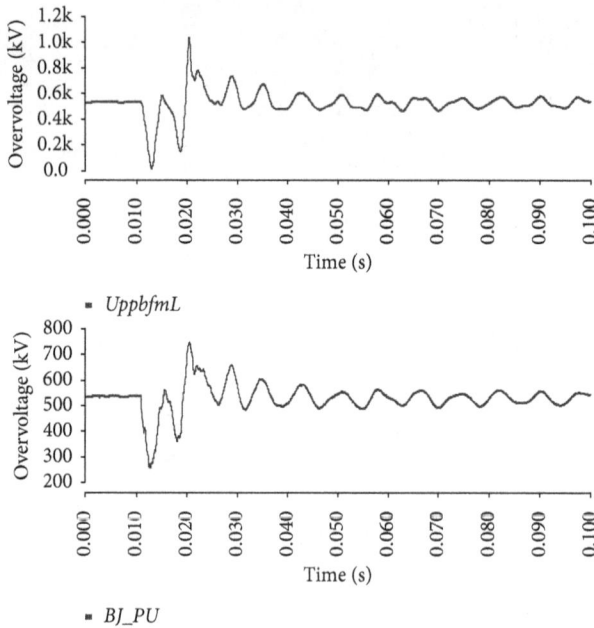

■ UppbfmL

■ BJ_PU

FIGURE 6: The waveforms of maximum overvoltage on C-D line and pole bus of D station under the mode of unconnected C station.

where H is the altitude above sea level measured in meters and the value of m depended on the type of voltage impulse.

The discharge curve of air gap of double circuit tower is provided by CEPRI tested on upper layer of ±500 kV DC double circuit transmission line, and the discharge curve of air gap of single circuit tower is tested on ±500 kV real type tower with V-type insulator string. The air gap flashover curve could be obtained by referring to relevant papers [18]. The calculated air gaps of single circuit and double circuit on one tower under switch impulse voltage are shown in Table 3. The required values of air gaps are considered under normal mode and abnormal mode (without C substation).

5. Conclusion

(1) It is different from circuit breaker adopted MMC-HVDC system and LCC-HVDC when confronted with overvoltage of transmission line. In LCC-HVDC, only the induced overvoltage on sound line caused by single-pole line fault is considered. Besides the induced overvoltage, the fault clearing overvoltage and reclosing overvoltage of transmission line have to

be taken into consideration in the case of multiterminal MMC-HVDC project.

(2) The statistics overvoltage on the sound line caused by ground fault presents the umbrella type distribution along the line. In the middle of line the overvoltage is high and to both ends of the line diminishing. In this paper, the simulated overvoltage of the project is up to 1.93 p.u., occurring during the DC circuit breaker clearing the DC side failure. The suppression measures of DC circuit breaker causing overvoltage need further study.

(3) The required minimal clearance of pole lines under switch impulse overvoltage can be selected as 2.79 m at an altitude of 0 m. Differential air gap selection for various transmission lines is also recommended.

(4) The mechanism for the generation of overvoltages has not been clearly explained. The quantitative influence of the line length, the grounding mode of the valve side of converter, and the network structure on the overvoltage level should be further studied to support the project design.

Conflicts of Interest

The authors declare that there are no conflicts of interest regarding the publication of this paper.

Acknowledgments

The authors are grateful for the support provided by the National Key Research and Development Program of China (Grant 2016YFB0900900) and the SGCC Project of Overvoltage and Insulation Coordination Technology Research of ±500 kV Flexible DC Grid (GY71-15-076).

References

[1] H. Zhiyuan, Y. Zhao, and T. Guangfu, "Key Technology Research and Application of the ±320 kV/1000 MW VSC-HVDC," Smart Grid, vol. 4, pp. 124–132, 2016.

[2] M. Weimin, W. Fangjie, Y. Yiming et al., "Flexible HVDC Transmission Technology's Today and Tomorrow," High Voltage Engineering, vol. 40, pp. 2429–2439, 2014.

[3] T. Guangfu, H. Zhiyuan, and P. Hui, "Discussion on Applying the VSC-HVDC Technology in Global Energy Interconnection," Smart Grid, vol. 4, no. 2, pp. 116–123, 2016.

[4] H. Alyami and Y. Mohamed, "Review and development of MMC employed in VSC-HVDC systems," in Proceedings of the

2017 IEEE 30th Canadian Conference on Electrical and Computer Engineering (CCECE), pp. 1–6, April 2017.

[5] L. Xingyuan, Z. Qi, W. Yuhong et al., "Control strategies of voltage source converter based direct current transmission system," *Gaodianya Jishu/High Voltage Engineering*, vol. 42, no. 10, pp. 3025–3037, 2016.

[6] N. Zhang, C. Kang, D. S. Kirschen et al., "Planning pumped storage capacity for wind power integration," *IEEE Transactions on Sustainable Energy*, vol. 4, no. 2, pp. 393–401, 2013.

[7] J. Lyu, X. Cai, and M. Molinas, "Frequency Domain Stability Analysis of MMC-Based HVdc for Wind Farm Integration," *IEEE Journal of Emerging and Selected Topics in Power Electronics*, vol. 4, no. 1, pp. 141–151, 2016.

[8] H. Chang, D. Chen, Y. Wu, and Y. Ma, "Analysis and optimization strategy of power disturbance on Xiamen flexible HVDC project," *Gaodianya Jishu/High Voltage Engineering*, vol. 42, no. 10, pp. 3045–3050, 2016.

[9] M. Weimin, J. Weiyong, and L. Yanan, "System design for dalian VSC-HVDC power transmission project," *Electric Power Construction*, vol. 34, no. 5, pp. 1–5, 2013.

[10] Z. Xu, G. Liu, and Z. Zhang, "Research on fault protection principle of DC grids," *Gaodianya Jishu/High Voltage Engineering*, vol. 43, no. 1, pp. 1–8, 2017.

[11] F. B. Ajaei and R. Iravani, "Cable surge arrester operation due to transient overvoltages under DC-side faults in the MMC-HVDC link," *IEEE Transactions on Power Delivery*, vol. 31, no. 3, pp. 1213–1222, 2016.

[12] M. Wang, Y. Hu, W. Zhao, Y. Wang, and G. Chen, "Application of modular multilevel converter in medium voltage high power permanent magnet synchronous generator wind energy conversion systems," *IET Renewable Power Generation*, vol. 10, no. 6, pp. 824–833, 2016.

[13] Z. Chengyong, *Modeling and Simulation of Flexible HVDC*, China Electric Power Press, Beijing, China, 2014.

[14] F. Zhang, J. Xu, and C. Zhao, "New control strategy of decoupling the AC/DC voltage offset for modular multilevel converter," *IET Generation, Transmission and Distribution*, vol. 10, no. 6, pp. 1382–1392, 2016.

[15] Y. Dong, W. Ling, J. Tian et al., "Control protection system for Zhoushan multi-terminal VSC-HVDC," *Electric Power Automation Equipment*, vol. 36, no. 7, pp. 169–175, 2016.

[16] K. Ou, H. Rao, Z. Cai et al., "MMC-HVDC simulation and testing based on real-time digital simulator and physical control system," *IEEE Journal of Emerging and Selected Topics in Power Electronics*, vol. 2, no. 4, pp. 1109–1116, 2014.

[17] Z. Gu, Y. Tang, W. Liu et al., "Electromechanical transient—electromagnetic transient hybrid simulation of double-fed induction generator," *Power System Technology*, vol. 39, no. 3, pp. 615–620, 2015.

[18] W. Liao, Y. Ding, Z. Sun, and Z. Su, "Altitude correction of switching impulse flashover voltage of tower gaps with V-shaped insulator strings for HVDC power transmission lines," *Dianwang Jishu/Power System Technology*, vol. 36, no. 1, pp. 182–188, 2012.

10

Verification of Thermal Models of Internally Cooled Gas Turbine Blades

**Igor Shevchenko, Nikolay Rogalev, Andrey Rogalev ⓘ,
Andrey Vegera, and Nikolay Bychkov**

National Research University Moscow Power Engineering Institute, 14 Krasnokazarmennaya Street, Moscow 111250, Russia

Correspondence should be addressed to Andrey Rogalev; r-andrey2007@yandex.ru

Academic Editor: Lei Tan

Numerical simulation of temperature field of cooled turbine blades is a required element of gas turbine engine design process. The verification is usually performed on the basis of results of test of full-size blade prototype on a gas-dynamic test bench. A method of calorimetric measurement in a molten metal thermostat for verification of a thermal model of cooled blade is proposed in this paper. The method allows obtaining local values of heat flux in each point of blade surface within a single experiment. The error of determination of local heat transfer coefficients using this method does not exceed 8% for blades with radial channels. An important feature of the method is that the heat load remains unchanged during the experiment and the blade outer surface temperature equals zinc melting point. The verification of thermal-hydraulic model of high-pressure turbine blade with cooling allowing asymmetrical heat removal from pressure and suction sides was carried out using the developed method. An analysis of heat transfer coefficients confirmed the high level of heat transfer in the leading edge, whose value is comparable with jet impingement heat transfer. The maximum of the heat transfer coefficients is shifted from the critical point of the leading edge to the pressure side.

1. Introduction

A gas turbine engine service life is mainly determined by lifespan of high-pressure turbine blades [1]. Creation of blade with effective cooling system is a complicated and labour-intensive process. It includes selection and design of cooling loop as well as its optimisation; gas-dynamic, thermal, and strength calculations; experimental study; development of production technology; and evaluation of reliability [2–4].

Hydraulic and thermal models of cooling passages providing the determination of boundary conditions from the side of gas and cooling air are used for the purpose of calculation of thermal state of a blade during operation. The adequacy of both hydraulic and thermal models of cooling passage, which are based on the values of hydraulic resistance of paths with turbulators and on criterion dependencies for calculation of heat transfer from the passage walls to cooling air, has an influence on the accuracy of obtained results. Possible inadequacy of thermal and hydraulic models of manufactured full-size blade prototype is related to the fact that the values of hydraulic resistance and criterion equations used for the design are usually obtained from models with constant cross section of a greater size than passages of a real blade while the modelling conditions significantly differ from the full-scale ones [5–8]. There is almost no data of calculation of heat transfer in transitional parts of passage junctions, turns, and bifurcations. All of the above decrease the accuracy of calculation of temperature fields of a blade and hence the safety factor.

The modelling of blade temperature field as a result of a simultaneous solution of a single package of problems of gas and coolant flows and heat transfer with conjugation of heat exchange conditions on the walls is an up-to-date design method. However, the accuracy of obtained results (and their matching to the experimental data) depend largely on a chosen turbulence model, an approach to three-dimensional modelling and a density of grid [9–11].

Figure 1 demonstrates an example of comparison results of calculated and measured temperature values in the middle section of a studied blade, which were obtained while

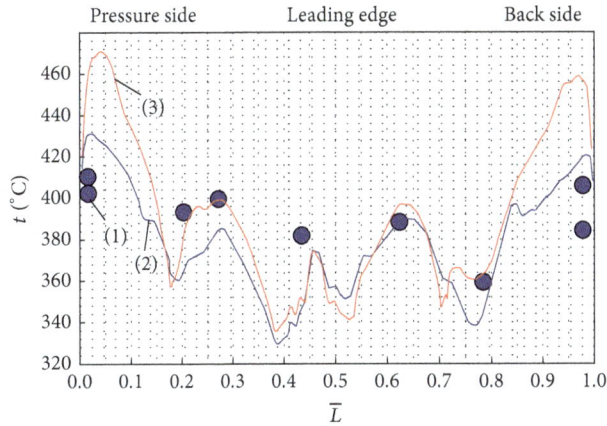

FIGURE 1: Temperature at the blade profile outline. (1) Experiment by Baranov Central Aircraft Engine Institute; (2) RKE model conjugated approach; (3) RKE model semiconjugated approach.

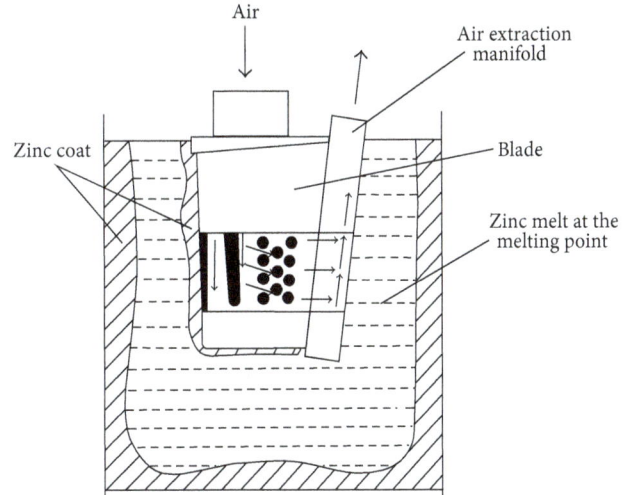

FIGURE 2: Method of calorimetric measurement in the molten metal thermostat.

selecting the most accurate calculating model. In particular areas, the discrepancy is up to 40°C while the gas temperature upstream the blade cascade is 614°C [12].

Verification of thermal and hydraulic models shall be performed by testing the full-size blade prototype manufactured according to a batch process by using the lost-wax casting [13–15]. Essential discrepancies of thermal and hydraulic parameters to calculated values determined during the test require correction of design and technological documentation, modification of a mould, or even production of a new one for pressing ceramic cores that hollow the blade during the casting [16, 17]. This results in significant additional costs and increases the time required for development of cooled blade as well as an engine as a whole.

During the process of reengineering of gas turbine hot section, a problem of recreation of thermal and hydraulic model of cooled blade for the purpose of calculation of its thermal state arises [18–20].

As for blades cooled by convection, a method of calorimetric measurement in a molten metal thermostat can help solve the problem of verification of thermal and hydraulic models of turbine blade cooling system [21].

This paper is dedicated to the development of verification method of thermal models of internally cooled gas turbine blades.

2. Method of Calorimetric Measurement in the Molten Metal Thermostat

Method of calorimetric measurement in the molten metal thermostat allows obtaining local values of heat flux in each point of blade surface experimentally [22]. An effect of phase transition of chemically pure metals is a physical basis of calorimetric method. The method proceeds as follows. A blade equipped with manifolds for supply and extraction of cooling air shall be submerged into the melt of pure zinc heated over its melting point. The blade shall be heated to the temperature of melt and then cooled along with it down to the equilibrium state at zinc melting point and below by

the cooling air within a particular period of time. After that the blade shall be taken out of the melt (Figure 2). A metal coat formed during the process of heat transfer to the coolant passing through internal paths of the blade is crystallized on its outer surface.

Experiments in the zinc melt are usually carried out in a wide range of pressure drops $\pi_a = P_b/P_0$, where the following parameters are measured: air consumption through the blade; incoming and outgoing air temperatures. After a series of experiments, zinc coats removed from the studied blade shall be weighted which is necessary for evaluation of the quality of experiments with respect to convergence of heat balance. The heat balance is defined as follows:

$$G_a \cdot C_p \cdot \left(T_a^* - T_{a\_out}^*\right) = \frac{m_c}{\tau} \cdot L_{cr}, \qquad (1)$$

where G_a represents air consumption through the blade; C_p represents heat capacity of air; T_a^* represents temperature of air incoming to the blade; $T_{a\_out}^*$ represents temperature of air outgoing from the blade; m_c represents coat mass; τ represents blowing time; L_{cr} represents zinc crystallization heat.

If the quality of experiment is high, then the difference in values of heat removed from the blade which were obtained by measuring of zinc coat masses and thermocouple indications should not exceed 5%.

After the convergence of heat balance is checked, coats shall be marked and cut with respect to sections accepted for the study. Usually these sections match constructive cross sections of the blade. Zinc coat images enlarged by 8 to 10 times shall be obtained by scanning. The use of enlarged images allows for essential improvement of accuracy of coat thickness measurement.

The heat flux for each calculation point q_i is defined as follows:

$$q_i = \frac{\rho_c \cdot L_{cr}}{\tau} \cdot \left[\left(\delta_i - \delta_d\right) \pm \frac{\left(\delta_i - \delta_d\right)^2}{2 \cdot R_i}\right], \qquad (2)$$

where R_i represents radius of outer surface of blade; δ_i represents coat thickness; $\delta_d = 0.1\,\text{mm}$ represents the thickness of dragged metal layer forming while the blade is taken out of the melt; ρ_c represents density of zinc coat; the "−" sign is used for points on the concave surface of blade.

In order to determine the values of local coefficients of heat elimination to the air, one shall know the temperature of cooling air T_{ai}^*. T_{ai}^* values shall be calculated with respect to distribution of coolant G_{ai} over internal passages for test conditions in thermostat and in order to solve the heat balance equations composed for particular areas of blade surface at which the values of heat flow are determined.

The local heat transfer coefficient h_{ai} is defined as follows:

$$h_{ai} = \frac{q_i}{K_{shi}} \cdot \left(T_m - T_{ai}^*\right)^{-1}, \tag{3}$$

where K_{shi} represents shape coefficient which considers the geometric difference between the blade wall and a flat thin wall; T_m represents zinc melting point.

The Nu_{ai} and Re_{ai} criteria shall be calculated for each point of blade surface; then they shall be approximated by the following dependencies using the least squares method:

$$\text{Nu}_{ai} = A_i \cdot \text{Re}_{ai}^{mi}. \tag{4}$$

The error of determination of local heat transfer coefficients δh_i consists of the error of determination of heat flux and local temperature of cooling air. The relative error of determination of heat flux δq_i is defined as follows:

$$\delta q_i = \sqrt{\left(\delta\delta_i\right)^2 + \left(\delta R_i\right)^2 + \left(\delta\tau_i\right)^2 + \left(\delta\rho_i\right)^2 + \left(\delta L_i\right)^2}, \tag{5}$$

where $\delta\delta_i$ represents the relative error of determination of crust thickness; δR_i represents the relative error of determination of curvature radius of the blade surface; $\delta\tau_i$ represents the relative error of blowing time determination; δL_i represents the relative error of specifying the latent value of zinc crystallization; $\delta\rho_i$ represents the relative error of setting the heat flux.

If we assume that the values of L_i and ρ_i are specified with high accuracy, the relative error δq_i does not exceed ±5%.

The error of determination of local heat transfer coefficient depends mainly on the error of determination of distribution of the air flow through the cooling channels δG_i and, accordingly, on air heating and local temperature T_{ai}^*. The value of δG_i is determined by the complexity of the blade cooling path.

The error of determination of local heat transfer coefficients usually does not exceed ±8% for blades with radial channels. An important feature of method is that all tests shall be carried out under the same external heat load and blade outer surface temperature T_b equals zinc melting point $T_m = 692.4\,\text{K}$.

3. Facility for Implementation of Method of Calorimetric Measurement in the Molten Metal Thermostat

Despite constructive and layout solutions, a facility for testing the cooled blades in molten metal thermostat shall include the following elements (Figure 3): a system of air purification and supply; a working area for connection of object to be studied; heat measuring system consisting of a crucible with zinc melt and an oven for its heating and melting; rotary lifting mechanism ejecting the crucible from the oven, moving it to the blade, allowing putting the blade into the melt and taking it out; a system of air consumption control and measurement and recording the experimental data.

It is possible to use facilities with various layouts which basically differ from one another by degree of mechanisation of test process. These constructive differences are determined by the amount of studies, the design of tested blade, its dimensions, allocation of holes for air ejection from the blade, and experimental modes, pressure of cooling air, and the need of creation of counterpressure at the blade output.

The thermostat is the crucible with zinc melt whose cross section is oval. The crucible shall be made of high-carbon steel; its internal surface shall be coated with a thin ceramic coat that prevents the contact of the crucible surface with the melt. Dimensions of crucible and the mass of zinc shall be determined by the mass and dimensions of blade equipped with manifolds for supply and extraction of cooling air. The control panel allows manipulating the rotary lifting mechanism and cooling air supply system, in both manual and automatic modes according to a signal of timer, and recording the experimental data.

The working area connects the blade to air supply mains and is a constructive unit, which also includes the blade prototype to be studied. The working area design shall be specially developed for each series of blades and is determined by test tasks. As the working area shall not influence the thermal state of blade during the test and shall provide the given parameters of incoming cooling air, it shall comply with the following requirements. The working area shall provide the following: ability of blowing of blade with coolant while it is dipped into the zinc melt, with coolant ejection to the atmosphere and with counterpressure; the minimum heat removal from the blade by working area; absence of massive elements of connections for the purpose of decrease of thermal accumulative ability of unit.

Figure 4 demonstrates the experimental test bench. The description of the main measurement equipment is presented in Table 1.

4. Blade Thermal Model Verification Technique

Method of calorimetric measurement in the molten metal thermostat can be effectively used for experimental verification of thermal models of internally cooled blades.

The thermal model shall be verified by comparison of distribution of heat flux q_c over the outer surface, which is calculated according to verified thermal model for test conditions, and distribution of heat flux q_z, which is determined from the results of experiments in the molten metal thermostat. Comparison of heat flows shall consider all parameters of thermal model that determine the thermal state of blade under the operating conditions: air consumption and temperature at the branches of equivalent hydraulic network,

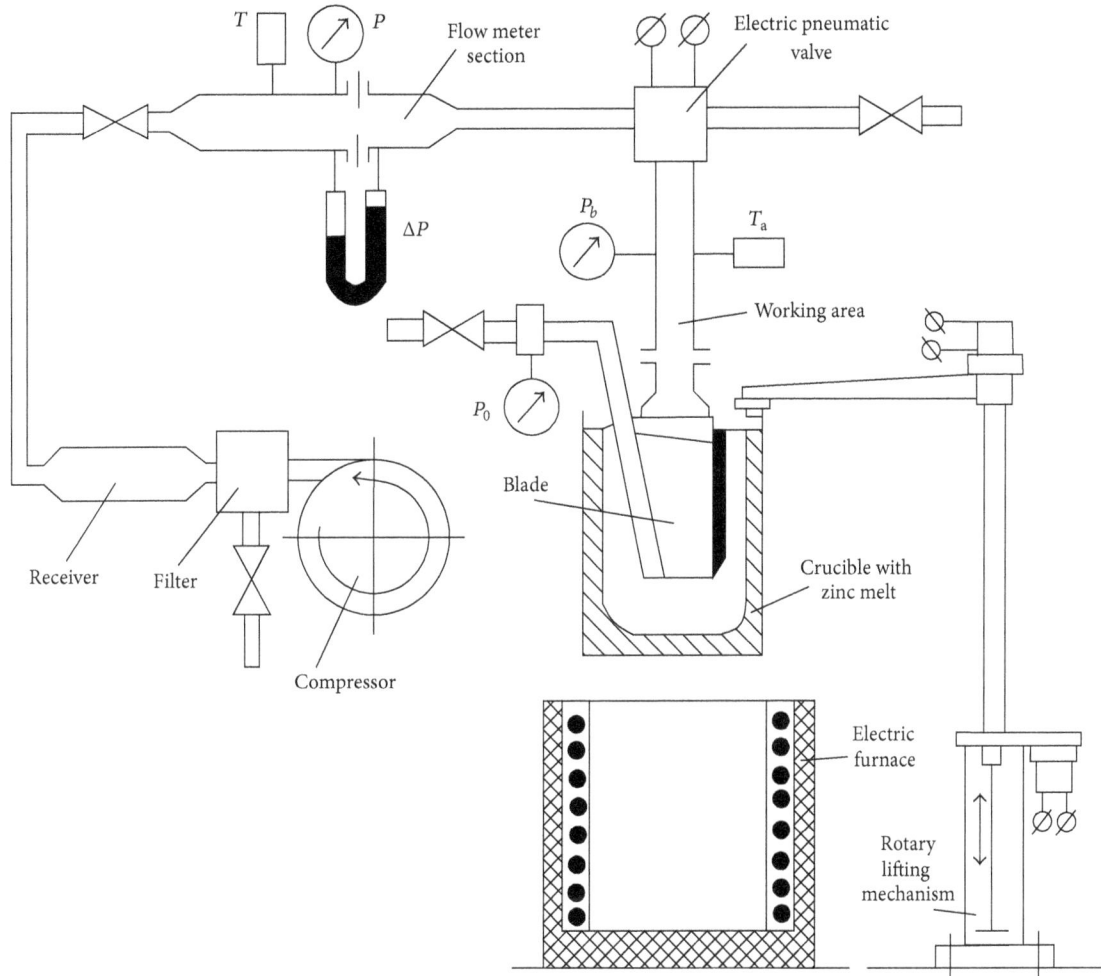

FIGURE 3: Experimental facility diagram.

TABLE 1: Accuracy of the measurement equipment.

Measurement equipment	Measurement error
Air flow meter EE741-A6D2DN20 + HAO79020	1.32%
Pressure transducer MBS 4500	0.5%
LED indicator-measuring microprocessor 2TRM0-N.U.	0.5%
Pressure transducer PD200-DI0, 1-315-0, 1-2-N	0.1%
High-precision industrial pressure sensor DMP 331i 111-2001-1-1-100-800-1-11R	0.1%
Thermoelectric converter DTPK454-05.200/2C.1	0,015 T

coefficients of heat transfer to the coolant, and blade wall shape coefficients.

While calculating the heat flow q_c under the modelling conditions of molten metal thermostat, the following boundary condition of the first kind shall be set: the surface temperature equals zinc melting point 692.4 K; the following boundary conditions of the third kind shall be set for the surface of cooling passages: temperature of cooling air T_{ai}^* and coefficients of heat transfer from the blade wall to cooling air h_{ai} obtained by using the verified hydraulic model of blade for conditions of blade test in the molten metal thermostat. The result of calculation is a two-dimensional temperature field of

blade cross section and a value of heat flux in given points of outer surface.

In order to make a decision regarding the adequacy of thermal model, dependence allowing evaluation of the influence of detected differences between q_z and q_c on the thermal state of blade is required. For this purpose, the use of dimensionless relative temperature (the intensity of blade cooling) is advisable:

$$\theta = \frac{T_g^* - T_b}{T_g^* - T_a^*}, \tag{6}$$

where T_g^* represents hot gas temperature.

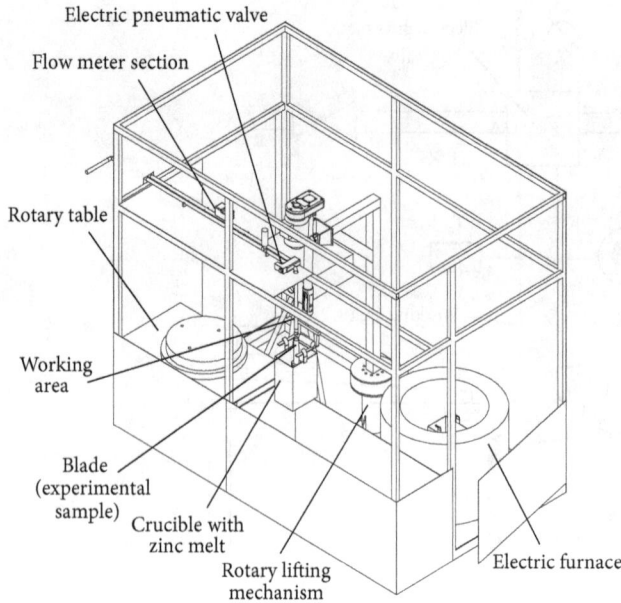

FIGURE 4: Test bench.

The following ratio is valid for each point of blade surface:

$$\frac{1/\theta_{bz} - 1}{1/\theta_{bc} - 1} = \frac{K_{sh} \cdot h_{ac}}{K_{sh} \cdot h_{az}} = \frac{q_z}{q_c}, \tag{7}$$

where θ_{bc} represents a dimensionless temperature of blade surface obtained from the results of calculation using the thermal model being verified; θ_{bz} represents a dimensionless temperature of blade surface obtained from the results of experiments in molten metal thermostat; h_{ac} represents a coefficient of heat transfer to cooling air applied in the thermal model being verified; h_{az} represents a coefficient of heat transfer to cooling air calculated on the basis of results of experiment.

As values of experimental and calculated air consumption coincide, the ratio is valid:

$$\frac{h_{ac}}{h_{az}} = \frac{\lambda_{az}}{\lambda_{ac}} \cdot \left(\frac{\mu_{az}}{\mu_{ac}}\right)^{-m}, \tag{8}$$

where λ_{az} represents thermal conductivity of air applied in the thermal model being verified; λ_{ac} represents thermal conductivity of air calculated on the basis of results of experiment; μ_{az} represents viscosity of air applied in the thermal model being verified; μ_{ac} represents viscosity of air calculated on the basis of results of experiment; m represents an exponent in the criterion equation for calculation of h_{ac}.

The exponential dependencies are used for calculation of variations of thermal conductivity λ_a and viscosity of air μ_a:

$$\lambda_a = f_1 \left(T_a\right)^{0.64};$$
$$\mu_a = f_2 \left(T_a\right)^{0.76}. \tag{9}$$

With regard to specific values of exponent at the Reynolds number $m = 0.6$–0.8, proceeding from (9), we shall obtain

$$\frac{h_{az}}{h_{ac}} = \left(\frac{T^*_{az}}{T^*_{ac}}\right)^{0.64-0.76 \cdot m}. \tag{10}$$

If $m = 0.8$, then the difference by 1.5 times in temperatures results in the difference $q_z/q_c = 1.013$. Using (7) and assuming θ_{bc} for particular area of blade as a permissible variation $\Delta\theta$, a permissive variation of q_c and q_z is defined as follows:

$$\left(\frac{q_z}{q_c}\right)_p = \frac{1/\left(\theta_{bc} - \Delta\theta\right) - 1}{1/\theta_{bc} - 1}. \tag{11}$$

If the obtained parameter value is $q_z/q_c \leq (q_z/q_c)_p$, then the thermal model describes the processes of internal heat exchange appropriately for this part of blade.

Comparison of q_z and q_c is advisable under several test modes, that is, for various pressure drops. The q_z/q_c value should not practically change depending on the P_b/P_0 pressure drop. In this case the major error in criterion dependencies of thermal model will relate to the value of coefficient at the Re number. If the comparison of experimental and calculated heat flows q_z/q_c, obtained for various pressure drops, gives the same value of variation in considered areas, one can conclude that the m exponent in the used criterion dependence of $Nu_i = A_i Re_i{}^m$ kind corresponds to the nature of cooling air flow.

In this case, criterion dependencies of thermal model for the calculation of heat transfer to the cooling air can be defined more precisely by using the ΔK_i correction factor:

$$\Delta K_i = \frac{q_{ci}}{q_{zi}}. \tag{12}$$

Then the revised value of A_{ci} in the criterion equation is equal to

$$A_{ci} = \Delta K_i \cdot A_i. \tag{13}$$

As various modes of cooling system operation with respect to pressure drops demonstrate essential difference of q_z/q_c, one can assume that the value of m exponent in the criterion equation m at this section of passage differs from that accepted in the thermal model. Usually turns of flow, stagnant zones in flow confluence points, areas of jet, and cyclone cooling can be these sections of passage [23–26].

In this case, the results of blade test in the molten metal thermostat are used for calculation of local heat transfer coefficients and derivation of criterion equation for the purpose of elaboration of thermal model.

5. Verification Technique Approbation

The developed technique was applied for verification of thermal model of blade of high-pressure turbine of a fixed gas turbine unit. The blade had an air distribution duct cooling system with ejection of air into the flow range of turbine through the slit in the exit edge and through the hole in

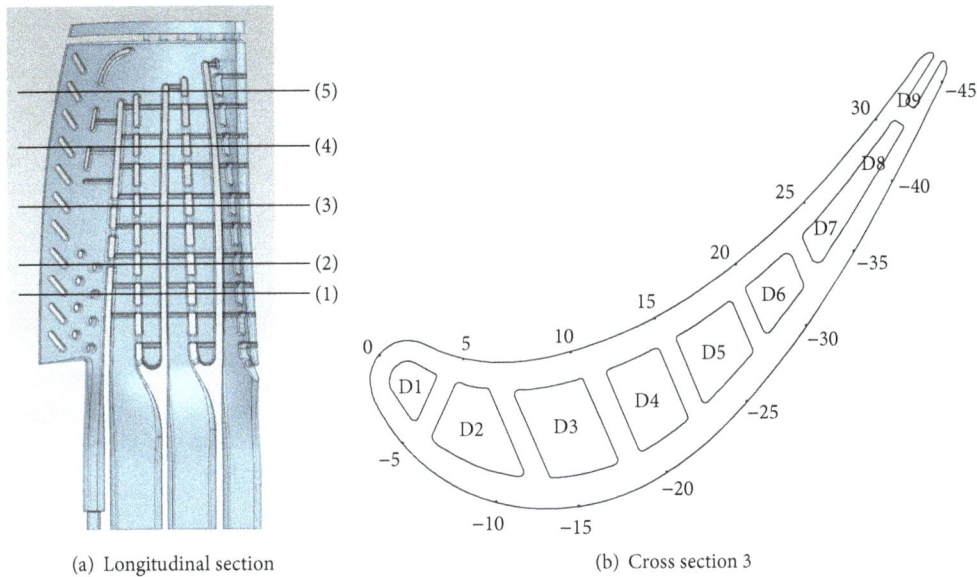

(a) Longitudinal section

(b) Cross section 3

FIGURE 5: Blade.

the end wall of blade. Figure 5 demonstrates a longitudinal section of internal space of the blade and cross section 3 in the middle throughout the blade length. A centrifugal flow of cooling air is realised in ducts D1–D6.

Figure 6 demonstrates that the internal channels of blade were modelled as an equivalent hydraulic network. The input node has number 20 and the output node 89. The pressure difference in the cooling path is modelled by setting pressure in these nodes. The calculation was carried out using the program "Gidra," developed by Baranov Central Aircraft Engine Institute.

The calculations were carried out for pressure differences corresponding to the experimental conditions. The wall temperature of the model was set equal to the crystallization temperature of zinc. The air heating in the channels was determined from the heat balance equation for each channel of the cooling path:

$$\Delta T_{ai}^* = \frac{\sum q_i}{C_p \cdot G_{ai}}. \tag{14}$$

The cooling air flow through the internal passages of the blade was measured under isothermal conditions: the wall temperature was equal to the air temperature. The difference between experimental and calculated consumption characteristics does not exceed 3%. Thus, the hydraulic model of the blade is adequate.

The blades were refined for tests: to reduce the thermal inertia the lock was abraded. The collectors for air extraction were made of a tube with a diameter of 20 mm and soldered with high-temperature solder in a vacuum furnace. The photo of the prepared blade is shown at Figure 7. To measure the temperature of the air at the inlet and outlet in section 3, the thermocouples were installed.

The values of the cooling air temperature, the blade inlet pressure, and the cooling air flow were recorded on the

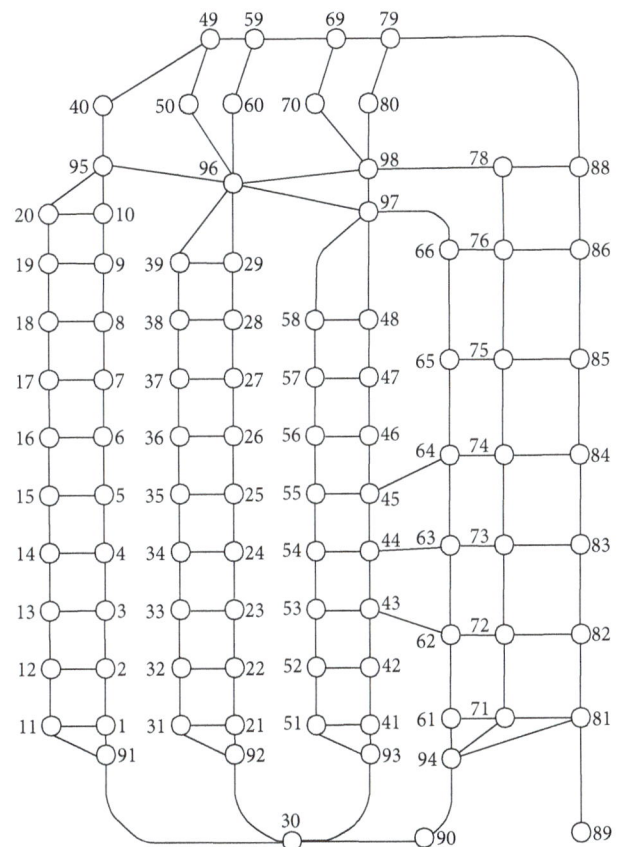

FIGURE 6: An equivalent hydraulic network of the blade cooling channels.

hard disk of the National Instruments industrial computer. Recording resolution was equal to 0.1 s. The accuracy of the temperature measurement was ±1°. The accuracy of the air flow measurement was ±0.1 g/s.

FIGURE 7: The blade model prepared for test.

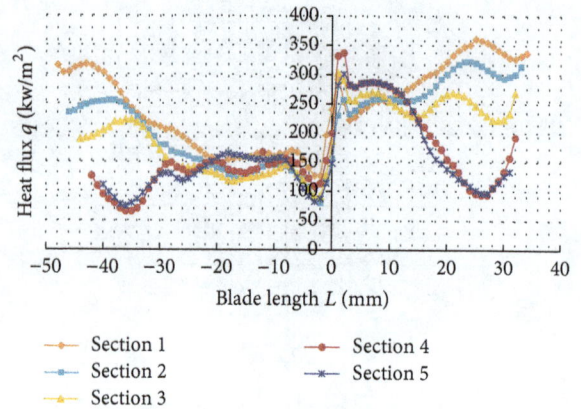

FIGURE 9: Heat flux in sections of blade obtained by averaging over three prototypes, $P_b/P_0 = 1.68$.

FIGURE 8: Photograph of zinc coat obtained under the pressure drop $P_b/P_0 = 1.68$.

A comparison between experimental and calculated values of the air outlet temperatures from the trailing edge showed that the difference does not exceed 9%. This allows us to conclude that the thermal model accurately calculates the total heating of air in the channels. Thus, the thermal model of the blade allows calculating the total air heating in the channels accurately.

The blade was blown in the zinc melt under pressure drops $P_b/P_0 = 1.48$, 1.68, 1.78, 1.97, and 2.37. P_b represents the cooling system inlet pressure, and P_0 represents the atmospheric one. For each working mode three experiments were carried out. Each thermogram (zinc crust) obtained during the tests was checked by the heat balance (see (1)).

Figure 8 demonstrates a photograph of zinc coat (from the back side) obtained under the pressure drop $P_b/P_0 = 1.68$. The thickness of coat unambiguously characterises the heat exchange intensity.

The qualitative analysis of coat demonstrates zones of duct flow along the blade. At the blade periphery, behind the barrier forming the flow turn into the duct of exit edge, an abrupt decrease of coat thickness was observed. It allows for assumption that there is a stagnant zone behind the barrier which causes the decrease of intensity of heat exchange.

Coats have been cut in calculated sections using the electric erosive method. Images of sections have been enlarged by 20 times and marked with the 1 mm step along the outer surface. Then the thickness of zinc coat was measured. An absolute error of coat thickness measurement did not exceed 0.01 mm. Then measured values of coat thickness δ_i have been averaged over the experiments. The obtained average value of δ_{iav} was used for calculation of heat flux along the outer surface of blade.

Graphs of distribution of heat fluxes along the outer surface in five sections depending on the pressure drop were drawn; a graph for the pressure drop $P_b/P_0 = 1.68$ is shown in Figure 9.

Heat flows from the back side are greater by 1.8–2.0 times. At the exit edge from the back side, q is greater by 10–20%. In sections 1, 2, and 3 the decrease of heat flows from the root sections to peripheral ones is observed in radial ducts. It is related to the heating of air along the duct length and the increase of its temperature. In sections 4 and 5 an abrupt decrease of heat flows after section length $L = 15$ mm from the pressure side and $L = -26$ mm from the back side is observed.

The local minima of heat flows ($q = 90$–100 kW/m^2) are located from the back side in the point $L = -36$ mm and from the pressure side in $L = 26$ mm. The heat flow in the local minima points decreases by 2.5–4 times compared to analogous points of sections 1, 2, and 3. Two air flows moving in the opposite directions meet in this area of internal space and form a stagnant zone. The stagnant zone can shrink during the rotation due to the action of centrifugal forces [27–31].

In order to verify the thermal model of blade, the distribution of heat flux in five sections along the blade length for two pressure drop values of 1.48 and 1.97 was calculated using special software developed for this purpose. The program allows calculating q_{zi} at specified points on the blade outer surface and providing the rapid formulation and solution of medium-sized computational grids (about 1000–1500 grid

(a) Computational grid

(b) Temperature field for test conditions in the molten metal thermostat, $P_b/P_0 = 1.97$

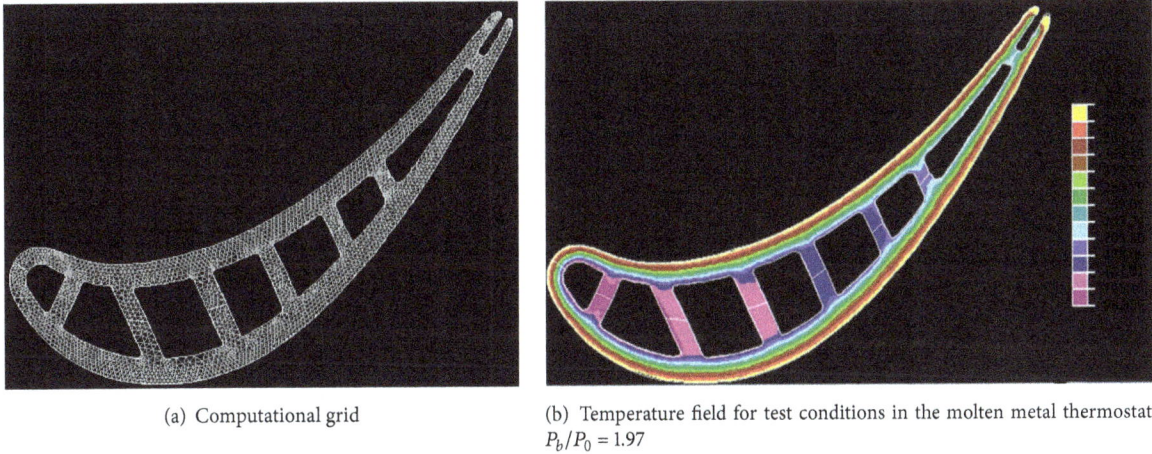

FIGURE 10: Section 3 of the blade.

nodes). It solves two-dimensional steady and unsteady state thermal conduction equations. The following mathematical formulation of the thermal conductivity process is used in program:

(i) plane-parallel steady flow:

$$\frac{d}{dx}\left(\lambda \cdot \frac{dT}{dx}\right) + \frac{d}{dy}\left(\lambda \cdot \frac{dT}{dy}\right) = Q; \qquad (15)$$

(ii) plane-parallel unsteady flow:

$$\frac{d}{dx}\left(\lambda \cdot \frac{dT}{dx}\right) + \frac{d}{dy}\left(\lambda \cdot \frac{dT}{dy}\right) + RC \cdot \frac{dT}{d\tau} = Q, \qquad (16)$$

where λ represents thermal conductivity; T represents temperature; Q represents power of the heat source/sink.

As an example, the computational grid for the section 3 is represented at Figure 10(a). The number of nodes was equal to 1400. The calculated temperature field in section 3 for the test conditions in zinc melt under the pressure drop of 1.97 is demonstrated at Figure 10(b). The obtained calculated values of heat flow q_{ci} have been compared to results of experiments in the molten metal thermostat q_{zi}.

Figure 11 demonstrates the distribution of heat flux q_z and heat flux q_c in middle section 3 of the model for two values of pressure drop equal to 1.48 and 1.97.

For each point of i on the surface, the permissive variation $(q_z/q_c)_p$ was determined based on the assumed permissive variation of blade surface temperature during the operation equal to ± 10 K. The use of $(q_z/q_c)_p$ allowed determining zones of discrepancy of thermal model and experimental results. These zones are the leading edge and ducts D5–D8 (the pressure side and the back side).

Comparison of q_z/q_c parameter obtained under two pressure drops (Figure 11) has revealed that the difference in all sections does not exceed 5%. It allows concluding that exponents in criterion equations of thermal model accepted as equal to $m = 0.8$ describe the specificity of heat exchange in ducts correctly.

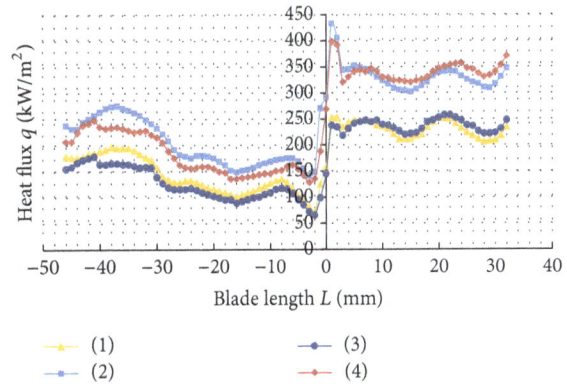

FIGURE 11: Distribution of heat flux along the surface of section 3: (1, 3) $P_b/P_0 = 1.48$; (2, 4) $P_b/P_0 = 1.97$; (1, 2) q_z; (3, 4) q_c.

As $q_z/q_c > (q_z/q_c)_p$ at particular areas of blade, the elaboration of thermal model using the dependency (11) and (12) was performed according to developed technique.

6. Conclusions

(1) An experimental method for thermal model verification of the blade with convective internal cooling based on the comparison of heat fluxes obtained by numerical modelling and experimental investigation in the molten metal thermostat was developed. The error of determination of local heat transfer coefficients using this method does not exceed 8% for blades with radial channels. An important feature of the method is that the heat load remains unchanged during the experiment and the blade outer surface temperature equals zinc melting point – 692.4 K. This experimental approach allows eliminating an influence of heat transfer on the results accuracy.

(2) The dependencies for estimation of calculation and experimental values of the heat flux influence on relative depth of the blade cooling under operation conditions were obtained.

(3) The verification of thermal-hydraulic model of high-pressure turbine blade with cooling allowing asymmetrical heat removal from pressure and suction sides was carried out using the developed method. The areas of blade requiring more accurate thermal model were identified.

(4) An analysis of heat flux distribution on the outer blade surface allows identifying more intensive cooling on the pressure side due to used heat transfer turbulators. The qualitative analysis of coat allows identifying the location of the area with low heat transfer. It is located in the peripheral blade sections behind the turning rib (sections 4 and 5) and associated with the formation of stagnation zone.

(5) An analysis of heat transfer coefficients confirmed the high level of heat transfer in the leading edge, whose value is comparable with jet impingement heat transfer. The maximum of the heat transfer coefficients is shifted from the critical point of the leading edge to the pressure side. A high intensification of heat transfer in the radial channels was detected and it was 1.5–2 times higher on the pressure side.

Conflicts of Interest

The authors declare that they have no conflicts of interest.

Acknowledgments

This study conducted by Moscow Power Engineering Institute has been sponsored financially by the Russian Federation through the Ministry of Education and Science of the Russian Federation under Subsidy Agreement no. 14.577.21.0210 of September 28, 2016, as a part of the Federal Targeted Programme for R&D in Priority Fields for the Development of Russia's S&T Complex for 2014–2020, with Applied Scientific Research Unique Identifier RFMEFI57716X0210.

References

[1] R. F. Hoeft, J. Janawitz, and R. Keck, *Heavy-duty gas turbine operating and maintenance considerations*, GE Energy Services, GA, USA, 1993.

[2] J.-C. Han, S. Dutta, and S. Ekkad, *Gas turbine heat transfer and cooling technology*, CRC Press, Boca Raton, FL, USA, 2nd edition, 2012.

[3] J. Hou, B. J. Wicks, and R. A. Antoniou, "An investigation of fatigue failures of turbine blades in a gas turbine engine by mechanical analysis," *Engineering Failure Analysis*, vol. 9, no. 2, pp. 201–211, 2002.

[4] B. L. Koff, "Gas turbine technology evolution: A designer's perspective," *Journal of Propulsion and Power*, vol. 20, no. 4, pp. 577–595, 2004.

[5] D. Chandran and B. Prasad, "Conjugate heat transfer study of combined impingement and showerhead film cooling near NGV leading edge," *International Journal of Rotating Machinery*, vol. 2015, Article ID 315036, 2015.

[6] M. E. Taslim and J. S. Halabi, "Heat transfer and friction studies in a tilted and rib-roughened trailing-edge cooling cavity with and without the trailing-edge cooling holes," *International Journal of Rotating Machinery*, vol. 2014, Article ID 710450, 14 pages, 2014.

[7] I. V. Shevchenko, A. N. Rogalev, I. V. Garanin, A. N. Vegera, and V. O. Kindra, "Research and development of asymmetrical heat transfer augmentation method in radial channels of blades for high temperature gas turbines," *Journal of Physics: Conference Series*, vol. 891, no. 1, Article ID 012142, 2017.

[8] I. V. Shevchenko, I. V. Garanin, A. N. Rogalev, V. O. Kindra, and V. P. Khudyakova, "Study of design and technology factors influencing gas turbine blade cooling," *Journal of Physics: Conference Series*, vol. 891, no. 1, Article ID 012253, 2017.

[9] D. C. Wilcox, *Turbulence modeling for CFD*, DCW Industries, Inc, La Canada, Calif, USA, 2nd edition, 1998.

[10] J. E. Bardina, P. G. Huang, and T. Coakley, "Turbulence modeling validation," in *Proceedings of the 28th Fluid Dynamics Conference*, Snowmass Village, CO, USA.

[11] C. G. Speziale, "Turbulence modeling for time-dependent RANS and VLES: A review," *AIAA Journal*, vol. 36, no. 2, pp. 173–184, 1998.

[12] M. I. Osipov and A. V. Veretelnik, *Modeling of conjugate problem of friction and heat exchange in transpiration cooling of blades of gas turbines*, vol. 1 of *Mechanical Engineering*, Herald of the Bauman Moscow State Technical University, 2007.

[13] C. Lane, *The Development of a 2D ultrasonic array inspection for single crystal turbine blades*, Springer International Publishing, Cham, Switzerland, 2014.

[14] G. J. S. Higginbotham, "Method of making gas turbine engine blades," Patent US 4417381 A, 1981.

[15] P. W. Schilke, *Advanced Gas Turbine Materials And Coatings*, E Energy, Schenectady, NY, USA, 2004.

[16] D. W. Richerson, *Modern ceramic engineering: properties, processing, and use in design*, CRC Press, Boca Raton, FL, USA, 3rd edition, 2006.

[17] H. Harada, "High temperature materials for gas turbines: the present and future," in *Proceedings of the International Gas Turbine Congress*, Tokyo, Japan, 2003.

[18] K. Mohaghegh, M. H. Sadeghi, and A. Abdullah, "Reverse engineering of turbine blades based on design intent," *The International Journal of Advanced Manufacturing Technology*, vol. 32, no. 9-10, pp. 1009–1020, 2007.

[19] L.-C. Chen and G. C. I. Lin, "Reverse engineering in the design of turbine blades - a case study in applying the MAMDP," *Robotics and Computer-Integrated Manufacturing*, vol. 16, no. 2, pp. 161–167, 2000.

[20] J. Gao, X. Chen, D. Zheng, O. Yilmaz, and N. Gindy, "Adaptive restoration of complex geometry parts through reverse engineering application," *Advances in Engineering Software*, vol. 37, no. 9, pp. 592–600, 2006.

[21] I. V. Shevchenko, A. N. Rogalev, M. I. Shevchenko, and A. N. Vegera, "Method of calorimetric measurements in molten metal thermostat and its application for developing blade cooling system of gas turbines," *International Journal of Applied Engineering Research*, vol. 12, no. 10, pp. 2382–2386, 2017.

[22] I. V. Shevchenko and Y. W. Kim, "Calorimetric heat transfer measurement in zinc bath and its application to blade internal cooling system development," in *Proceedings of the in Proceedings of the 1997 International Mechanical Engineering Congress Exposition*, Dallas, TX, USA, November 1997.

[23] A. Khalatov, N. Syred, P. Bowen, R. Al-Ajmi, A. Kozlov, and A. Schukin, "Innovative cyclone cooling scheme for gas turbine blade: Thermal-hydraulic performance evaluation," in *Proceedings of the ASME Turbo Expo 2000: Power for Land, Sea, and Air, GT 2000*, Germany, May 2000.

[24] A. A. Khalatov, N. Syred, P. J. Bowen, and R. Al-Ajmi, "Quasi two-dimensional cyclone-jet cooling configuration: Evaluation of heat transfer and pressure losses," in *Proceedings of the ASME Turbo Expo 2001: Power for Land, Sea, and Air (GT'01)*, New Orleans, LA, USA, 2001.

[25] N. Winter and H. Schiffer, "Effect of rotation on the cyclone cooling method mass transfer measurements," in *Proceedings of International Symposium on Heat Transfer in Gas Turbine Systems*, pp. 1–17, Antalya, Turkey, 2009.

[26] A. A. Khalatov, N. Syred, P. J. Bowen, and R. Al-Ajmi, "Enhanced cyclone cooling technique for high performance gas turbine blades," in *Proceedings of the 12-th International Heat Transfer Conference*, Grenoble, France, 2002.

[27] J. W. Wagner, B. V. Johnson, and F. C. Kopper, "Heat transfer in rotating serpentine passages with smooth walls," *Journal of Turbomachinery*, vol. 113, no. 3, pp. 321–330, 1991.

[28] W. D. Morris and T. Ayhan, "Observations on the influence of rotation on heat transfer in the coolant channels of gas turbine rotor blades," *Proceedings of the Institution of Mechanical Engineers*, vol. 193, pp. 303–311, 1979.

[29] B. V. Johnson, J. H. Wagner, G. D. Steuber, and F. C. Yeh, "Heat transfer in rotating serpentine passages with trips skewed to the flow," in *Proceedings of the ASME 1992 International Gas Turbine and Aeroengine Congress and Exposition, GT 1992*, Germany, June 1992.

[30] M. E. Taslim, A. Rahman, and S. D. Spring, "An experimental investigation of heat transfer coefficients in a spanwise rotating channel with two opposite rib-roughened walls," in *Proceedings of the ASME International Gas Turbine and Aeroengine Congress and Exposition (GT'89)*, Toronto, Canada, 1989.

[31] T. Lei, Y. Zhiyi, X. Yun, L. Yabin, and C. Shuliang, "Role of blade rotational angle on energy performance and pressure fluctuation of a mixed-flow pump," *Proceedings of the Institution of Mechanical Engineers, Part A: Journal of Power and Energy*, vol. 231, no. 3, pp. 227–238, 2017.

11

Numerical Simulation of Wind Turbine Rotors Autorotation by using the Modified LS-STAG Immersed Boundary Method

Ilia K. Marchevsky and Valeria V. Puzikova

Applied Mathematics Department, Bauman Moscow State Technical University (BMSTU), 5 2nd Baumanskaya, Moscow 105005, Russia

Correspondence should be addressed to Ilia K. Marchevsky; iliamarchevsky@mail.ru

Academic Editor: Lin Zhu

A software package is developed for numerical simulation of wind turbine rotors autorotation by using the modified LS-STAG level-set/cut-cell immersed boundary method. The level-set function is used for immersed boundaries description. Algorithm of level-set function construction for complex-shaped airfoils, based on Bézier curves usage, is proposed. Also, algorithm for the level-set function recalculation at any time without reconstructing the Bézier curve for each new rotor position is described. The designed second-order Butterworth low-pass filter for aerodynamic torque filtration for simulations using coarse grids is presented. To verify the modified LS-STAG method, the flow past autorotating Savonius rotor with two blades was simulated at Re = $1.96 \cdot 10^5$.

1. Introduction

The rotor is the first element in the chain of functional elements of a wind turbine. Its aerodynamic and dynamic properties, therefore, have a decisive influence on the entire system in many respects. The designer faces the problem of finding the relationship between the actual shape of the rotor, for example, the number of rotor blades or the airfoil of its blades, and its aerodynamic properties.

To simulate rotor's dynamics [1–6] and, in particular, its autorotation, there is a need to solve coupled aeroelastic problems. Such problems are complicated for numerical solution, since it is necessary to take into account interference between the flow and moving immersed body. So, the aim of this research is to develop an efficient numerical method for simulation of wind turbine rotors autorotation.

In case of massive body, when its average density is significantly higher than density of the flow, coupled aeroelastic problems can be solved using "step-by-step" weak-coupling numerical algorithm, firstly simulating flow around the body, which moves according to known parameters, and then computing the dynamics of the body under known hydrodynamic loads. Such case is considered in this research.

Immersed boundary methods [7] are suitable for numerical simulation in coupled aeroelastic problems, since they do not require a coincidence of cell edges and boundaries of the computational domain and allow solving problems when domain shape is irregular or it changes significantly in the simulation process due to aeroelastic body motion. The main advantage of these methods is that the mesh should not be reconstructed at every time step.

In the present study, the LS-STAG cut-cell immersed boundary method [8] is used for rotors autorotation simulation. This method permits solving problems on the Cartesian grid. The immersed boundary is represented with the level-set function [9]. Linear systems resulting from the LS-STAG discretization of the Navier-Stokes or Reynolds-averaged Navier-Stokes equations are solved using the BiCGStab method [10] with the ILU and Multigrid [11, 12] preconditioning. An original algorithm for the solver cost-coefficient estimation [13] is used for the optimal parameters of the multigrid preconditioner choice.

2. Governing Equations

The problem is considered in 2D unsteady case when the flow around a rigid airfoil is assumed to be viscous and

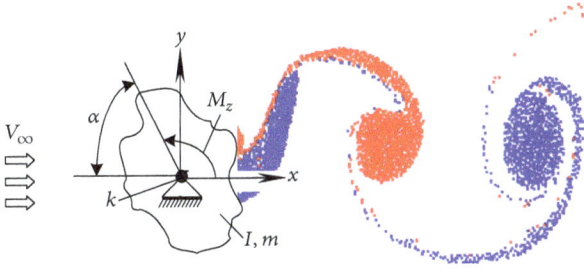

FIGURE 1: The airfoil of irregular shape with one rotational degree of freedom and schematic view of vortex wake behind it.

incompressible. The continuity and momentum equations are as follows:

$$\nabla \cdot \mathbf{v} = 0,$$

$$\frac{\partial \mathbf{v}}{\partial t} + (\mathbf{v} \cdot \nabla)\,\mathbf{v} = \frac{1}{\rho}\nabla p + \nu \Delta \mathbf{v}. \tag{1}$$

Here $\mathbf{v}$ is the velocity vector, p is the pressure, t is the time, ρ is the flow density, and ν is the flow kinematic viscosity. The boundary conditions on the outer boundaries of the computational domain are as follows:

$$\mathbf{v}|_{inlet} = \mathbf{v}_\infty,$$

$$\left.\frac{\partial \mathbf{v}}{\partial \mathbf{n}}\right|_{outlet} = 0, \tag{2}$$

$$\left.\frac{\partial p}{\partial \mathbf{n}}\right|_{inlet\&outlet} = 0.$$

Here $\mathbf{v}_\infty$ is the velocity vector on the inlet boundary and $\mathbf{n}$ is the unit outer normal vector. The boundary conditions on the surface line of the airfoil are no-slip conditions:

$$\mathbf{v}|_{airfoil} = \mathbf{v}^{ib},$$

$$\left.\frac{\partial p}{\partial \mathbf{n}}\right|_{airfoil} = 0. \tag{3}$$

Here $\mathbf{v}^{ib}$ is the velocity of the immersed boundary.

To simulate the rotation of wind turbine rotors, the following dynamics equation is being solved:

$$I\ddot{\alpha} + k\dot{\alpha} = M^{flow}. \tag{4}$$

Here α is the rotation angle of the rotor, I is the polar inertia moment of the rotor, k is the viscous friction coefficient in the axis, and M^{flow} is the aerodynamic torque. Two-dimensional problem is considered, so $M^{flow} = M_z$ (Figure 1).

3. Numerical Method

The Cartesian mesh with cells $\Omega_{i,j} = (x_{i-1}, x_i) \times (y_{j-1}, y_j)$ is introduced in the rectangular computational domain. It is considered that $\Gamma_{i,j}$ is the face of $\Omega_{i,j}$ cell and $\mathbf{x}_{i,j}^c = (x_i^c, y_j^c)$ is its center. Unknown components $u_{i,j}$ and $v_{i,j}$ of velocity

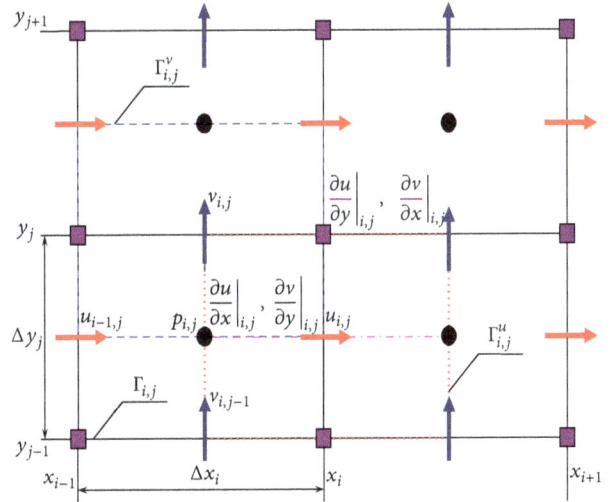

FIGURE 2: Staggered arrangement of the variables on the modified LS-STAG mesh.

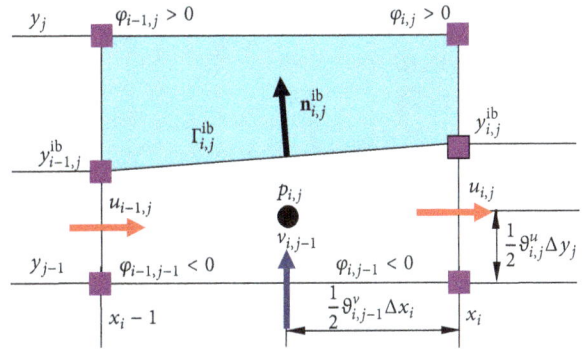

FIGURE 3: Example of a cut-cell.

vector $\mathbf{v}$ are computed in the middle of fluid parts of the cell faces. These points are the centers of control volumes $\Omega_{i,j}^u = (x_i^c, x_{i+1}^c) \times (y_{j-1}, y_j)$ and $\Omega_{i,j}^v = (x_{i-1}, x_i) \times (y_j^c, y_{j+1}^c)$ with faces $\Gamma_{i,j}^u$ and $\Gamma_{i,j}^v$, respectively (Figure 2). The main idea of this numerical method is described in our previous papers, for example, in [14].

Cells which the immersed boundary intersects are the so-called cut-cells (Figures 3 and 4). These cells contain the solid part together with the liquid one. The level-set function φ is used for immersed boundary Γ^{ib} description. The boundary Γ^{ib} is represented by a line segment on the cut-cell $\Omega_{i,j}$. Locations of this segment's endpoints are defined by a linear interpolation of the variable $\varphi_{i,j} = \varphi(x_i, y_j)$.

A few notes should be mentioned about construction of the level-set function. The level-set function cannot be constructed analytically for rotor with complex shape, that is, Darrieus rotor, as for circular airfoil and for simple rotor shapes, for example, for Savonius rotor (Figures 5 and 6). For this reason, it is necessary to approximate the rotor surface line with a curve, which would make it possible to simulate both smooth parts of the boundary and the sharp edges. Moreover, it is desirable that the distance from an arbitrary

(a)

(b)

(c)

(d)

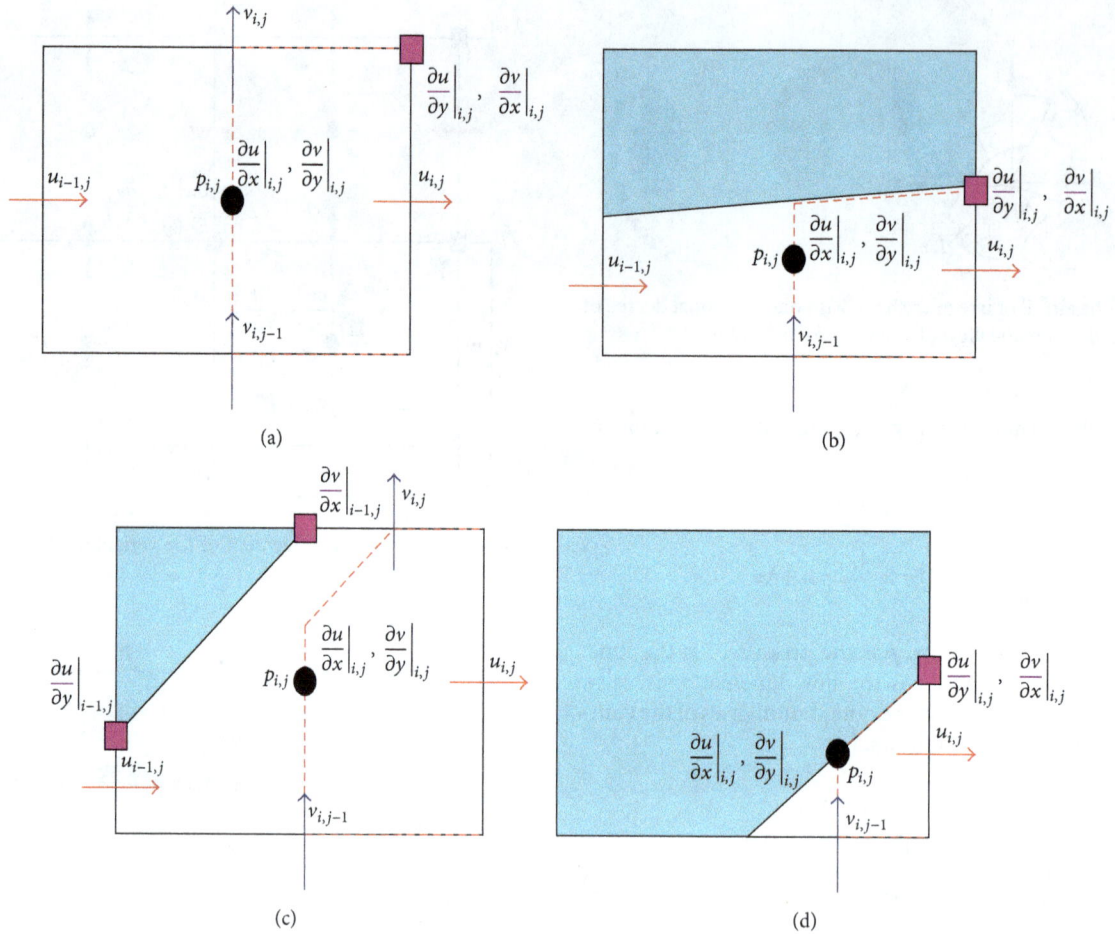

FIGURE 4: Locations of the variables discretization points in case of generic cells on the modified LS-STAG mesh: (a) Cartesian fluid cell; (b) north trapezoidal cell; (c) northwest pentagonal cell; (d) northwest triangle cell.

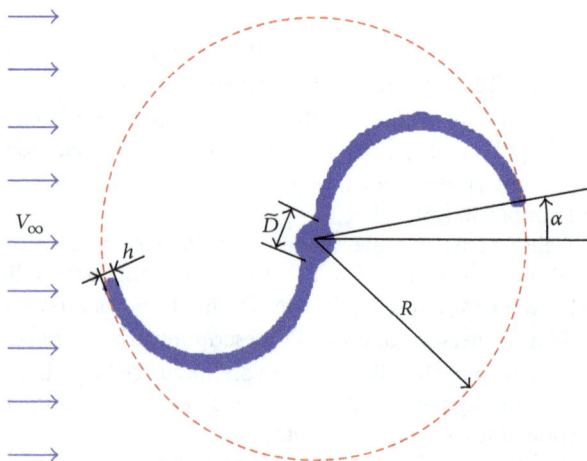

FIGURE 5: Schematic of the two-bucket Savonius rotor.

FIGURE 6: Flow past autorotating Savonius rotor with two blades.

point to the boundary can be calculated easily. An efficient approach for the level-set function construction for an airfoil of arbitrary shape, which corresponds to the mentioned requirements, is described in [15]. The airfoil's surface line

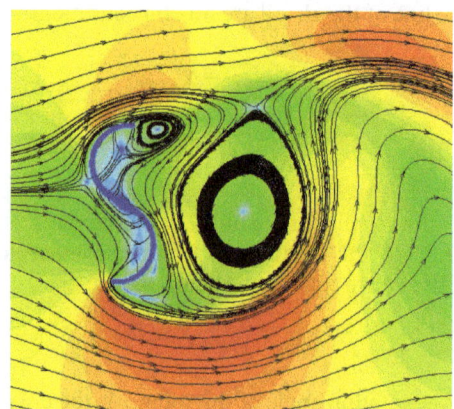

approximation by using Bézier curve is taken as a basis of the developed algorithm.

In order to avoid the Bézier curve reconstruction at every time step for the airfoil surface line, the following approach can be used. At the beginning of the computation, the Bézier curve and its derivative should be built for the rotor blade airfoil at position $\alpha = 0$. Then, the level-set function φ^n in point (x_i, y_j) can be computed as $\varphi^n(x_i, y_j) = \varphi^0(\widetilde{x}_i, \widetilde{y}_j)$ for the rotor rotated by angle α^n counterclockwise. Here φ^0 is the level-set function built for the rotor surface line at position $\alpha = 0$ by using the Bézier curve and $(\widetilde{x}_i, \widetilde{y}_j)$ is the image of the point (x_i, y_j) after clockwise rotation on the angle α^n. Thus, the level-set function can be recalculated at any time without reconstructing the Bézier curve for each new rotor position.

The cell-face fraction ratios $\vartheta^u_{i,j}$ and $\vartheta^v_{i,j}$ are introduced. They take values in interval $[0, 1]$ and represent the fluid parts of the east and north faces of $\Gamma_{i,j}$, respectively. One-dimensional linear interpolation of $\varphi(x_i, y)$ on the segment $[y_{j-1}, y_j]$ and $\varphi(x, y_j)$ on the segment $[x_{i-1}, x_i]$ is used for the cell-face fraction ratios computing:

$$\vartheta^u_{i,j} = \frac{\min\left(\varphi_{i,j-1}, \varphi_{i,j}\right)}{\min\left(\varphi_{i,j-1}, \varphi_{i,j}\right) - \max\left(\varphi_{i,j-1}, \varphi_{i,j}\right)},$$

$$\vartheta^v_{i,j} = \frac{\min\left(\varphi_{i-1,j}, \varphi_{i,j}\right)}{\min\left(\varphi_{i-1,j}, \varphi_{i,j}\right) - \max\left(\varphi_{i-1,j}, \varphi_{i,j}\right)}. \tag{5}$$

According to the concept of the LS-STAG method, normal Reynolds stress components are sampled on the base mesh (similar to pressure discretization) and shear ones are sampled in the upper right corners of the base mesh cells.

The time integration of the differential algebraic system resulting from continuity and momentum equations sampling in space is performed with a semi-implicit Euler scheme [8]. Predictor step leads to discrete analogues of the Helmholtz equation for velocities prediction at the next time layer. Corrector step leads to discrete analogue of the Poisson equation for pressure correction. After computation of the flow variables, the dynamics equation for the airfoil motion (4) should be solved.

The rotor angular velocity is $\omega = \dot{\alpha}$. So difference analogue of (4) can be written in the following form:

$$I\frac{\omega^{n+1} - \omega^n}{\Delta t} + k\omega^n = M^n_z. \tag{6}$$

The aerodynamic torque at the n-th time step can be computed as follows:

$$M^n_z$$

$$= \sum_{\text{Cut-cells }\Omega^{ib}_{i,j}} \left[\left(x^C_i - X_C\right)F^h_{ya}\big|^n_{i,j} - \left(y^C_j - Y_C\right)F^h_{xa}\big|^n_{i,j}\right]. \tag{7}$$

Here (x^C_i, y^C_j) are coordinates of the center of $\Omega_{i,j}$ cell, (X_C, Y_C) are coordinates of the point around which the airfoil rotates, and $F^h_{xa}\big|^n_{i,j}$ and $F^h_{ya}\big|^n_{i,j}$ are the drag and lift forces acting on the solid part of the cut-cell $\Omega_{i,j}$ at the n-th time step:

$$F^h_{xa}\big|_{i,j} = [n_x \Delta S]^{ib}_{i,j}\left(p_{i,j} - \nu\frac{\partial u}{\partial x}\Big|_{i,j}\right)$$

$$- \nu\text{Quad}^{ib}_{i,j}\left(\frac{\partial u}{\partial y}\overrightarrow{e}_y \cdot \overrightarrow{n}\right),$$

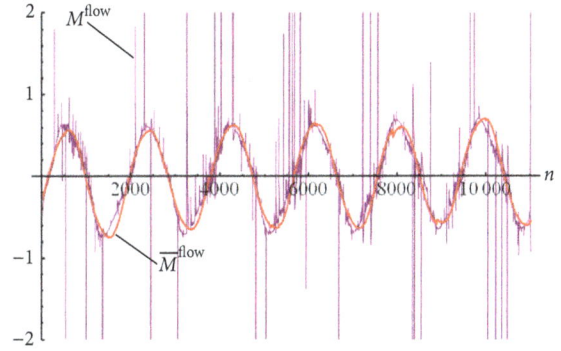

FIGURE 7: The time dependency for the torque $M^{\text{flow}} = M_z(t)$ (on the coarse grid 272×292 with time step $\Delta t = 0.0001$) and filtered time dependency $\overline{M}^{\text{flow}} = \overline{M}_z(t)$ for the torque.

$$F^h_{ya}\big|_{i,j} = -\nu\text{Quad}^{ib}_{i,j}\left(\frac{\partial v}{\partial x}\overrightarrow{e}_x \cdot \overrightarrow{n}\right)$$

$$+ [n_y \Delta S]^{ib}_{i,j}\left(p_{i,j} - \nu\frac{\partial v}{\partial y}\Big|_{i,j}\right). \tag{8}$$

Here $[n_x\Delta S]^{ib}_{i,j} = (\vartheta^u_{i-1,j} - \vartheta^u_{i,j})\Delta y_j$, $[n_y\Delta S]^{ib}_{i,j} = (\vartheta^v_{i,j-1} - \vartheta^v_{i,j})\Delta x_i$, $\Delta y_j = y_j - y_{j-1}$, $\Delta x_i = x_i - x_{i-1}$, and $\text{Quad}^{ib}_{i,j}$ are the quadratures of the shear stress which depend on the cut-cells [8].

The value of the rotor angular velocity at the next time step is computed from (6). New rotor position and the immersed boundary velocity can be defined by using this value:

$$\alpha^{n+1} = \alpha^n + \omega^n\Delta t,$$

$$\mathbf{v}^{ib,n+1}_{i,j} = \left\{-\frac{y^C_j - Y_C}{|y^C_j - Y_C|}\cdot\omega^n\left|x^C_i - X_C\right|, \frac{x^C_i - X_C}{|x^C_i - X_C|}\right.$$

$$\left.\cdot\omega^n\left|y^C_j - Y_C\right|\right\}. \tag{9}$$

Such approach allows reconstructing the level-set function and all matrices and the source terms required for the next time step.

When investigating some complicated physical phenomenon, preliminary qualitative estimations for the considered construction behavior are very important. They allow finding domains for grid refinement, prediction of the structure's dynamic, and estimation of the CFL number. But large fluctuations in the aerodynamic loads can occur on the coarse grid. So the values of the torque acting on the airfoil should be filtered by the low-pass filter (Figure 7). For this purpose, a second-order Butterworth low-pass filter [16] is designed.

It is necessary to explain the reasons for choosing this filter. Filters with infinite impulse response are less expensive from a computational point of view than filters with finite impulse response. The frequency response of the Butterworth filter is maximally flat (i.e., has no ripples) in the passband

and rolls off towards zero in the stopband [16]. Compared with a Chebyshev type I/type II filter or an elliptic filter, the Butterworth filter has a slower roll-off and thus will require a higher order to implement a particular stopband specification, but Butterworth filters have a more linear phase response in the passband than Chebyshev type I/type II and elliptic filters can achieve [16].

The transfer function of the Butterworth second-order filter is as follows:

$$H(s) = \frac{1}{\left(s - (i-1)/\sqrt{2\varepsilon_p}\right)\left(s + (i+1)/\sqrt{2\varepsilon_p}\right)}$$

$$= \frac{\varepsilon_p}{1 + \sqrt{2\varepsilon_p}s + \varepsilon_p s^2}. \tag{10}$$

Here $s \in \mathbb{C}, i = \sqrt{-1}, \varepsilon_p = \sqrt{10^{R_p/10} - 1}$, and R_p is a distortion level in the passband. As practice shows, filters of higher order can lead to the appearance of numerical instability in the filtered signal. For this function, $H(0) = \varepsilon_p$. Therefore the following transfer function corresponds to the normalized Butterworth second-order filter:

$$H^{\text{norm}}(s) = \frac{1}{1 + \sqrt{2\varepsilon_p}s + \varepsilon_p s^2}. \tag{11}$$

In order to control the filter cutoff frequency, it is necessary to use the following transfer function:

$$H_{\text{LP}}(s) = H^{\text{norm}}\left(\frac{s}{\Omega_x}\right). \tag{12}$$

The sampling frequency of the torque is equal to $f_d = 1/\Delta t$ in the numerical simulation (Δt is a time discretization step). It is necessary that the filter cut-off frequency is equal to $f_s = 5$ Hz and the suppression level on the cut-off frequency is equal to $R_s = 3$ dB. So, the following condition imposed on the required filter frequency response function $|H_{\text{LP}}(i \cdot \omega)|$ (Figure 8) must be satisfied for the required filter transfer function $H_{\text{LP}}(s)$:

$$|H_{\text{LP}}(i \cdot \Omega_s)| = \gamma = 10^{-R_s/20},$$

$$\Omega_s = \tan\left(\frac{\pi f_s}{f_d}\right) = \tan(\pi f_s \Delta t). \tag{13}$$

Solution of (13) leads to the following result:

$$\frac{1}{\sqrt{1 + \left(\sqrt{\varepsilon_p}/\Omega_x\right)^4 \Omega_s^4}} = \gamma \iff$$

$$\frac{\sqrt{\varepsilon_p}}{\Omega_x} = \frac{\sqrt[4]{1 - \gamma^2}}{\sqrt{\gamma}\Omega_s} = \xi. \tag{14}$$

Thus, the desired filter transfer function can be written as follows:

$$H_{\text{LP}}(s) = \frac{1}{1 + \sqrt{2}\xi s + (\xi s)^2}. \tag{15}$$

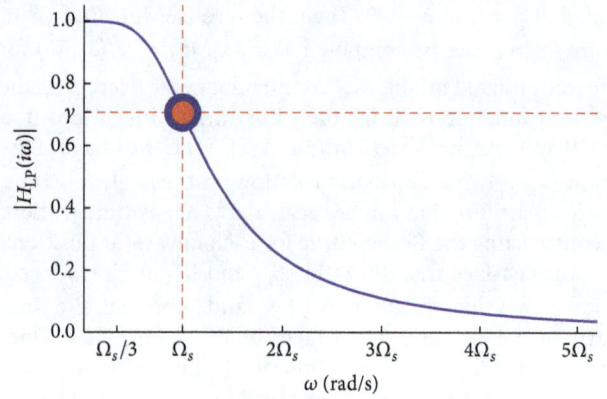

FIGURE 8: The required filter frequency response function.

To obtain the corresponding digital filter coefficients, there is a need to use the bilinear transformation [16]:

$$H_{\text{LP}}\left(\frac{1 - z^{-1}}{1 + z^{-1}}\right)$$

$$= \frac{1 + 2z^{-1} + z^{-2}}{\left(1 + \sqrt{2}\xi + \xi^2\right) + 2\left(1 - \xi^2\right)z^{-1} + \left(1 - \sqrt{2}\xi + \xi^2\right)z^{-2}}. \tag{16}$$

Here $z \in \mathbb{R}$. It is known that

$$H_{\text{LP}}\left(\frac{1 - z^{-1}}{1 + z^{-1}}\right) = \frac{b_0 + b_1 z^{-1} + b_2 z^{-2}}{1 + a_1 z^{-1} + a_2 z^{-2}} \tag{17}$$

for the second-order low-pass filter with infinite impulse response. Therefore, it can be obtained from formulae (16) and (17) that the designed digital filter coefficients are the following:

$$b_0 = b_2 = \frac{1}{1 + \sqrt{2}\xi + \xi^2},$$

$$b_1 = \frac{2}{1 + \sqrt{2}\xi + \xi^2},$$

$$a_1 = \frac{2\left(1 - \xi^2\right)}{1 + \sqrt{2}\xi + \xi^2}, \tag{18}$$

$$a_2 = \frac{1 - \sqrt{2}\xi + \xi^2}{1 + \sqrt{2}\xi + \xi^2}.$$

So, the filtered value of the torque acting on the airfoil from the flow at the time $t_{n+1} = (n+1)\Delta t$ is

$$\overline{M}_z^{n+1} = b_0 \cdot M_z^{n+1} + b_1 \cdot M_z^n + b_2 \cdot M_z^{n-1}$$

$$- \left(a_1 \cdot \overline{M}_z^n + a_2 \cdot \overline{M}_z^{n-1}\right). \tag{19}$$

4. Numerical Experiments

We considered autorotation of Savonius rotor with two blades (Figure 5) at Re $= 1.96 \cdot 10^5$. The radius of rotation R was measured from the axis of rotation to the outer edge of the buckets. The turbine-swept area A_s is equal to πR^2. Reynolds number is based on rotor diameter. The following parameters were used in the simulation:

$$V_\infty = 1.0,$$

$$R = \frac{D}{2} = 1.0,$$

$$\widetilde{D} = 0.1D,$$

$$h = 0.05R, \qquad (20)$$

$$k = 0.0,$$

$$I = 10.0,$$

$$\alpha_0 = \alpha(0) = \frac{\pi}{18}.$$

It should be noted that the above-proposed algorithm for level-set reconstruction can be easy to apply for rotors of other shapes, that is, Darrieus rotor.

The dimensionless average rotor angular velocity is in the following range:

$$\overline{\omega} = 0.4, \ldots, 1.6. \qquad (21)$$

Averaging was carried out over 16 dimensionless time units. This time is enough for the rotor to make full turn with the smallest value of average rotor angular velocity $\overline{\omega} = 0.4$. At the chosen values of the parameters, tip speed ratio is equal to

$$X_\infty = \frac{R\overline{\omega}}{V_\infty} = \overline{\omega}; \qquad (22)$$

torque coefficient value $\overline{C}_Q$ is obtained by averaging of nonstationary dependency $C_Q(t)$ over time, where

$$C_Q(t) = \frac{2\overline{M}_z(t)}{\rho V_\infty^2 R A_s} = \frac{2\overline{M}_z(t)}{\pi}. \qquad (23)$$

Similarly, the power coefficient $\overline{C}_P$ is obtained by averaging of nonstationary dependency $C_P(t)$ over time, where

$$C_P(t) = \frac{2\overline{M}_z(t)\overline{\omega}}{\rho V_\infty^3 A_s} = \frac{2\overline{M}_z(t)\overline{\omega}}{\pi}. \qquad (24)$$

A number of computations have been performed on non-uniform grid 544×496. The uniform mesh block with space discretization step $\Delta h = D/128$ was used in the vicinity of the rotor. Time discretization step was equal to $\Delta t = 10^{-5}$. Computations were performed on a server based on the Intel C610 platform using the Intel Xeon E5-1620 V3 4-core processor (3.5 GHz) with Hyper-Threading support (8 logical cores). The server is equipped with 16 GB of ECC DDR4-2133 RAM and two hard drives (2 TB), united in a RAID1 disk volume. This server is running Windows Server 2012 R2 operating system. To simulate 1 dimensionless time unit, about 24 hours are required.

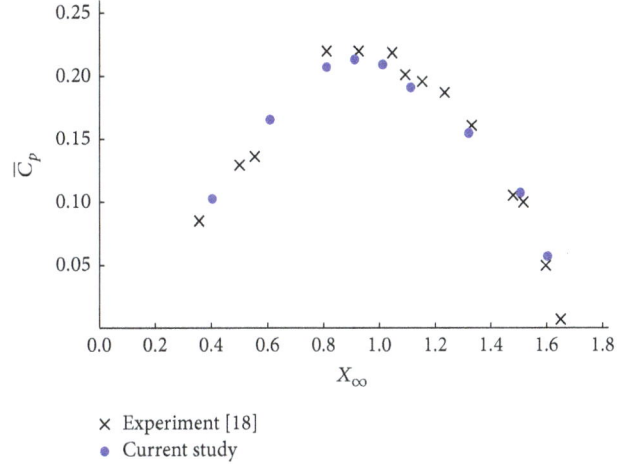

FIGURE 9: Comparison of computed power coefficients $\overline{C}_P$ for a two-bucket Savonius rotor at Re $= 1.96 \cdot 10^5$ with experimental data (Sheldahl et al. [17] at Re $= 3.9 \cdot 10^5$ and Shankar [18] at Re $= 1.96 \cdot 10^5$).

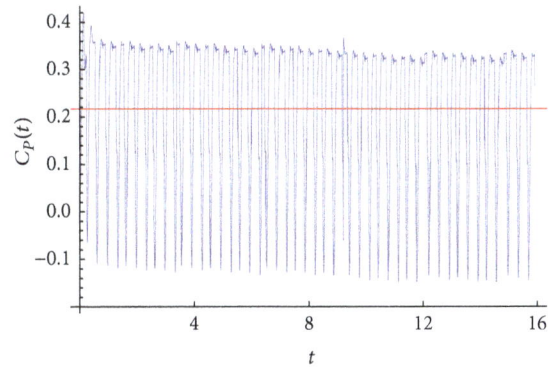

FIGURE 10: Time evolution of the power coefficient $C_P(t)$ at Re $= 1.96 \cdot 10^5$ and $\overline{\omega} = 0.9$. The red line corresponds to the average value $\overline{C}_P$.

The computed values of the power coefficient $\overline{C}_P$ were compared with the experimental data [17, 18] presented in [18]. As can be seen from Figure 9, the computational results are in good agreement with experiment. An example of simulated nonstationary dependency $C_P(t)$ and computed value of the corresponding power coefficient $\overline{C}_P$ at Re $= 1.96 \cdot 10^5$ and $\overline{\omega} = 0.9$ is shown in Figure 10.

5. Conclusions

We have the following conclusions

(i) A software package is developed for numerical simulation of wind turbine rotors autorotation by using the modified LS-STAG level-set/cut-cell immersed boundary method. This software package can be used for rotors of other shapes (Savonius rotor, Darrieus rotor, etc.).

(ii) Algorithms for level-set function construction and recalculation are described.

(iii) A second-order Butterworth low-pass filter is designed for aerodynamic torque filtrating at simulation on the coarse grid.

(iv) Simulation of flow past autorotating Savonius rotor with two blades is considered at Re $= 1.96 \cdot 10^5$.

(v) Computational results are in good qualitative agreement with the experimental data.

Conflicts of Interest

The authors declare that there are no conflicts of interest regarding the publication of this paper.

Acknowledgments

This research is partially supported by Russian Ministry of Education and Science (Project no. 9.2422.2017/PP) and Russian Federation President Grant for Young Russian Ph.D. Scientists (Project no. MK-7431.2016.8).

References

[1] P.-A. Krogstad and J. A. Lund, "An experimental and numerical study of the performance of a model turbine," *Wind Energy*, vol. 15, no. 3, pp. 443–457, 2012.

[2] N. N. Sorensen, J. A. Michelsen, and S. Schreck, "Navier-stokes predictions of the NREL phase VI rotor in the NASA Ames 80-by-120 wind tunnel," in *Proceedings of the 2002 ASME Wind Energy Symposium*, pp. 94–105, Reno, Nev, USA, January 2002.

[3] K. Rogowski and R. Maroński, "CFD computation of the savonius rotor," *Journal of Theoretical and Applied Mechanics*, vol. 53, no. 1, pp. 37–45, 2015.

[4] R. Gupta and K. K. Sharma, "Flow physics of a combined Darrieus-Savonius rotor using computational fluid dynamics (CFD)," *Int. Research J. Eng. Science, Technology and Innovation*, vol. 1, pp. 1–13, 2012.

[5] L. Oggiano, "CFD Simulations on the NTNU Wind Turbine Rotor and Comparison with Experiments," in *Proceedings of the Renewable Energy Research Conference, RERC 2014*, pp. 111–116, June 2014.

[6] M. H. Mohamed, A. M. Ali, and A. A. Hafiz, "CFD analysis for H-rotor Darrieus turbine as a low speed wind energy converter," *Engineering Science and Technology, an International Journal*, vol. 18, no. 1, pp. 1–13, 2015.

[7] R. Mittal and G. Iaccarino, "Immersed boundary methods," in *Annual review of fluid mechanics. Vol. 37*, vol. 37 of *Annu. Rev. Fluid Mech.*, pp. 239–261, Annual Reviews, Palo Alto, CA, 2005.

[8] Y. Cheny and O. Botella, "The LS-STAG method: A new immersed boundary/level-set method for the computation of incompressible viscous flows in complex moving geometries with good conservation properties," *Journal of Computational Physics*, vol. 229, no. 4, pp. 1043–1076, 2010.

[9] S. Osher and R. Fedkiw, *Level Set Methods and Dynamic Implicit Surfaces*, vol. 153, Springer, New York, NY, USA, 2003.

[10] H. A. van der Vorst, "BI-CGSTAB: a fast and smoothly converging variant of BI-CG for the solution of nonsymmetric linear systems," *SIAM Journal on Scientific and Statistical Computing*, vol. 13, no. 2, pp. 631–644, 1992.

[11] P. Wesseling, *An Introduction to Multigrid Methods*, John Wiley & Sons, Chichester, UK, 1992.

[12] J. van Kan, C. Vuik, and P. Wesseling, "Fast pressure calculation for 2D and 3D Time dependent incompressible flow," *Numerical Linear Algebra with Applications*, vol. 7, no. 6, pp. 429–447, 2000.

[13] I. Marchevsky and V. Puzikova, "OpenFOAM iterative methods efficiency analysis for linear systems solving," *Proceedings of the Institute for System Programming of RAS*, vol. 24, pp. 71–86, 2013.

[14] V. V. Puzikova and I. K. Marchevsky, "Application of the LS-STAG immersed boundary method for numerical simulation in coupled aeroelastic problems," in *Proceedings of the Joint 11th World Congress on Computational Mechanics, WCCM 2014, the 5th European Conference on Computational Mechanics, ECCM 2014 and the 6th European Conference on Computational Fluid Dynamics, ECFD 2014*, pp. 1995–2006, esp, July 2014.

[15] V. V. Puzikova, "Construction of Level-Set Function for an Airfoil of Arbitrary Topology when Modelling a Flow past It Using the LS-STAG Method," in *Herald of the Bauman Moscow State Technical University. Natural Sciences*, pp. 163–173, 2012.

[16] R. G. Lyons, *Understanding Digital Signal Processing*, Pearson Education, 2010.

[17] R. E. Sheldahl, B. F. Blackwell, and L. V. Feltz, "Wind tunnel performance data for two- and three-bucket savonius rotors," *J Energy*, vol. 2, no. 3, pp. 160–164, 1978.

[18] P. N. Shankar, *National Aeronautical Laboratory*, AE-TM-3-76, National Aeronautical Laboratory, Bangalore, India, 1976.

12

Research on Aerodynamic Characteristics of Straight-Bladed Vertical Axis Wind Turbine with S Series Airfoils

Fang Feng ⓘ,[1,2] **Shouyang Zhao,**[3] **Chunming Qu,**[3] **Yuedi Bai,**[3] **Yuliang Zhang,**[1] **and Yan Li** ⓘ[2,3]

[1]*College of Science, Northeast Agricultural University, Harbin 150030, China*
[2]*Heilongjiang Provincial Key Laboratory of Technology and Equipment for Utilization of Agricultural Renewable Resources in Cold Region, Harbin 150030, China*
[3]*College of Engineering, Northeast Agricultural University, Harbin 150030, China*

Correspondence should be addressed to Fang Feng; fengfang@neau.edu.cn and Yan Li; liyanneau@163.com

Academic Editor: Ahmad Sedaghat

Background. In order to investigate the effect of aerodynamic characteristics of S series airfoils on the straight-bladed vertical axis wind turbine (SB-VAWT), numerical simulations and wind tunnel experiments were carried out using a small SB-VAWT model with three kinds of blade airfoils, which are asymmetric airfoil S809, symmetric airfoil S1046, and NACA0018 used for performance comparison among S series. The aerodynamics characteristics researched in this study included static torque coefficient, out power coefficient, and rotational speed performance. The flow fields of these three kinds of blade under static and dynamic conditions were also simulated and analyzed to explain the mechanism effect of aerodynamic performance. According to the results, the SB-VAWT with airfoil S1046 has better dynamic aerodynamic characteristics than other two airfoils, while the SB-VAWT with airfoil S809 is better in terms of the static characteristics. As the most suitable airfoil for SB-VAWT, the S series airfoil is worth researching deeply.

1. Introduction

In recent years, as one of vertical axis wind turbines (VAWT), the straight-bladed vertical axis wind turbine (SB-VAWT) has developed rapidly and attracted attention of scientists due to its advantages such as wind direction independence, simple structure, and unique shape [1]. The selection of the blade airfoil was found one of the main factors which has influenced the output characteristics of SB-VAWT. Normally, the airfoils of horizontal axis wind turbine (HAWT) are NACA series airfoils, SERL series airfoils studied by America, S8xx series airfoils, FFA-W series airfoils manufactured by FOI of Sweden, RisΦ-A-XXX series airfoils developed by Denmark, DU series airfoils developed by Delft University, and so on [2]. Common VAWT airfoils refer to NACA series, among which airfoil l0018 is widely used in SB-VAWT due to its better wind energy utilization coefficients. Researchers have great interests in the airfoils comparative analysis and selection. In 2008, Canadian researchers found that airfoils which are suitable for HAWT are not necessarily suitable for

VAWT [3]. Zhang et al. analyzed six different kinds of airfoils including NACA, FFA, and FX series and found that the maximum lift coefficient and the maximum lift-drag ratio of FFA series airfoils which are more suitable for SB-VAWT working environment are better than those of the NACA series airfoils [4]. Liao et al. researched the aerodynamic performance of small SB-VAWT based on different airfoils [5]. In 2012, Mohamed analyzed the performance investigation of VAWT using new airfoil shapes [6]. Xu et al. used nonsymmetrical airfoils DU06-W-200 as the research object to investigate the influence of installation on the performance of SB-VAWT [7]. Xu et al. analyzed the aerodynamic performance of thickened DU series airfoils [8]. Yang and Li studied an improved VAWT airfoil on the basis of airfoil 4418 of NACA aeries [9]. In 2016, Jia et al. studied three kinds of DU series airfoils including DU25, DU30, and DU35 and found the influence of relative thickness on aerodynamic performance [10].

Based on the past researches mentioned above, it can be found that the researches on SB-VAWT blade airfoils mainly focus on NACA series, DU series, FFA series, and

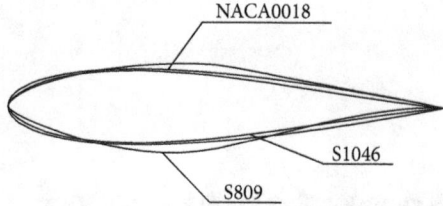

FIGURE 1: Three kinds of airfoil.

TABLE 1: Main features of the analyzed rotor.

Denomination	Value
Blade number (N)	3
Blade shape	Straight
Blades airfoil (attended)	NACA0018, S809, S1046
Height (H) [m]	0.5
Radius (R) [m]	0.4
Chord (c) [m]	0.125
Solidity (σ)	0.149
Texture	FRP

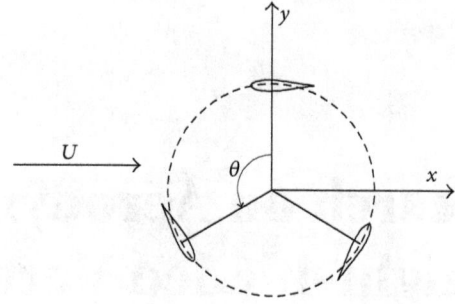

FIGURE 2: Structural parameters of wind turbine.

FX series. However, there is a lack of research on S series airfoils which is widely used in HAWT. Therefore, in this study, two typical kinds of S series including S1046 and S809 were selected to investigate whether this kind of blade airfoil is suitable for SB-VAWT or not. NACA0018 airfoil was also selected for the comparison study. The aerodynamics characteristics researched in this study included static starting torque coefficient, out power coefficient, and rotational speed performance. The flow fields of three kinds of blade under both static and dynamic conditions were also simulated and analyzed. This study can provide a good reference for the blade airfoil research on SB-VAWT.

2. Model Design

2.1. Airfoils. The cross sections of airfoils NACA0018, S809, and S1046 are shown in Figure 1. NACA0018 and S1046 are symmetrical airfoils, and S809 is an asymmetric airfoil. As is shown in the figure, the leading edge of airfoil S809 is thinner than the other two. However, the middle of airfoil S809 is thicker than the other two. NACA0018 and S1046 are symmetrical airfoils. Both the leading edge and the trailing edge of airfoil NACA0018 are thicker than those of airfoil S1046 and the trailing edge of airfoil S1046 is the inner convergence.

2.2. Main Structural Parameters. According to the theory of hydrodynamics and the design method of VAWT, a kind of small blade SB-VAWT is designed [11–13]. For the experimental validation, the geometry considered in the study is summarized in Table 1 based on the size of experimental segment (1 m × 1 m) of the wind tunnel in the laboratory. The blade is rotating counterclockwise; the angle between blade beam and axis Y is azimuth angle θ. The azimuth angle is shown in Figure 2.

3. Research Methods

In order to explore the aerodynamic characteristics of SB-VAWT with S series airfoils, numerical simulations and wind tunnel experiments were used in this paper.

3.1. Numerical Simulation. A finite volume CFD solver by ANSYS was used in this work, implementing Reynolds averaged Navier-Stokes equations. Because the cross sections of wind turbine blade are exactly the same, this study adopts two-dimensional model numerical simulation method. The calculation of the structure is simplified. The turbulence model based on the pressure solver is the k-ε RNG model. The transport equations for k and ε are shown in (1). The pressure velocity coupling is the SIMPLEC algorithm, and the flow is unsteady. The turbulent kinetic energy dissipation rate epsilon, K equation, and the momentum equation are the two-order upwind scheme.

$$\frac{\partial (\rho k)}{\partial t} + \frac{\partial (\rho k u_i)}{\partial x_i} = \frac{\partial}{\partial x}\left(\alpha_k \mu_{\text{eff}} \frac{\partial k}{\partial x_j}\right) + G_k + G_b - \rho\varepsilon$$
$$- Y_M + S_k,$$
$$\frac{\partial (\rho \varepsilon)}{\partial t} + \frac{\partial (\rho \varepsilon u_i)}{\partial x_i} = \frac{\partial}{\partial x_j}\left(\alpha_\varepsilon \mu_{\text{eff}} \frac{\partial \varepsilon}{\partial x_j}\right)$$
$$+ C_{1\varepsilon}\frac{\varepsilon}{k}\left(G_k + C_{3\varepsilon}G_b\right)$$
$$- G_{2\varepsilon}\rho\frac{\varepsilon^2}{k} - R. \tag{1}$$

The calculation area of the outside wind turbine is a rectangle as is shown in Figure 3, which is 10 times the width and 15 times the length of the radius of wind turbine. The flow is relatively stable before the wind turbine, and after the wind turbine, the flow field changes a lot, so the position of the wind turbine in the computation domain is far from the exit boundary, thus helping to observe the variation of the flow field behind the wind turbine. The flow from the left entry of the graph is the velocity inlet, with the calculation wind speed being 10 m/s, the right side being the pressure outlet, the rectangular upper and lower sides being stationary wall surfaces, and the blade wall moving on the wall surface. In order to guarantee the accuracy of the calculated results and

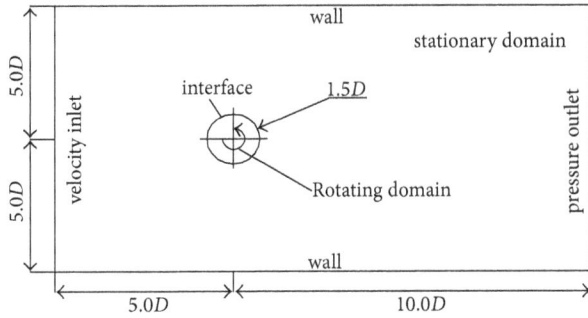

FIGURE 3: Computational region diagram and boundary conditions.

TABLE 2: Component of the experimental equipment.

Equipment	Function	Precision
Wind tunnel	Provide a stable source of wind	3%
Model of wind turbine	Experimental model	
Torque sensor	Acquisition data	±0.2%
Torque analyzer	Display the data	
Motor	Provide load, stable model speed	3%
Power supply	Control motor speed	
Computer	Process the data	

control the appropriate calculation time, the grid of VAWT was refined. The grid settings are shown in Figure 4.

In order to prove the independence of the mesh number, the static torque coefficient of wind turbine was studied when the azimuth is 5 degrees. Five different mesh numbers are 73765, 110647, 156971, 200432, and 248956, respectively. The fluid computational domain model is chosen for trial calculation. It clearly appears in Figure 5 that using RNG k-ε as the turbulence model for the model meshed by ANSYS gives unstable solution over a wide range of a number of nodes. When the number of mesh increases from 156971 to 248956, the numerical results are not much different. Taken into consideration, the mesh number selected in this study is 156971.

3.2. Wind Tunnel Experiment. In order to verify the influence of eccentricity on the power characteristics of the SB-VAWT, the experiments were conducted at Northeast Agricultural University using the large-scale low-speed opening wind tunnel which is 9.1 m long and 2.3 m wide, its exit size being 1 m × 1 m wide, and the wind speed being constant at the exit from 1–20 m/s. The wind tunnel experiment system consisted of a low-speed wind tunnel, an experimental model, a speed torque tester, an induction motor, a frequency converter, and a computer. The system diagrams are shown in Figures 6 and 7 and the function of them is shown in Table 2.

In the experiment, the height of wind turbine center is consistent with that of the wind tunnel exit center. The wind turbine is driven by an induction motor, with its speed controlled by a frequency converter. The speed and torque are detected by the speed torque sensor. The speed torque sensor measuring range is 5 N·m, the accuracy is ±0.2%, and the original data sampling interval is 0.1 s. The wind generated in the wind tunnel pushes the blades and applies rotation power

to rotor axis, the rotational power can be transferred to the torque meter, and the rotational speed is measured in real time by the topic sensor. When the experiment is carried out, the signal is converted to the voltage level which is used to acquire the power of the wind turbine using a data program.

4. Results and Discussion

The aerodynamic characteristics of the wind turbine investigated in this study are mainly output power coefficients at different tip speed ratio and static torque coefficients at different azimuth angles for SB-VAWT. The definitions of power coefficient, torque coefficient, and tip speed ratio are shown in the following equation:

$$C_P = \frac{P}{(1/2)\,\rho A U^3},$$
$$C_M = \frac{M}{(1/2)\,\rho A U^2 R}, \quad (2)$$
$$\lambda = \frac{V}{U} = \frac{\omega R}{U},$$

where C_P is the power coefficient, P is the power absorbed by a wind turbine, ρ is air density, A is the swept area of the wind turbine relative to the current, U is inflow velocity, C_M is the torque coefficient, M is the torque absorbed by a wind turbine, R is the radius of wind turbine, λ is the tip speed ratio, V is the linear velocity of wind turbine, and ω is the rotation speed of wind turbine.

4.1. Dynamic Characteristics

4.1.1. Power Coefficient. In this paper, numerical simulations and wind tunnel tests were carried out, respectively, to reflect the output characteristics of the SB-VAWT with different airfoils. Figure 8 shows that power coefficients vary with the tip speed ratio when wind speed is 10 m/s. Figure 9 shows the maximum power coefficients of the three kinds of SB-VAWT at different wind speeds.

As shown in Figure 8, for simulation results, when λ is in the range from 0.2 to 0.8, the power coefficients are slowly increasing and the growth rate is low. When λ is in the range from 0.8 to 1.8, the power coefficients are obviously increasing, and the value of power coefficients reaches the maximum when λ is up to 1.8. The power coefficient of SB-VAWT with airfoil S809 is the lowest among the three airfoils. When λ is in the range from 0.2 to 1.0, the power coefficients of the SB-VAWT with airfoil S1046 are much close to NACA0018. When the tip speed ratio is in the range from 1.0 to 2.0, the power coefficient of airfoil S1046 is better than those of S809 and NACA0018. The wind tunnel test results and numerical results are in good agreement with the overall trend, except a bit difference in the value. The reasons for the difference between numerical and experimental results are mainly as below: first, there are no shafts, beams, flanges, and other components in the normal numerical simulation model, and when these factors are taken into account in the experiment, the experimental error cannot be ignored. The

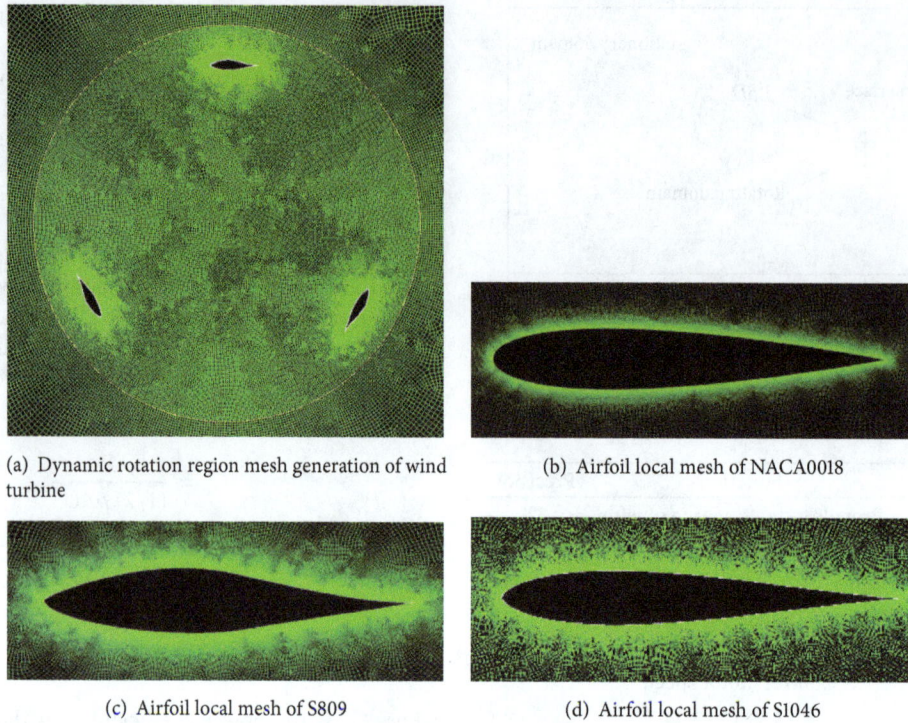

(a) Dynamic rotation region mesh generation of wind turbine

(b) Airfoil local mesh of NACA0018

(c) Airfoil local mesh of S809

(d) Airfoil local mesh of S1046

FIGURE 4: Grid mesh of computing model.

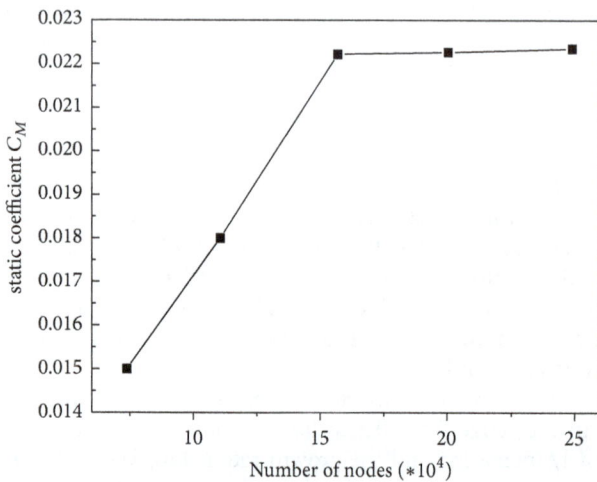

FIGURE 5: Grid independent verification.

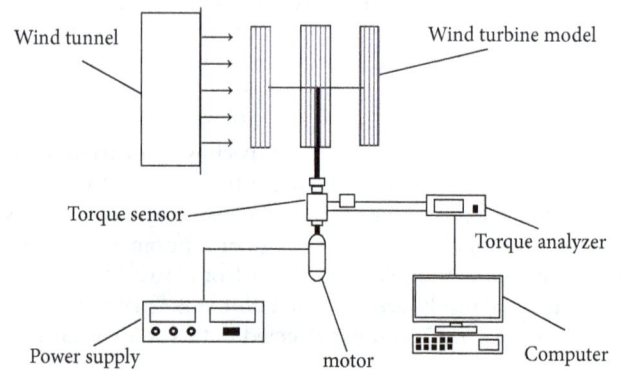

FIGURE 6: Schematic diagram of experimental system.

S1046 airfoil SB-VAWT are better than those of the other two airfoils.

second is that the fluid field of numerical calculation is large, and the range of the experimental wind field is relatively small, which will produce certain errors.

As can be seen from Figure 9, the maximum power coefficient of SB-VAWT increases with the increasing of wind speed. Under the same wind speed, the order of the power coefficient from high to low is S1046, NACA0018, and S809. When wind speed is 10 m/s, the maximum power coefficient of airfoil S1046 wind turbine is 0.32, which is 6.7% higher than that of NACA0018 and 7.1% higher than that of S809. Therefore, the results of Figures 8 and 9 show that, under the experimental conditions, the output characteristics of the

4.1.2. Dynamic Flow Field. In order to further analyze the influence of blade airfoil selection on the output characteristics for SB-VAWT, the torque characteristics and the flow field of the blade were compared and analyzed.

It can be seen from Figure 8 that under the experiment condition of $U = 10$ m/s the power coefficients of the three airfoils wind turbine reach the maximum value when C_P is up to 1.8 according to the results listed above. The power characteristics of the S1046 airfoil wind turbine are better than the other two. Under the experimental conditions, the torque coefficients of SB-VAWT are different from the azimuth angles when the wind turbine rotates in one circle.

FIGURE 7: Physical diagram of experimental system.

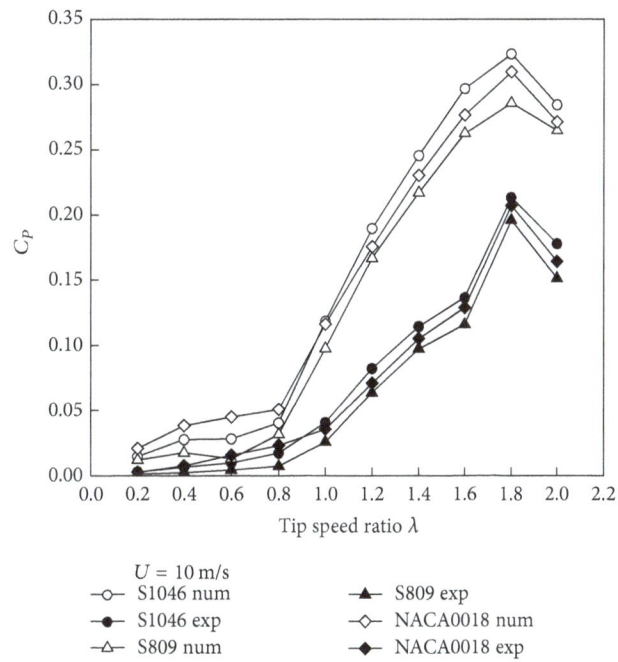

FIGURE 8: Power coefficient.

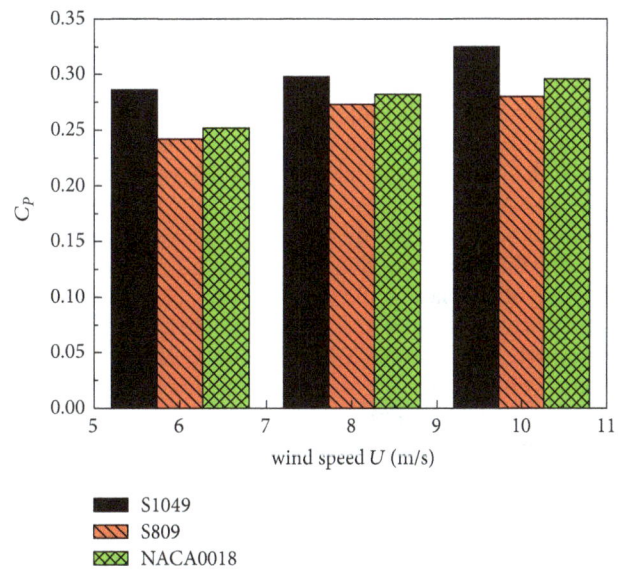

FIGURE 9: Maximum power coefficient.

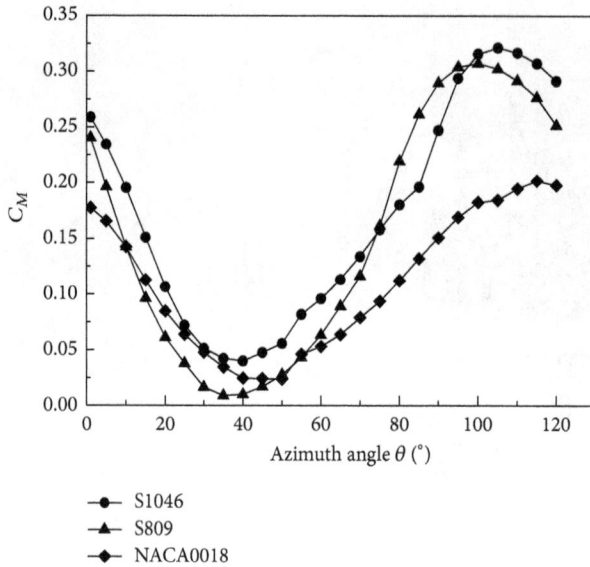

FIGURE 10: Torque coefficient at different azimuth angle ($\lambda = 1.8$).

Figure 10 shows the curves of the torque coefficients at different azimuth angles when λ is 1.8. As we can see from Figure 10, the torque coefficients of the SB-VAWT with three airfoils have a peak value and a valley value in a circle. Airfoils S1046 and S809 have the lowest torque coefficient at 35 degrees, and the airfoil NACA0018 has the lowest torque coefficient at 45 degrees. The maximum torque coefficients of the SB-VAWT with airfoils S1046 and S809 are near 100 degrees, and the airfoil NACA0018 is near 115 degrees. In conclusion, the torque characteristics of SB-VAWT with S series airfoils are better than NACA series airfoils, the average value of the torque coefficients of airfoil S1046 is the largest, and the output characteristics is the best.

In order to analyze the reasons for the difference in output aerodynamic characteristics of three airfoils SB-VAWT in detail, velocity and pressure distribution flow charts of the blades with three different airfoils are compared when λ is 1.8 and the azimuth angles of 35 and 100 degrees are selected. The influence of airfoils on the flow field of the blade is shown in Figures 11 and 12.

As is shown in Figure 11, for blade, the SB-VAWTs with airfoils S1046 and NACA00018 have a small range of high pressure zone on the windward side of the leading edge. In comparison, the high pressure zone of SB-VAWT with airfoil S809 is larger and there is a wide range of negative pressure zone on the ventral of the blade. The difference on the pressure between the ventral and the back of the blade could generate more aerodynamic forces, thus driving the forward rotation of the wind turbine. Furthermore, the SB-VAWT with airfoil NACA00018 produces a large vortex at the back of the blade, which leads to the loss of energy and lower output characteristics. For blade (b), a large zone of negative pressure is found in the ventral of blades, where the low value area of S809 is more obvious. Therefore, it generates lower power coefficient. For blade (c), three wind turbines with different airfoils all have vortex in the ventral side, among which NACA00018 is the largest, and the energy loss is also greater

than those of other two turbines. In conclusion, among the SB-VAWTs with three different airfoils, the performance of airfoil S1046 is better in the aspects of the pressure difference between the ventral and back of the blade and the energy utilization, and the torque coefficient is slightly higher than the other two airfoils at 35 degrees.

As can be seen from Figure 12, when the azimuth angle is 100 degrees, the variation of the flow field of the blade is obvious. For blade (a), SB-VAWT with three kinds of airfoils has large pressure zone at the leading edge of the blade, and the vortex appears on the ventral, among which NACA00018 has the largest pressure, causing energy loss, producing a great influence on the output characteristics and leading to the smallest torque coefficient. For blade (b), both NACA00018 and S1046 (b) have vortex. However, the blade of airfoil S809 does not appear. The pressure difference between the ventral and the back of the blade is obvious, so there is a great contribution to the output characteristics of airfoil S809. For blade (c), the pressure difference of three kinds of airfoils is not obvious, so the contribution to the output characteristics is small. Comparing with S809 and S1046, the high pressure area of the blade of S809 is smaller, which causes smaller aerodynamic force, so that the torque is slightly smaller.

4.1.3. Rotational Speed Performance.
In order to study the rotational speed characteristics and the starting characteristics of SB-VAWT with different airfoils, the change of rotational speed under different wind speeds was investigated by wind tunnel experiments, and the tested wind speed is 6–10 m/s with the interval of 1 m/s. The steady speed of the SB-VAWT and the time required to reach the steady speed were tested at each wind speed. The speed change curve with different wind speeds is shown in Figure 13. When the wind speed is under 7 m/s, the SB-VAWT with airfoil NACA0018 cannot start, so the wind speed change curves are not given.

As can be seen from Figure 13, the steady speed of three different airfoils wind turbines is on the rise with the increase of wind speed, and the time required to reach a steady rotating speed is increasing gradually. When the wind speed is certain, the steady speed from high to low is followed by S1046, NACA0018, and S809. The time required to reach a steady speed from long to short is followed by S1046, NACA0018, and S809. When the wind speed is 10 m/s, three different wind turbine airfoils steady speeds are the highest. For airfoil S1046, the steady rotational speed is 140 r/min, for NACA0018 is 120 r/min, and for S809 is 45 r/min. When the wind speed is less than or equal to 7 m/s, the SB-VAWT with airfoil NACA0018 cannot start itself, which shows that the starting performance is not good. This is one of the important reasons which restricts the development of current NACA series of SB-VAWT. This problem can be overcome to some extent by using S series airfoils.

4.2. Static Characteristics

4.2.1. Static Torque Coefficient.
In order to further study different influence of blade on the starting characteristics of SB-VAWT with different airfoils, the static torque of SB-VAWT at different wind speeds was investigated by numerical

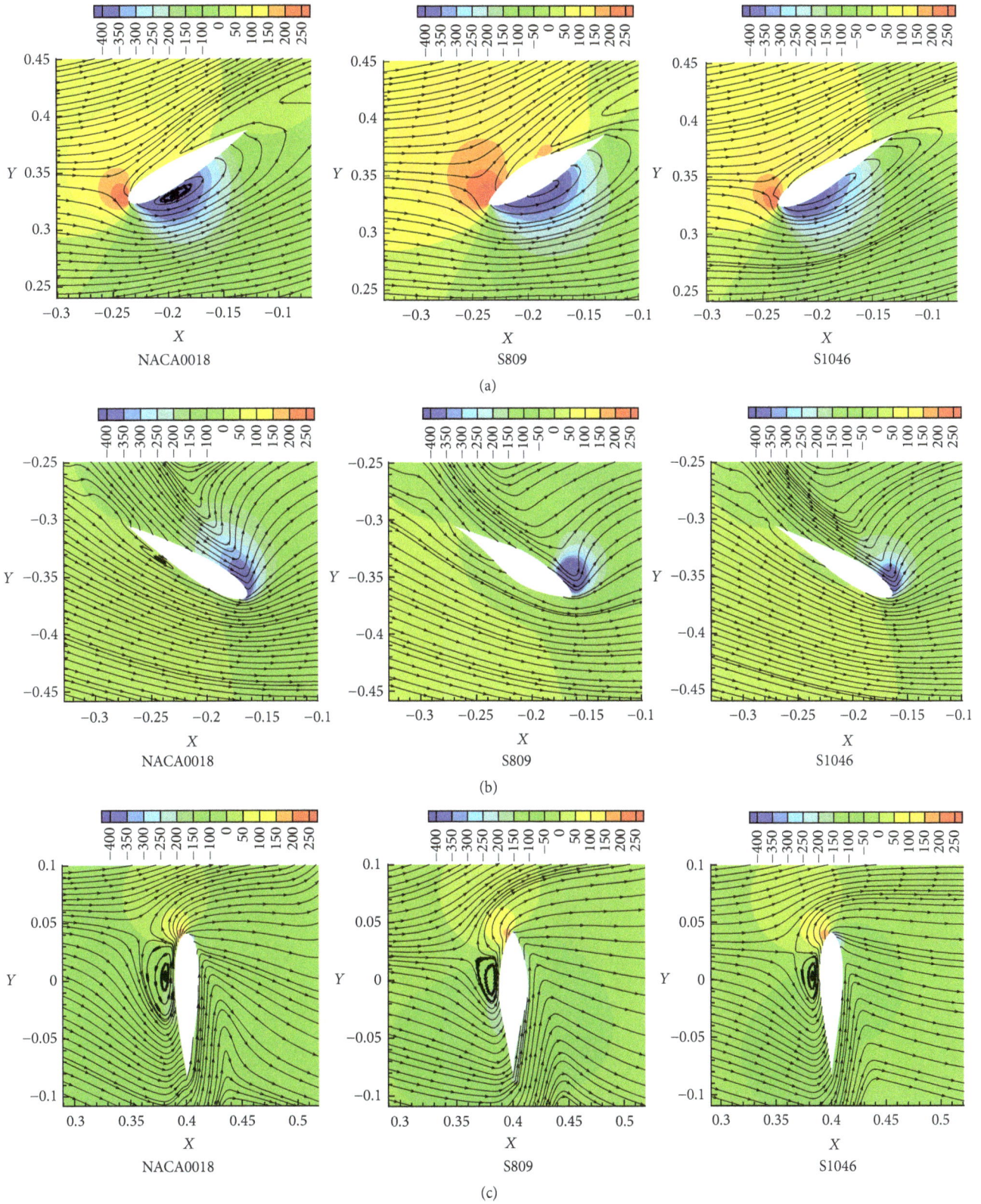

FIGURE 11: Flow fields around blade with different airfoils ($\theta = 35°$).

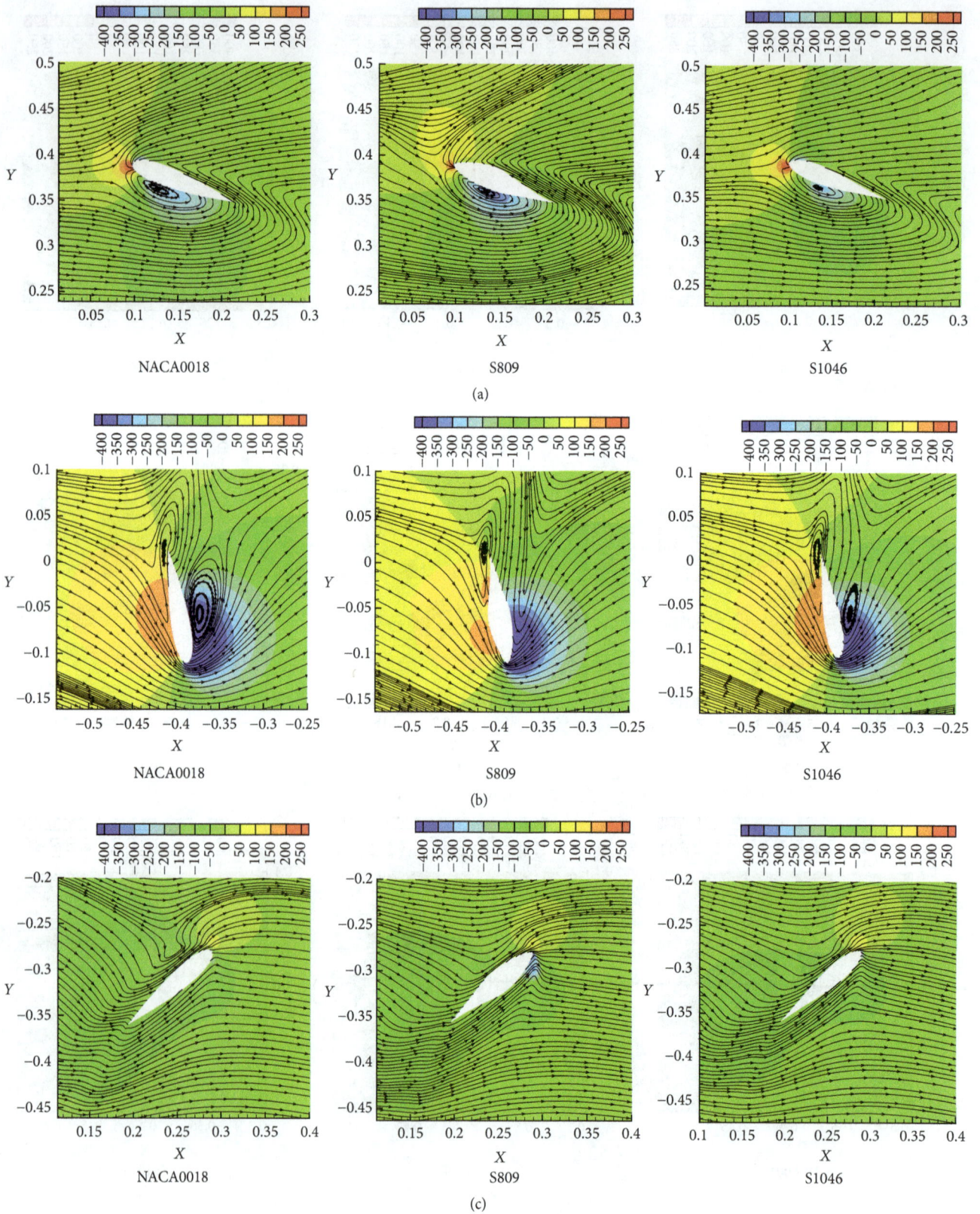

FIGURE 12: Flow fields around blade with different airfoils ($\theta = 100°$).

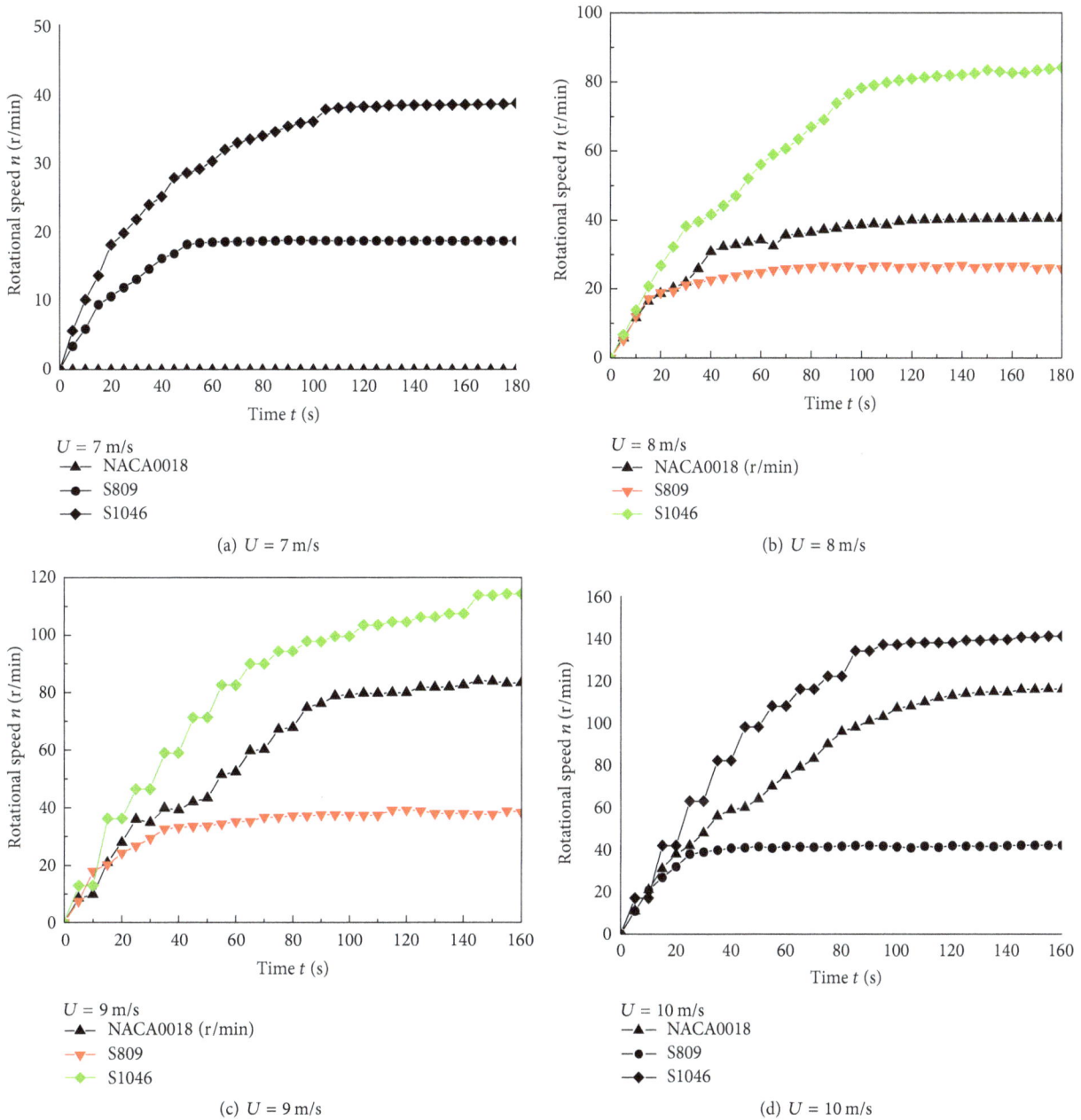

(a) $U = 7$ m/s

(b) $U = 8$ m/s

(c) $U = 9$ m/s

(d) $U = 10$ m/s

FIGURE 13: Rotational speed under different wind speeds.

simulation and wind tunnel experiments. Figure 14 shows the simulation and tests results of the curves of static torque coefficients (C_{ts}) change with the azimuth angles when the wind speed is 10 m/s.

As can be seen from Figure 14, concerning simulation results, the torque coefficient of SB-VAWT with three different airfoils in a rotation period namely 0 to 120 degrees shows two peaks and a trough. The value of the static torque coefficient reaches the maximum when the azimuth angle is 15 degrees, the static torque coefficient of SB-VAWT with airfoil S809 is 0.062, airfoil NACA0018 is 0.054, and airfoil S1046 is 0.049. When the azimuth angle is 45 degrees, the minimum value reaches at this moment, and the static torque coefficient of airfoil S809 wind turbine is −0.0075, airfoil

NACA0018 is −0.0014, and airfoil S1046 is −0.0091. The average static torque coefficient of SB-VAWT with airfoil S809 is 0.032, the highest among the three airfoils. For experimental results, the results of wind tunnel experiments and numerical simulations have good consistency in the overall trend except being slightly lower in value, because in the process of torque transmission, there will be some friction loss between the flange and the beam, and the influence of the experimental error. In general, the asymmetric airfoil S809 has bigger static torque coefficients comparing with the other two kinds of symmetric airfoils.

4.2.2. Static Flow Field. In order to analyze the reason of the difference of starting characteristics in SB-VAWT with

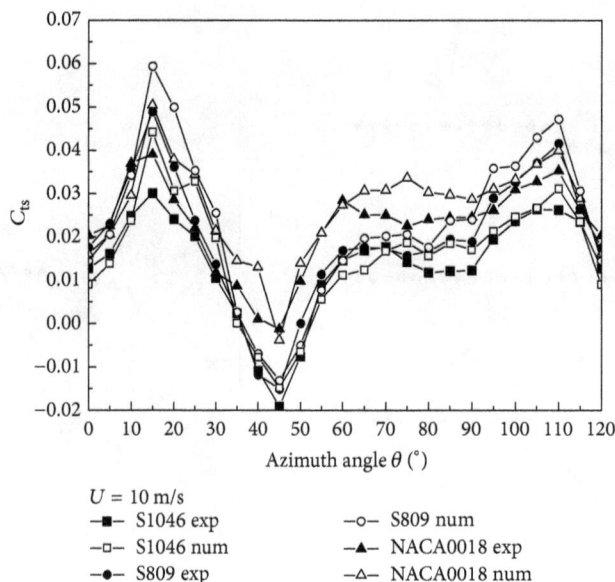

FIGURE 14: Static torque coefficients ($U = 10$ m/s).

three kinds of airfoils in detail, we selected streamlines and pressure distribution of three kinds of airfoil wind turbine at the azimuth angle of 15 and 45 degrees which are shown in Figures 15 and 16. The influence of airfoils on the flow fields of the blade was analyzed.

As can be seen in Figure 15, for blade (a), the force acting on the blades of the three airfoils is mainly at the leading edge. The negative pressure zone is found at the ventral of the blade, while the positive pressure zone is concentrated at the leading edge of the windward point. The aerodynamic force of the pressure difference causes the wind turbine to rotate counter-clockwise. At the same time, the vortex appears at the trailing edge of airfoils S809 and S1046, which makes a certain energy loss on the upper blade. For blade (b), the trailing edge of the blade of the three airfoils is windward; the positive pressure zone is about 1/2 of the blade size; the back of the blade generates low pressure area; the pressure difference is more obvious than the upper blade; the aerodynamic force is greater, which is the main reason for starting coefficients of the SB-VAWT under this angle. The back of the three airfoils produces varying degrees of vortex, which also results in energy loss. The negative pressure zone of the blade in airfoil S1046 is larger than those of S809 and NACA0018, but its aerodynamic force is relatively small, which is an important coefficient for a minimum starting torque at 15 degrees. For blade (c), the positive pressure zone of three kinds of airfoil appears at the trailing edge of the blade, and the pressure difference is small, so the aerodynamic force is slightly smaller. In general, when the azimuth angle is 15 degrees, the high pressure area of non-symmetrical airfoil S809 is bigger than those of the other two kinds of airfoils. The aerodynamic force is relatively larger, and the starting torque is bigger, so the starting performance is better than those of the other two kinds of SB-VAWT.

As can be seen in Figure 16, for blade (a), three airfoils have a large positive pressure area at the trailing edge of the blade, while the negative pressure zone which generates

from the ventral of airfoil has obvious difference. The negative pressure zone around airfoil S809 accounts for 2/3 of the leeward side which results in great vortex and the energy loss is larger. For blade (b), the flow field of the three airfoils is basically the same. The pressure difference between the windward and leeward surfaces of the trailing edge is found, but the torque is small because of the aerodynamic position. For blade (c), when the positive pressure area is larger, the leeward vortex appears on the different positions of the three airfoils. The vortex appearing in the airfoil NACA0018 is smaller comparing with the other two airfoils. We found that the vortex range of the ventral side of blade S1046 is the largest and the energy loss is the largest, so the torque coefficient is the lowest at this azimuth angle. Overall, when the azimuth angle is 45 degrees, the pressure difference between ventral and the back of the three airfoils is small. The aerodynamic force is small so that the starting torque is small, which leads to the appearance of trough.

5. Conclusions

The main conclusions obtained under the condition of this study are as follows:

(1) For dynamic characteristics, the power coefficient of the wind turbine model with airfoil S1046 is higher than those of the model with S809 and NACA0018 airfoils. Furthermore, the rotational speed performance of the test model with airfoil S1046 is also better than the other two kinds of airfoils.

(2) For static characteristics, the static torque coefficient of SB-VAWT with airfoil S809 is higher than the S1046 and NACA0018. The asymmetry of the blade allows the turbine to obtain the forward torque that causes the turbine to rotate under the wide azimuth angle. Therefore, it can be said that the S series airfoils can

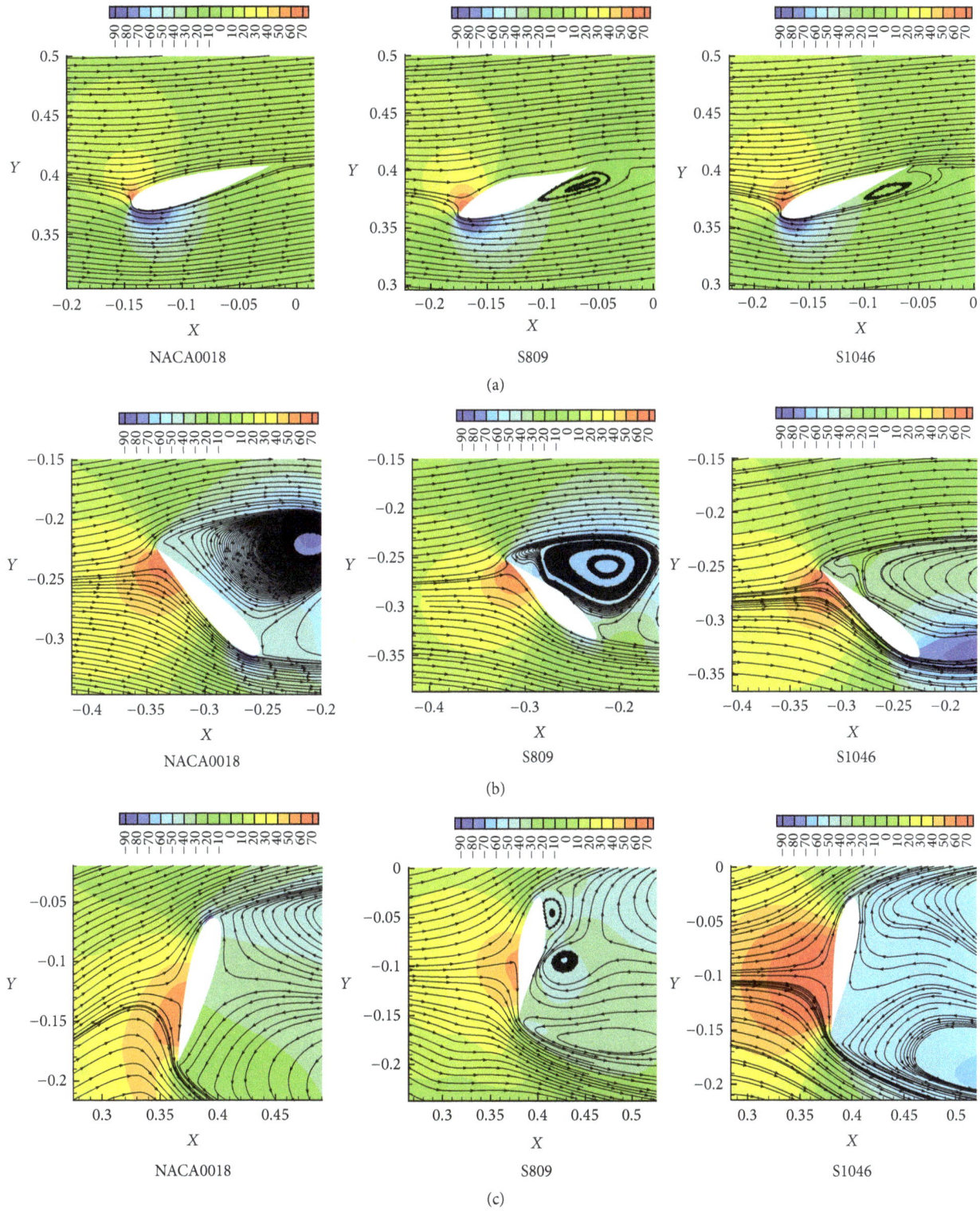

FIGURE 15: Flow fields around blade with different airfoils.

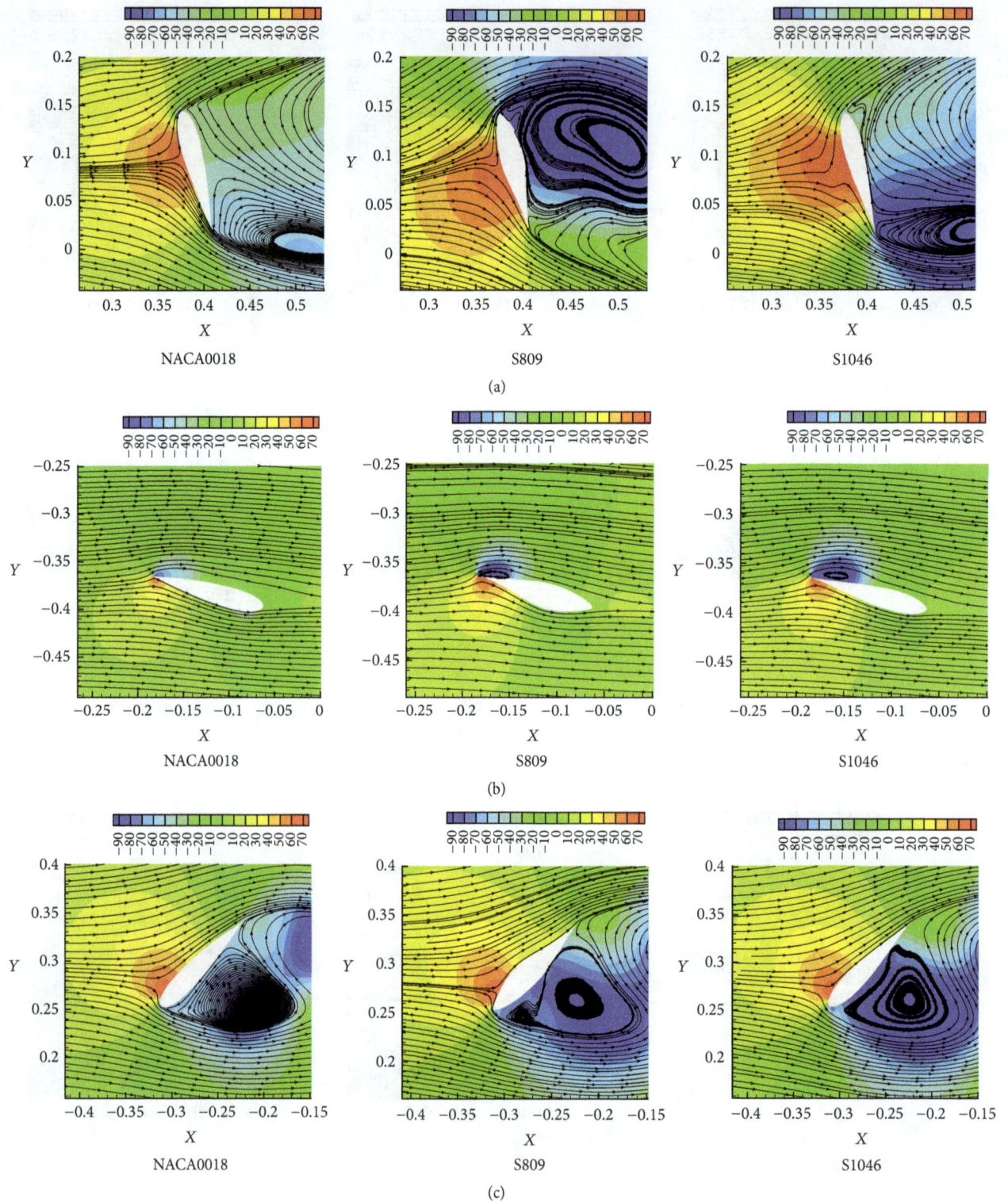

(a)

(b)

(c)

FIGURE 16: Flow fields around blade with different airfoils ($\theta = 45°$).

be used for SB-VAWT and as the most suitable airfoil for SB-VAWT, the S series airfoil is worth researching deeply.

Conflicts of Interest

The authors declare that they have no conflicts of interest.

Acknowledgments

This research is sponsored by the Project 51576037 supported by National Natural Science Foundation of China (NSFC) and Project 12541012 supported by Science and Technology Research Project of Heilongjiang Provincial Department of Education. The authors give thanks to their supporters.

References

[1] Y. Li, Y. Zheng, S. Zhao et al., "A review on aerodynamic characteristics of straight-bladed vertical axis wind turbine," *Acta Aerodynamica Sinica*, vol. 35, no. 6, pp. 368–382, 2017.

[2] P. Chen, M.-Y. Du, and J.-P. Liu, "Development status and key aerodynamic problems of wind turbine dedicated airfoils," *Power System and Clean Energy*, vol. 25, no. 2, pp. 36–42, 2009.

[3] M. Islam, M. R. Amin, R. Carriveau, and A. Fartaj, "Designing straight-bladed vertical axis wind turbine using the cascade theory," in *12th AIAA/ISSMO Multidisciplinary Analysis and Optimization Conference*, American Institute of Aeronautics and Astronautics, Reston, Va, USA, 2008.

[4] G.-Y. Zhang, W.-M. Feng, and C.-L. Liu, "Numerical simulation on the aerodynamic performance of six kinds of aerofoil of wind turbine blade," in *Renewable Energy Resources*, vol. 2, 2009.

[5] S.-X. Liao, C. Li, J.-B. Nie et al., "The analysis of aerodynamic performance for small H type VAWT based on different airfoils," in *Machine Design and Research*, vol. 3, 2011.

[6] M. H. Mohamed, "Performance investigation of H-rotor Darrieus turbine with new airfoil shapes," *Energy*, vol. 47, no. 1, pp. 522–530, 2012.

[7] Z.-H. Xu, Y. Zhang, B. Yang et al., "Effects of installation method of asymmetric airfoils on performance of H-type wind turbine," *Acta Energy Solaris Sinica*, vol. 34, no. 6, pp. 933–937, 2013.

[8] H. Xu, H. Yang, and C. Liu, "Numerical value analysis on aerodynamic performance of DU series airfoils with thickened trailing edge," *Transactions of the Chinese Society of Agricultural Engineering*, vol. 17, 2014.

[9] C.-X. Yang and S.-T. Li, "Study of post stalled airfoil of a H-type vertical axis wind turbine," *Journal of Lanzhou University of Technology*, vol. 41, no. 1, pp. 51–54, 2015.

[10] Y.-L. Jia, P. Peng, Q.-J. Li et al., "Effects of relative thickness of airfoil on aerodynamics of DU airfoil," *Machinery Design & Manufacture*, vol. 3, 2016.

[11] F. Feng, Y. Li, L. Chen, W. Tian, and Y. Zhang, "A simulation and experimental research on aerodynamic characteristics of combined type vertical axis wind turbine," *Acta Energiae Solaris Sinica*, vol. 35, no. 5, pp. 855–860, 2014.

[12] I. Paraschivoiu, C. Li et al., *The vertical axis wind turbine principle and design*, Shanghai Science and Technology Press, 2013.

[13] Q.-B. He, *Study on calculation of structure and aerodynamic characteristics for vertical axis wind turbine with double-layer retractile blades*, Northeast Agricultural University, Harbin, China, 2015.

13

Interpreting Aerodynamics of a Transonic Impeller from Static Pressure Measurements

Fangyuan Lou ⓘ, John Charles Fabian, and Nicole Leanne Key ⓘ

Purdue University, West Lafayette, IN 47907, USA

Correspondence should be addressed to Fangyuan Lou; louf@purdue.edu

Academic Editor: Jingyin Li

This paper investigates the aerodynamics of a transonic impeller using static pressure measurements. The impeller is a high-speed, high-pressure-ratio wheel used in small gas turbine engines. The experiment was conducted on the single stage centrifugal compressor facility in the compressor research laboratory at Purdue University. Data were acquired from choke to near-surge at four different corrected speeds (Nc) from 80% to 100% design speed, which covers both subsonic and supersonic inlet conditions. Details of the impeller flow field are discussed using data acquired from both steady and time-resolved static pressure measurements along the impeller shroud. The flow field is compared at different loading conditions, from subsonic to supersonic inlet conditions. The impeller performance was strongly dependent on the inducer, where the majority of relative diffusion occurs. The inducer diffuses flow more efficiently for inlet tip relative Mach numbers close to unity, and the performance diminishes at other Mach numbers. Shock waves emerging upstream of the impeller leading edge were observed from 90% to 100% corrected speed, and they move towards the impeller trailing edge as the inlet tip relative Mach number increases. There is no shock wave present in the inducer at 80% corrected speed. However, a high-loss region near the inducer throat was observed at 80% corrected speed resulting in a lower impeller efficiency at subsonic inlet conditions.

1. Introduction

High-pressure-ratio centrifugal compressors have been widely used in turbochargers and turboshaft engines because of their compact size, high efficiency, and wide operating range. Modern turbochargers and turboshaft engines are continuously pushing the boundary of pressure ratio and flow capacity. Size limitations on the outer diameter lead to larger rotational speeds and result in transonic flow conditions at the compressor inlet. Compared to the impeller flow with subsonic inlet conditions, the flow in transonic impellers is more complex due to the presence of shock waves. It is believed that, in a transonic centrifugal compressor, additional losses could be generated due to the interaction between the shock waves and blade surface boundary layer/tip clearance flow. Investigations by Rodgers and Sapiro [1] showed a drop in impeller peak efficiency (PE) with the increase of impeller inlet shroud relative Mach number, and this has prompted several investigations on transonic impellers.

The flow inside a centrifugal impeller with transonic inlet conditions was studied by Senoo et al. [2] with static pressure measurements along the impeller shroud and later by Hayami et al. [3] with velocity measurements inside the inducer using a laser-2-focus velocimeter (L2FV). At subsonic flow conditions, the impeller stalled at low flow rates, and flow separation was observed. However, the flow separation was limited to the semi-open area between the inducer leading edge (LE) and the throat. Thus, it allowed for a stable operation range without deterioration of impeller efficiency. At supersonic inlet conditions, the results revealed two shock waves inside the inducer: one detached shock wave in front of the main blade LE and the other on the pressure surface in the flow passage. Additionally, the highest polytropic efficiency was observed at conditions near impeller stall, indicating that shock waves do not severely deteriorate impeller performance.

Krain et al. [4, 5] conducted detailed measurements of the flow field inside a transonic backswept impeller first using L2FV and later using 3-component laser Doppler velocimetry

(LDV). The relative inducer tip Mach number was 1.3. Results showed that the first flow instability occurred near the shroud in the inducer due to the interference of shock waves with the tip leakage vortex. The tip leakage flow extended over the full passage as the flow moved towards the impeller exit, and the flow pattern at the impeller exit seemed to be dominated by the wake formed by the tip leakage flow.

Ibaraki et al. [6–8] and Marconcini et al. [9, 10] studied the flow field inside a transonic turbocharger impeller at both design and off-design conditions. The impeller features a double-splitter design. The relative Mach number at the impeller leading edge tip was about 1.3, the same as the case studied by Krain et al. [4, 5], and the results showed similar flow patterns as well. The low-velocity regions were first observed near the shroud/suction-side area in the inducer due to the interaction between the shock wave and the tip leakage flow. This interaction enhances the total pressure loss in the inducer. At the exit of impeller, the low-velocity regions formed by the tip leakage vortex dominated the flow field.

Higashimori et al. [11] carried out detailed flow measurements on an impeller with a relative tip Mach number of 1.6 at the inducer, and a CFD simulation was also conducted for comparison. The research was mainly focused on the inducer flow, and measurements at four traverse planes were acquired. Results showed that an oblique shock was formed at the leading edge of the inducer as the flow entered the impeller, and a passage shock appeared at the inducer throat. Reversed flow near the shroud in the inducer was present because of the interaction between the shock wave and the tip leakage flow. Buffaz and Trébinjac [12] investigated the flow in the inducer of a transonic centrifugal compressor from both velocity measurements using LDV and time-resolved static pressure measurements along the impeller shroud at the inducer section. It was concluded that the change in flow pattern from choke to surge is mainly due to the change in the tip leakage flow.

In summary, flow in transonic impellers is more complex due to the presence of shock waves; reversed flow near the shroud in the inducer may occur due to the presence of shock waves; the tip leakage flow plays a more important role both within the impeller and at the exit of impeller; flow instabilities may start further upstream at the throat of inducer due to the interaction of shock waves and tip leakage flow.

2. Scope of the Paper

There are conflicting results on transonic impellers due to the limited impeller geometries on which the previous research was performed. For example, results from Senoo et al. [2] showed that the effect of the interaction between the shock wave and tip leakage flow is confined within the semi-open area of the impeller near the LE. However, results from Krain et al. [4, 5] and Ibaraki et al. [6–8] showed an extended effect of the interaction between shock wave and tip leakage flow at the impeller exit.

This paper presents detailed analysis of transonic impeller aerodynamics using data acquired from both steady and time-resolved static pressure measurements along

the impeller shroud. The impeller is a high-speed, high-pressure-ratio wheel used in aerogas turbine engines. The results help in understanding the aerodynamics in transonic impellers and also enrich the database of research in impeller aerodynamics in the open literature.

3. Facility and Instrumentation

The present study was performed on the single stage centrifugal compressor (SSCC) facility with an APU-style inlet in the compressor research laboratory at Purdue University. The facility includes a 1400 horsepower AC electric motor, a 30.46 : 1 ratio gearbox, an exhaust plenum, and a Honeywell experimental compressor. Details of the SSCC facility were documented by Lou et al. [13]. Figure 1 shows the cross-section of the compressor and the location of performance instrumentation. The entire stage includes an inlet housing, an unshrouded impeller, a vaned diffuser, a bend, and deswirl. The impeller features 17 main blades and 17 splitter blades. The compressor performance is calculated based on the total pressure and total temperature measurements at compressor inlet and deswirl exit, as indicated in Figure 1. The design operating speed for the compressor is about 45,000 rpm, and the total pressure ratio for the entire stage is on the order of 6.5.

Figure 2 shows the pressure instrumentation along the impeller shroud. Static pressure taps were equally spaced along the impeller shroud from the LE to trailing edge (TE) at 10 meridional locations. Additionally, 10 fast-response pressure transducers were installed along the impeller chord from impeller LE to 40% meridional position.

The static pressures were measured using Scanivalve pressure modules of high accuracy, and a high-speed data acquisition system was used for acquiring the signals from the fast-response transducers. The analog millivolt signals from the fast-response pressure transducers are amplified using Precision Filter Cards 28118. The amplified signals are then digitized using National Instruments cards of model 6358. Since the blade passing frequency is on the order of 25 KHz at design speed, a sampling rate of 1 MHz was used for digitizing the unsteady pressure measurements. This gives more than 1300 data points per rotor revolution, and, thus, it provides very good spatial resolution in the time domain that enables detailed flow features in the impeller to be captured.

Experimental data were acquired from choke to near-surge at four different corrected speeds (Nc) from 80% to 100% design speed. A constant 5.1% exducer clearance in the axial direction relative to impeller tip blade height was maintained throughout the test. The impeller-diffuser match bleed flow was closed, and a constant 1% back face bleed flow relative to the compressor inlet flow rate was maintained throughout the test.

4. Results and Discussions

Compressor steady performance results and a detailed analysis of impeller performance using steady and time-resolved static pressure measurements are presented. The steady

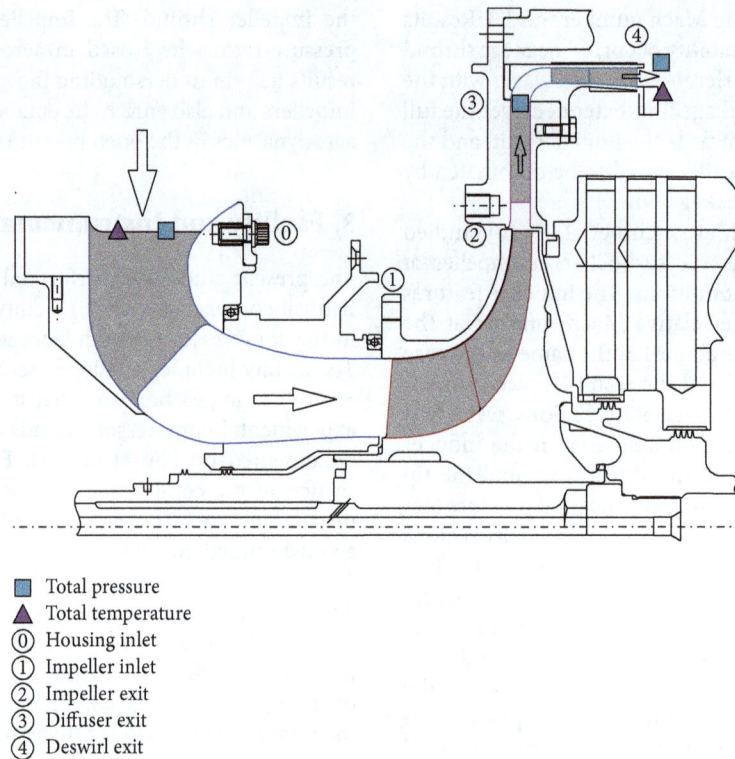

- ■ Total pressure
- ▲ Total temperature
- ⓪ Housing inlet
- ① Impeller inlet
- ② Impeller exit
- ③ Diffuser exit
- ④ Deswirl exit

FIGURE 1: Cross-section of compressor and performance instrumentation.

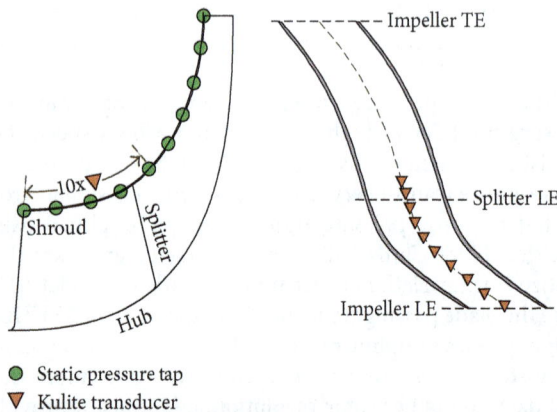

- ● Static pressure tap
- ▼ Kulite transducer

FIGURE 2: Pressure instrumentation along impeller shroud.

performance is characterized by the total pressure ratio, total temperature ratio, and efficiency. The performance of the entire compressor stage is calculated from the area-averaged inlet and exit conditions measured at stations 0 and 4. The impeller performance is evaluated using the area-averaged inlet condition measured at station 0 and the impeller exit condition at station 2, where static pressure measurements are available. The total temperature at the impeller exit is assumed to be the same as that measured at the deswirl exit (station 4) based on the adiabatic assumption. The impeller exit total pressure is derived from the measured total temperature at the deswirl exit, the inlet mass flow rate, and the area-averaged static pressure measured at the impeller trailing

edge using the continuity and the turbomachinery Euler equation [14]. The compressor corrected conditions (speed and mass flow rate) [15] and efficiency [16] are calculated using properties for humid air retrieved from REPROP [17]. Results presented in this section are normalized using the operating condition at design point. The compressor inlet pressure is measured using highly accurate 2.5 psid modules with an uncertainty less than 0.12%. Compressor exit pressure is measured using the 100 psid modules with uncertainty less than 0.05% of full scale. This renders an uncertainty in the total pressure ratio less than 0.2% from 80% to 100% corrected speed. The mass flow rate is measured using a calibrated bell mouth with an uncertainty less than 0.5%.

Figure 3 shows the normalized total pressure versus the normalized corrected mass flow rate. Results for both the impeller and the entire stage from 80% to 100% corrected speed are presented. The peak efficiency (PE) conditions are represented by the solid green symbols. The low loading (LL) conditions are represented by the solid blue symbols. The choke conditions are shown as solid red symbols. Those color schemes are consistent in Figures 4–8. Regardless of the variation in the total pressure ratio for the entire compressor stage relative to the changes in the loading conditions, the total pressure ratio for the impeller stays very constant along the choke line due to the choked flow in the diffuser.

Figure 4 shows the performance of the impeller and entire compressor stage in terms of isentropic efficiency. Compared to the entire stage, the impeller operates more efficiently over the entire operating range from choke to near-surge. There is no obvious deterioration in impeller efficiency along the

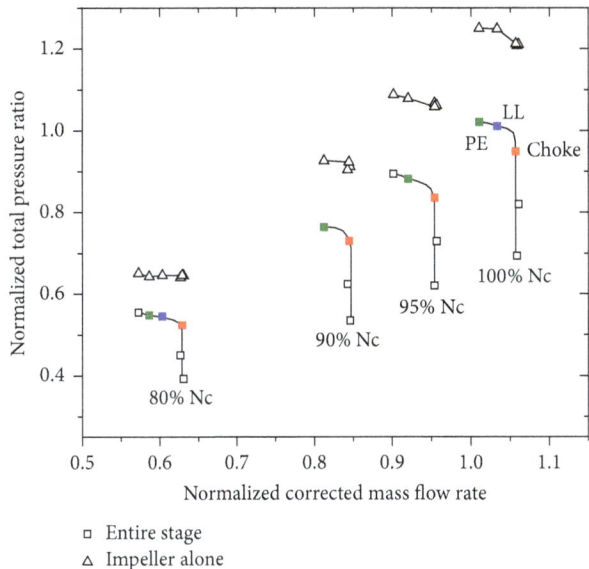

FIGURE 3: Total pressure ratio for impeller and entire stage.

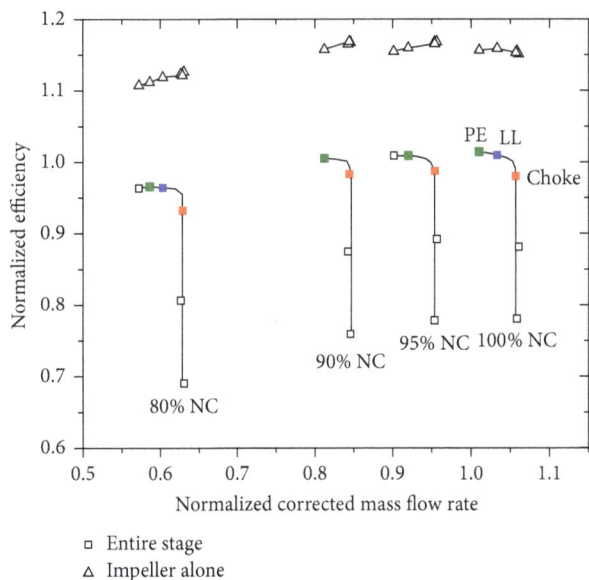

FIGURE 4: Isentropic efficiency for impeller and entire stage.

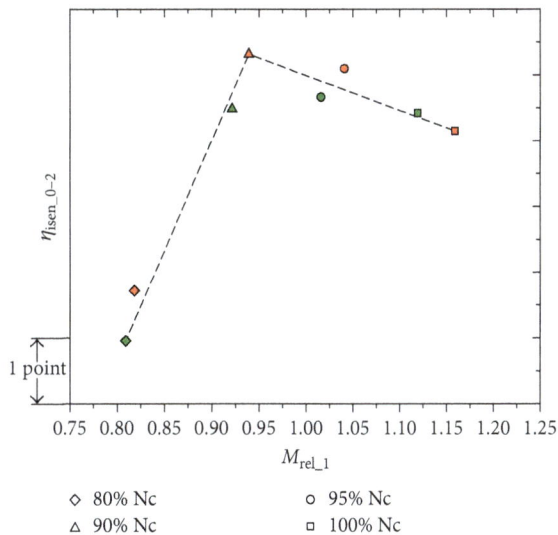

FIGURE 5: Effect of inlet tip Mach number on impeller efficiency.

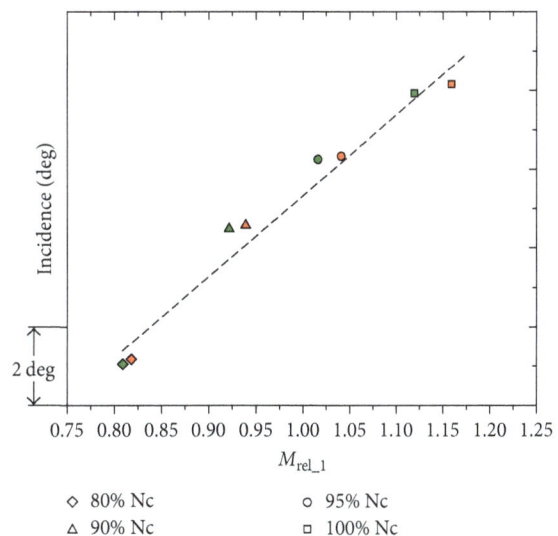

FIGURE 6: Relationship between impeller inlet incidence angle and impeller inlet tip relative Mach number.

choke line as loading decreases. In fact, at subsonic inlet conditions from 80% to 95% corrected speed, the impeller efficiency increases as loading decreases. At design speed with supersonic inlet conditions, the trend in impeller efficiency lines up with the trend for the entire compressor stage in that they both increase from choke to LL condition.

Figure 5 shows the impeller efficiency versus the inlet tip relative Mach number. The inlet tip relative Mach number is mainly determined by the inlet mass flow rate and prewhirl angle. At supersonic inlet conditions, the impeller efficiency drops with the increase of inlet tip relative Mach number, and there is about a 0.9-point drop in the efficiency from 95% to 100% corrected speed. This trend agrees with the observation from Rodgers [18]. However, at subsonic inlet

conditions, the impeller peak efficiency increases as the inlet tip relative Mach number increases, and there is a 3.5-point improvement in the impeller efficiency from 80% to 90% corrected speed.

Figure 6 shows the relationship between inlet tip relative Mach number and impeller inlet incidence angle. The results show a fairly linear correlation between the tip relative Mach number and incidence at the impeller inlet from 80% to 100% corrected speed. The incidence angle increases with the increased tip relative Mach number, and there is about a 7.5-degree change in the impeller incidence angle from 80% to 100% corrected speed.

Furthermore, the detailed distribution of relative Mach number was evaluated from both the steady and time-resolved static pressure measurements along the impeller shroud at each operating condition. Figure 7 shows the

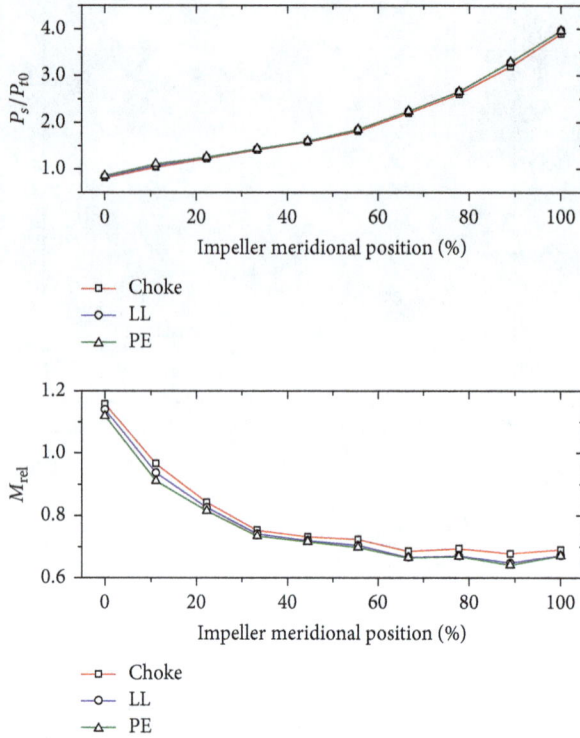

FIGURE 7: Distribution of static pressure and isentropic relative Mach number along impeller shroud at 100% N_c.

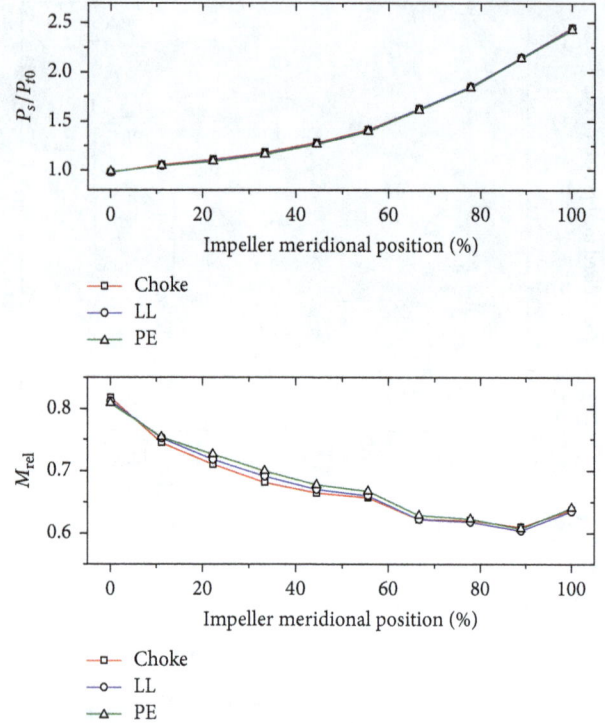

FIGURE 8: Distribution of static pressure and isentropic relative Mach number along impeller shroud at 80% N_c.

distribution of static pressure and isentropic relative Mach number along the impeller shroud from LE to TE at 100% corrected speed. The static pressure is normalized by the area-averaged total pressure measured at station 0. The isentropic relative Mach number along the impeller shroud is derived from the equation for conservation of rothalpy and the isentropic flow assumption [3, 8]. It is calculated as follows:

$$M_{\mathrm{rel}} = \frac{\sqrt{2\left(I_1 - h_s + U^2/2\right)}}{a}, \qquad (1)$$

where I_1 represents rothalpy at impeller inlet, h_s is static enthalpy, U is tangential velocity, and a is speed of sound. Results at three different loading conditions from choke to peak efficiency (as indicated in Figures 3 and 4) are presented. The inlet tip relative Mach number is about 1.15. The inlet tip relative Mach number is obtained from static pressure at the impeller LE and the total pressure and total temperature measured at station 0, assuming isentropic flow through the inlet housing.

At 100% corrected speed, the majority of static pressure rise (nearly 81%) is achieved in the knee and exducer because of the centrifugal effect, with only about 19% of the static pressure rise achieved in the inducer. In contrast, over 86% of relative diffusion occurs in the inducer. Since all the losses occur in the relative diffusion process, the inducer strongly influences impeller efficiency. Additionally, the variation in the shape of the static pressure distribution from choke to peak efficiency is small. In contrast, the distributions of isentropic relative Mach number deviate starting at the

impeller leading edge. The relative Mach number distribution at LL and PE conditions matches those at the knee and exducer, but they are different at the inducer, which leads to the differences in impeller efficiency.

Distributions of static pressure and isentropic relative Mach number along the impeller shroud at subsonic inlet conditions are shown in Figure 8 using the data at 80% corrected speed. The inlet tip relative Mach number is around 0.82, and results at three different loading conditions from choke to PE are presented. The normalized static pressure distributions are similar from choke to peak efficiency. At 80% corrected speed, about 12% of the static pressure rise is achieved in the inducer, which is 7 points lower compared to that at 100% corrected speed. This indicates more loss in the impeller inducer at 80% corrected speed. Approximately 65% of the relative diffusion occurs in the inducer, which is 21 points less than the value at 100% corrected speed. At both 80% and 100% corrected speeds, the majority of diffusion is achieved in the inducer while the majority of the static pressure rise occurs in the knee and exducer. This once again highlights the merit of the relative diffusion effectiveness as an indicator of design quality since it removes the bias of using static pressure rise, which is mostly achieved via centrifugal effects.

Figure 9 shows the distribution of static pressure and isentropic relative Mach number at PE conditions from 80% corrected speed to 100% corrected speed. The inlet tip relative Mach number covers both subsonic and supersonic flow conditions. The relative Mach numbers are normalized by the values at the leading edge. The normalized static pressure

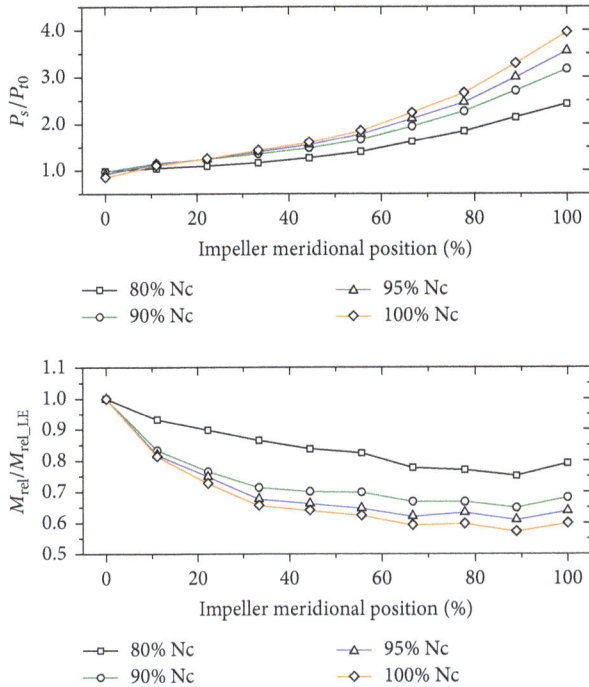

FIGURE 9: Distribution of static pressure and isentropic relative Mach number along impeller shroud at PE conditions.

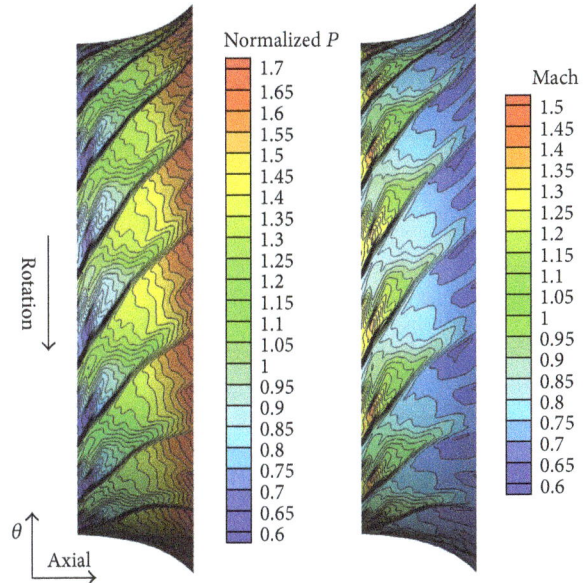

FIGURE 10: Contour of static pressure and isentropic relative Mach number at peak efficiency condition.

distributions from 90% to 100% corrected speed agree well in the inducer, but they deviate starting at the beginning of the radial turn. However, differences in static pressure distributions start at the LE for 80% corrected speed, with 69% less static pressure rise in the inducer. Thus, there is more loss generated in the inducer at 80% corrected speed and, therefore, lower impeller efficiency (by 3.6 points as shown in Figure 5). The same conclusion is drawn from the isentropic relative Mach number distributions from 80% to 100% corrected speed. Additionally, the majority of the loss occurs in regions near the inducer throat.

Furthermore, time-resolved static pressure along the blade chord is processed to reveal more flow features in the inducer. The static pressure is ensemble-averaged using data from 500 rotor revolutions. Figure 10 shows the contours of static pressure and relative Mach number on the shroud surface at the peak efficiency condition. The static pressure is normalized by the inlet total pressure measured at compressor inlet, and the relative Mach number is derived from static pressure and compressor inlet conditions assuming isentropic flow.

The presence of the main blade is indicated by the large gradient between the pressure surface (PS) and the suction surface (SS). In contrast, the presence of the splitter blade is noticeable but less obvious, as indicated by the discontinuity in static pressure contour at mid-passage. A deficit in static pressure corresponds to increased isentropic relative Mach number and vice versa. In addition, both contours show very repeatable flow features in each blade passage with negligible variability due to blade-to-blade differences. Therefore, contours for a single blade are used in Figures 11–14. The

contours of a single blade passage were obtained by passage-averaging the ensemble-averaged properties across the entire annulus.

Contours of isentropic relative Mach number derived from time-resolved static pressure measurements at 100% corrected speed from choke to peak efficiency conditions are shown in Figure 11. The isentropic relative Mach number is calculated using (1). The inlet conditions at 100% corrected speed are supersonic. There is a significant gradient in Mach number in the pitchwise direction, with high Mach numbers near the SS and low Mach numbers near the PS. Compared to flow on the SS, the flow on the PS gets diffused better, with lower Mach number near the PS, and this applies to both the main and splitter blades. Shock waves near the inducer throat are observed at all three loading conditions, and their positions are shown in the sketch. The position of shock wave is represented by the isolines with isentropic relative Mach number equal to unit. The shock wave forms on the pressure surface near the impeller LE, which further extends into the blade passage. The position of the shock wave changes with loading condition: it moves towards the impeller leading edge from choke to peak efficiency as the inlet tip relative Mach number decreases. Additionally, the corresponding contours of normalized static pressure at 100% corrected speed are shown in Figure 12. At all three loading conditions, there exists a low-pressure zone in the semi-open region near the suction surface. This is because of the high velocity flow upstream of the shock wave, as shown in the sketch. Both contours get skewed near the entrance of the inducer because of the presence of the shock wave and its associated interactions, with contour lines parallel to the trajectory of the shock waves.

Figure 13 shows contours of isentropic relative Mach number from choke to peak efficiency at 80% corrected speed

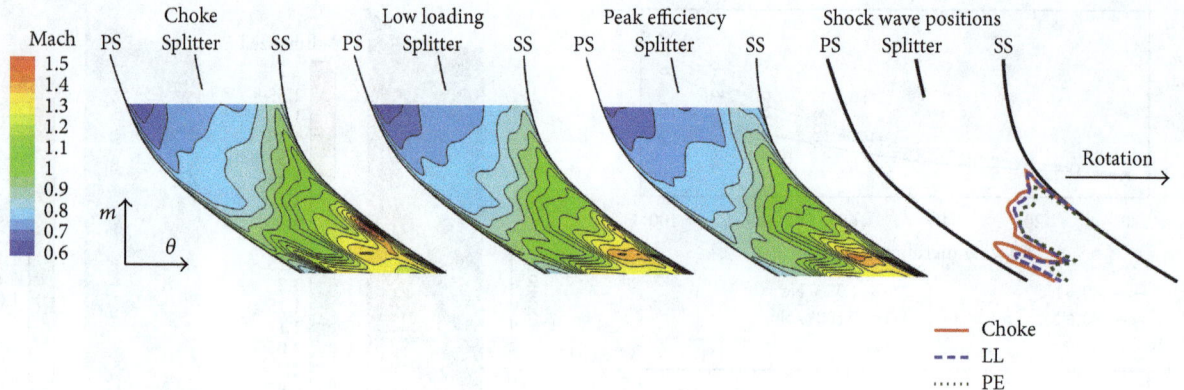

FIGURE 11: Contour of isentropic relative Mach number along the impeller shroud at 100% Nc.

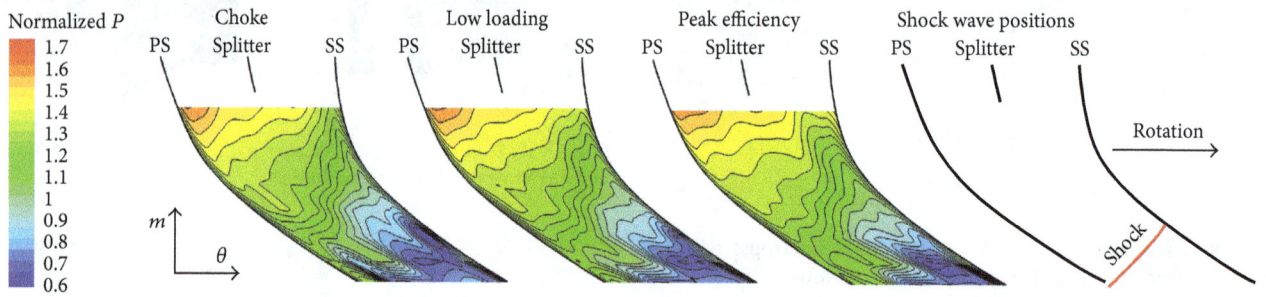

FIGURE 12: Contour of normalized static pressure along the impeller shroud at 100% Nc.

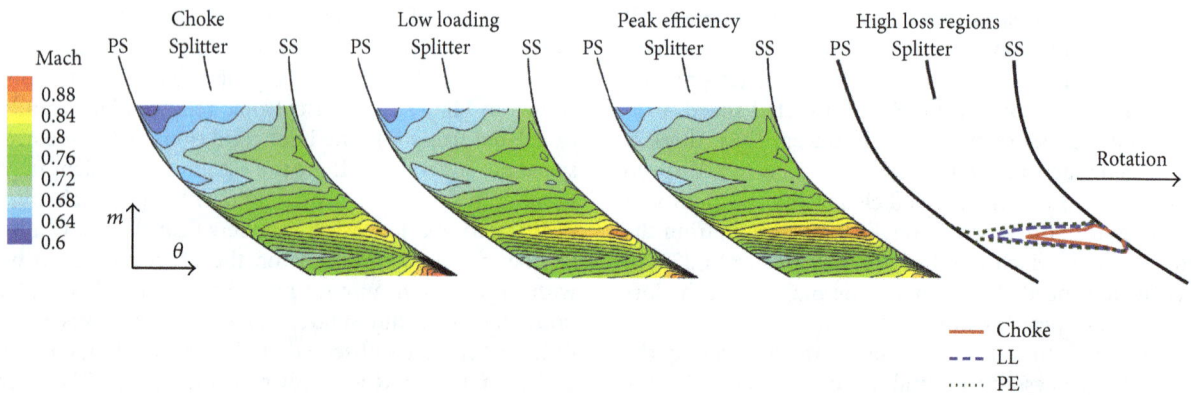

FIGURE 13: Contour of isentropic relative Mach number along the impeller shroud at 80% Nc.

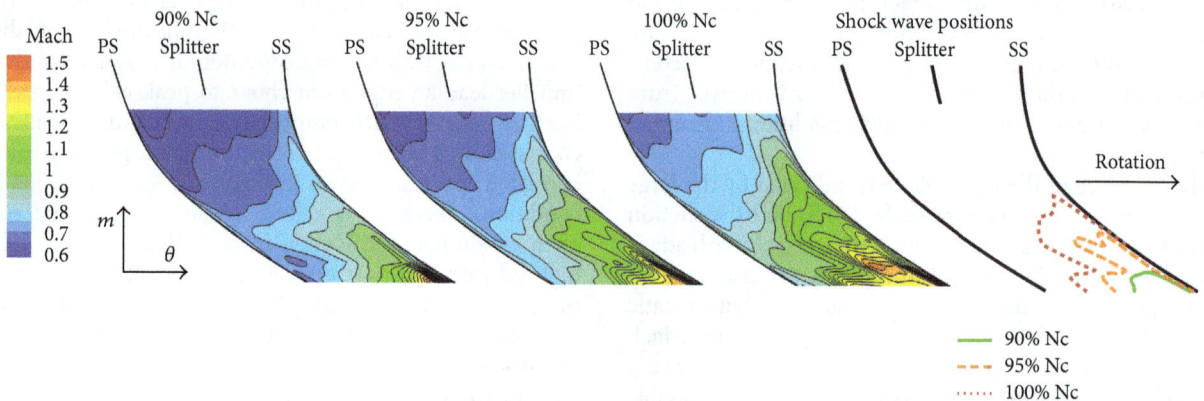

FIGURE 14: Contour of isentropic relative Mach number along the impeller shroud at PE conditions.

for subsonic inlet conditions. The relative Mach number gradient in the pitchwise direction is much smaller compared to that at 100% corrected speed. The contour lines are not skewed but are parallel to the impeller leading edge. However, there is a high Mach number zone observed near the inducer throat area at all three loading conditions. The high Mach number represents a deficit in static pressure and additional loss generation. The high-loss region stems from the suction surface and stretches towards the pressure surface. The area of this high-loss region, as shown in the sketch, grows as the loading increases from choke to peak efficiency. It covers 60% pitch at choke and reaches the pressure surface at peak efficiency. The existence of this high-loss region near the throat in the inducer explains the low efficiency of the impeller at 80% corrected speed. Additionally, the increased area for the high-loss region from choke to peak efficiency explains the decrease in impeller efficiency as the loading of the entire compressor stage increases. Despite the growth of this high-loss region with increased loading, it is limited to the rear portions of the impeller throat, which agrees with observations from Senoo et al. [2].

Figure 14 shows the contours of isentropic relative Mach number at peak efficiency conditions from 90% to 100% corrected speed. The inlet conditions are supersonic at all three loading conditions, and there are shock waves observed in the inducer, as illustrated in the sketch. The subsonic inlet conditions at 90% corrected speed calculated from the steady measurements were underpredicted. This is caused by the steady measurements, which do not include the pressure gradient between the SS and PS. There is strong pressure gradient from PS to SS with lower static pressure near impeller SS. This potential field redistributes the flow and enters into impeller eye with flow of higher relative velocity near the SS and lower relative velocity near the PS, which renders a transonic inlet condition at 90% corrected speed. However, the relative velocity derived from the time-averaged pressure from the steady measurements represents the mean relative velocity, the value of which is between the lower relative velocity on PS and higher relative velocity on the SS. In the present study, it calculates a subsonic inlet condition at 90% corrected speed. Additionally, the shock wave position in the inducer is closely related to the inlet tip relative Mach number. As the inlet tip Mach number increases, the shock wave moves downstream.

5. Conclusions

This paper presents a detailed analysis of transonic impeller aerodynamics using data acquired from both steady and time-resolved static pressure measurements along the impeller shroud. The impeller is a high-speed, high-pressure-ratio wheel used in aerogas turbine engines. The experiment was conducted on the single stage centrifugal compressor (SSCC) facility in compressor research laboratory at Purdue University. Data were acquired from choke to near-surge at four different corrected speeds (Nc), from 80% to 100% design speed, which covers both subsonic and supersonic inlet conditions. By using the parameter isentropic relative Mach number derived from shroud static

pressure measurements, the relative diffusion together with shock wave position could be characterized. The information provides a measure of the flow inside the impeller.

Inside the impeller, the isentropic relative Mach numbers along the impeller shroud show that the majority of diffusion is achieved in the inducer at both subsonic and supersonic inlet conditions. The inducer diffuses the flow more efficiently when the inlet tip relative Mach numbers are close to unity at 90% and 95% corrected speed, and it becomes less efficient as the inlet tip relative Mach number departs from unity at 80% and 100% corrected speed, particularly in the subsonic region.

Shock waves emerging near the LE of the pressure surface were observed from 90% to 100% corrected speed with fast-response pressure measurements. The position of the shock wave changes relative to the inlet tip Mach number. It moves towards the impeller exit as the inlet tip relative Mach number increases. The shock wave at 100% corrected speed is near the inducer throat, and it moves slightly upstream towards the impeller LE with the increase in loading from choke to peak efficiency. While there is no shock for the subsonic conditions at 80% corrected speed, there is a high-loss region around the inducer throat. It stems from the suction surface and expands towards the pressure surface as the loading increases from choke to PE.

Static pressure measurements acquired along the impeller shroud are critical in evaluating the impeller performance. However, the relative Mach number calculated from steady static pressure measurements is underestimated because of the pressure gradient from pressure side to suction side in the blade passage. A subsonic value of inlet tip relative Mach number obtained from steady static pressure measurements does not guarantee shock-free flow in the inducer. Compared to the steady static pressure measurements, time-resolved pressure provides more insight for characterizing the impeller flow. They can be used to identify the high-loss regions and shock wave positions.

Nomenclature

a: Speed of sound
h: Enthalpy
I: Rothalpy
M: Mach number
P: Pressure
U: Blade speed.

Subscripts

isen: Properties derived from isentropic assumption
rel: Properties in relative frame coordinate
s: Static properties
t: Stagnation properties.

Conflicts of Interest

The authors declare that there are no conflicts of interest regarding the publication of this paper.

Acknowledgments

This research has been sponsored by Honeywell, Inc., and this support is most gratefully acknowledged. The authors also wish to thank Honeywell for granting permission to publish this work. Additionally the guidance and advice offered by Mr. Darrell James and Dr. Rakesh Srivastava of Honeywell were very valuable to this project. Assistance from Herbert Harrison and Amelia Brooks at the Purdue Compressor Research Laboratory during data acquisition was also very much appreciated.

References

[1] C. Rodgers and L. Sapiro, "Design considerations for high-pressure- ratio centrifugal compressors," in *Proceedings of the ASME Gas Turbine Conference*, 72-GT-91 pages, 1972.

[2] Y. Senoo, H. Hayami, Y. Kinoshita, and H. Yamasaki, "Experimental study on flow in a supersonic centrifugal impeller," *Journal of Engineering for Power-Transactions of the ASME*, vol. 101, no. 1, p. 32, 1979.

[3] H. Hayami, Y. Senoo, and H. Ueki, "Flow in the inducer of a centrifugal compressor measured with a laser velocimeter," *Journal of Engineering for Gas Turbines and Power*, vol. 107, no. 2, pp. 534–540, 1985.

[4] H. Krain, B. Hoffmann, and H. Pak, "Aerodynamics of a centrifugal compressor impeller with transonic inlet conditions," in *Proceedings of the ASME 1995 International Gas Turbine and Aeroengine Congress and Exposition, GT 1995*, USA, June 1995.

[5] H. Krain, G. Karpinski, and M. Beversdorff, "Flow analysis in a transonic centrifugal compressor rotor using 3-component laser velocimetry," in *Proceedings of the ASME Turbo Expo 2001: Power for Land, Sea, and Air, GT 2001*, USA, June 2001.

[6] S. Ibaraki, H. Higashimori, and T. Matsuo, "Flow investigation of a transonic centrifugal compressor for turbocharger," in *Proceedings of the 23nd CIMAC Congress Proceedings*, pp. 339–346, 2001.

[7] S. Ibaraki, T. Matsuo, H. Kuma, K. Sumida, and T. Suita, "Aerodynamics of a transonic centrifugal compressor impeller," *Journal of Turbomachinery*, vol. 125, no. 2, pp. 346–351, 2003.

[8] S. Ibaraki, K. Sumida, and T. Suita, "Design and off-design flow fields of a transonic centrifugal compressor impeller," in *Proceedings of the 2009 ASME Turbo Expo*, pp. 1375–1384, USA, June 2009.

[9] M. Marconcini, F. Rubechini, A. Arnone, and S. Ibaraki, "Numerical investigation of a transonic centrifugal compressor," *Journal of Turbomachinery*, vol. 130, no. 1, Article ID 011010, 2008.

[10] M. Marconcini, F. Rubechini, A. Arnone, and S. Ibaraki, "Design and off-design numerical investigation of a transonic double-splitter centrifugal compressor," in *Proceedings of the 2008 ASME Turbo Expo*, Germany, June 2016.

[11] H. Higashimori, K. Hasagawa, K. Sumida, and T. Suita, "Detailed flow study of mach number 1.6 high transonic flow with a shock wave in a pressure ratio 11 centrifugal compressor impeller," *Journal of Turbomachinery*, vol. 126, no. 4, pp. 473–481, 2004.

[12] N. Buffaz and I. Trébinjac, "Detailed analysis of the flow in the inducer of a transonic centrifugal compressor," *Journal of Thermal Science*, vol. 21, no. 1, pp. 1–12, 2012.

[13] F. Lou, H. M. Harrison, J. C. Fabian, N. L. Key, D. K. James, and R. Srivastava, "Development of a centrifugal compressor facility for performance and aeromechanics research," in *Proceedings of the ASME Turbo Expo 2016: Turbomachinery Technical Conference and Exposition, GT 2016*, Republic of Korea, June 2016.

[14] H. Simon, T. Wallmann, and T. Moenk, "Improvements in performance characteristics of single-stage and multistage centrifugal compressors by simultaneous adjustments of inlet guide vanes and diffuser vanes," in *Proceedings of the ASME Gas Turbine Conference*, vol. 109, 1986.

[15] R. A. Berdanier, N. R. Smith, J. C. Fabian, and N. L. Key, "Humidity effects on experimental compressor performance-corrected conditions for real gases," *Journal of Turbomachinery*, vol. 137, no. 3, Article ID 031011, 2015.

[16] F. Lou, J. Fabian, and N. L. Key, "The effect of gas models on compressor efficiency including uncertainty," *Journal of Engineering for Gas Turbines and Power*, vol. 136, no. 1, Article ID 012601, 2014.

[17] E. W. Lemmon, M. L. Huber, and M. O. McLinden, "NIST Standard Reference Database 23: Reference Fluid Thermodynamic and Transport Properties—REFPROP, Version 9.1, National Institute of Standards and Technology, Standard Reference Data Program," Gaithersburg, MD, USA, 2013.

[18] C. Rodgers, "Centrifugal compressor inducer," in *Proceedings of the ASME Gas Turbine Conference*, 98-GT-032 pages, 1998.

14

Fault Tolerant Control of Internal Faults in Wind Turbine: Case Study of Gearbox Efficiency Decrease

Younes Ait El Maati ⓘD, Lhoussain El Bahir ⓘD, and Khalid Faitah

National School of Applied Sciences, LGECOS Laboratory, Cadi Ayyad University, Marrakech, Morocco

Correspondence should be addressed to Younes Ait El Maati; aitelmaati.younes@gmail.com

Academic Editor: Ali Mostafaeipour

This paper presents a method to control the rotor speed of wind turbines in presence of gearbox efficiency fault. This kind of faults happens due to lack of lubrication. It affects the dynamic of the principal shaft and thus the rotor speed. The principle of the fault tolerant control is to find a bloc that equalizes the dynamics of the healthy and faulty situations. The effectiveness decrease impacts on not only the dynamics but also the steady state value of the rotor speed. The last reason makes it mandatory to add an integral term on the steady state error to cancel the residual between the measured and operating point rotor speed. The convergence of the method is proven with respect to the rotor parameters and its effectiveness is evaluated through the rotor speed.

1. Introduction

The wind turbine is an electromechanical device to extract the energy from the wind and feed it to the customer through the grid. The wind turbine is composed of several interconnected components. First, the rotor transforms the aerodynamic torque defined by (1) into mechanical torque. The latter is transformed into electricity through a conventional generator. The bond between the rotor and the generator is performed by the bias of a mechanical gearbox. The role of the gearbox is to maintain the same power from rotor to generator through a transformation ratio Ng [1]. Figure 1 summarizes the different components of a modern wind turbine.

$$T_a(t) = \frac{1}{2}\rho\pi R^3 V^2 C_q(V, \Omega_r, \beta).\tag{1}$$

The wind turbine (WT) operates in two distinct regions. The first region is called moderate winds (<7 s/m). In this interval, the WT is controlled through the generator torque to maximize the extraction of the energy contained in the wind. The second region is called high winds region (>7 m/s) in which the objective is to maintain a constant power at the nominal value. This is achieved through controlling three actuators existing in each blade. Those actuators are called

pitch because they let the blade turn a pitch angle about its longitudinal axis. By this pitching movement, the blade is exposed (0° pitch) or not (90° pitch) to the wind and then the rotor speed is accelerated or decelerated [2]. An overview of the modelling and control of the wind turbine systems could be found in [3].

However, the dusty and wet environment induces degradation in some critical components such as the blades, the shafts, the sensors, the generator, and the mechanical components such as the gearbox. Moreover, the challenging situations in which the wind turbines operate (high winds and turbulences, faults on the sensors, and actuators [4]) require highly available systems. For this reason, fault tolerant control strategies [5] are elaborated to prevent the damaging effect of faults and failures on the turbine structure [6]. Most of the FTC methods are composed of two blocs. The first bloc estimates the fault and provides information about the amplitude and the shape of the fault. The last information is then provided to the FTC bloc to build a new control law suitable for the faulty situation. The production of the new control law could be performed either by changing the regulator's parameters or by adding a new term to the old control law to compensate the fault term. More details on faults in the wind turbines and their FTC could be found in [7], where different types of faults and their corresponding

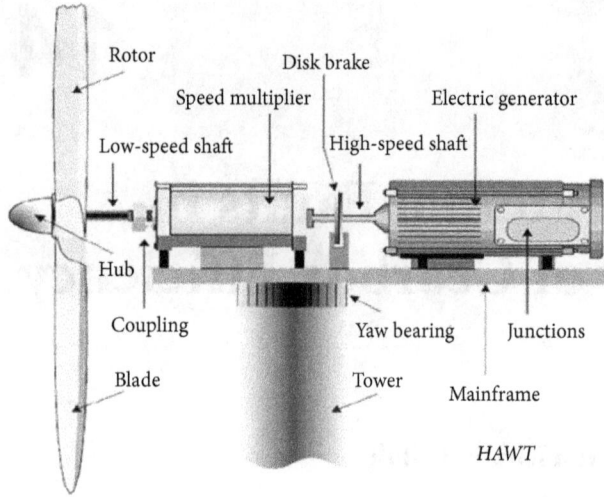

FIGURE 1: The composition of a wind turbine system.

severities are cited. Leakage fault in the hydraulic actuators which is of high severity could not be resolved and the only solution is to shut down the wind turbine for possible maintenance.

In the present paper, the considered fault is the degradation of the efficiency of the gearbox linking the rotor to the generator. It is a fault of medium severity which impacts the dynamics of the rotor and then deteriorates the result of the speed regulation. It will be demonstrated that the fault could be considered an internal fault and a suitable fault tolerant strategy is then applied.

2. The Wind Turbine Model

The considered nominal objective of the wind turbine control is to regulate the rotor speed about the operating value of 40 rpm. The chosen operating wind speed is 18 m/s with 30% of variations according to Kaimal distribution.

The blade pitch operating angle is 9°. This operating point belongs to the high wind region where the only objective is to regulate the power by regulating the rotor speed. This prevents the wind turbine from exceeding the nominal values and from being damaged due to high winds. The parameters of the wind turbine are extracted through linearization from the software FAST [8]. FAST is industrial software developed by the National Renewable Energy Laboratory in Colorado to test and validate the control laws on the wind turbines before physical implementation.

The considered control objective requires considering the rotor and the gearbox models.

2.1. The Rotor Speed Model. The rotor model of the wind turbine is extracted by applying the first law of mechanics to the turbine and is given by

$$\delta\dot{\Omega}_r = \frac{\gamma}{J_t}\delta\Omega_r + \frac{\xi}{J_t}\delta\beta + \frac{\alpha}{J_t}\delta\omega, \tag{2}$$

where $\partial T_a/\partial\omega = \alpha$; $\partial T_a/\partial\Omega_r = \gamma$ et $\partial T_a/\partial\beta = \xi$. T_a is the aerodynamic torque applied by the wind on the blades and

defined by (1). $\delta\omega$ is the variation of the wind speed, $\delta\beta$ is the variation of the pitch angle, and $\delta\Omega_r$ is the variation of the rotor speed, about the operating point. J_t is the total inertia of the wind turbine. The linearization about the chosen operating point gives $\gamma = -0.1039J_t$; $\xi = -2.5727J_t$; and $\alpha = 0.61141J_t$.

2.2. The Gearbox Model. The gearbox is used to adapt the low speeds (40 rpm) of the rotor to the high speeds of the generator (1500 rpm) while maintaining the same power between the two nodes. This relationship could be modelled by the following equation:

$$J_g\dot{\Omega}_g = \eta_{\text{gbx}}T_{\text{lss},e} - N_gT_g. \tag{3}$$

Ω_g is the generator speed, J_g is the generator inertia, T_g is the generator torque, $T_{\text{lss},e}$ is the principle shaft torque, and η_{gbx} and N_g are, respectively, the efficiency and the multiplication ratio of the gearbox. The variation of the parameter η_{gbx} induces a variation of the generator speed. This variation is also transmitted to the rotor due to N_g:

$$\Omega_r = \frac{\Omega_g}{N_g}. \tag{4}$$

The torque $T_{\text{lss},e}$ is a picture of the aerodynamic torque. The torque should be carefully estimated as in [9].

2.3. Aerodynamic Torque Estimation. The aerodynamic torque could be estimated from the drive train model.

2.3.1. Drive Train Model. The drive train is composed of a low speed shaft (rotor side) interconnected with a high-speed shaft (generator side) through a mechanical gearbox with a ratio Ng. The drive train is modelled by the following differential equations:

$$\dot{\Omega}_r = \frac{D_{\text{lss},e}}{J_r}\Omega_r - \frac{X}{J_r} + \frac{D_{\text{lss},e}}{J_r}\Omega_g + \frac{T_a}{J_r}$$

$$\dot{X} = K_{\text{lss},e}\Omega_r - K_{\text{lss},e}\Omega_g \tag{5}$$

$$\dot{\Omega}_g = \frac{D_{\text{lss},e}}{J_g}\Omega_r - \frac{X}{J_g} + \frac{D_{\text{lss},e}}{J_g}\Omega_g + \frac{N_gT_g}{J_g}.$$

J_r is the rotor inertia, J_g is the generator inertia, and X is the restoring force applied on the low speed shaft. The shaft is driven by the aerodynamic torque. The generator torque about the equivalent low speed shaft N_gT_g is used to accelerate or decelerate the shaft. $D_{\text{lss},e}$ and $K_{\text{lss},e}$ are the damping and stiffness coefficients of the principle shaft.

2.3.2. Torque Estimation Loop. In literature, many approaches have been proposed for aerodynamic torque estimation. Authors in [10] proposed a PI based observer to estimate aerodynamic torque. In [11] authors used the Kalman filter to reconstruct the aerodynamic torque Fourier coefficients. In this paper, the proposed method is based on the transfer function between the aerodynamic torque T_a and the rotor

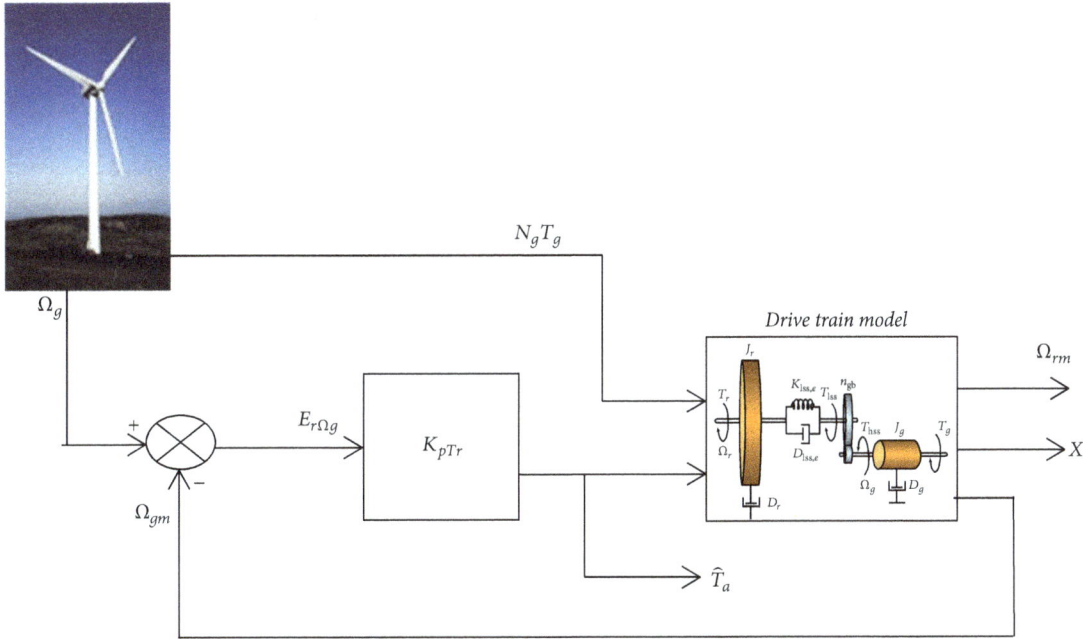

FIGURE 2: Aerodynamic torque feedback estimation loop.

speed Ω_r. However, in most cases, the rotor speed cannot be measured; the measured generator speed about low speed shaft could be considered as a good approximation to the rotor speed. In fact, after transients, the rotor and generator speed about the equivalent low speed shaft are the same, in the case of a rigid equivalent shaft, or when damping the torsional modes of the mechanical shaft.

The transfer function $F(s)$ between the aerodynamic torque T_a and generator speed Ω_r is obtained after manipulations of (5) as follows:

$$F(s) = \frac{1}{s} \frac{as + b}{(s^2 + cs + d)}, \quad (6)$$

where $a = D_{lss,e}/J_r J_g$; $b = K_{lss,e}/J_r J_g$; $c = ((J_r + J_g)/J_r J_g)D_{lss,e}$; and $d = K_{lss,e}((J_r + J_g)/J_r J_g)$.

The idea is to keep the model speed Ω_{gm} sufficiently close to the measured one, Ω_g, by acting on the model with an adequate aerodynamic torque $\hat{T}_a$. This could be performed through a feedback estimation loop as presented in Figure 2. In contrast to authors in the previous works, and since the transfer function $F(s)$ already contains an integrator term, and assuming that the mean of Ω_g is sufficiently low frequency, only a proportional action is needed to estimate T_a. After transients, the model output Ω_{gm} converges to Ω_g and its input $\hat{T}_a$ converges to the actual torque T_a. Finally, $\hat{T}_a$ can be considered as an estimation of the actual aerodynamic torque T_a. The proportional torque estimator gain is chosen in such way that the slowest pole of the closed loop of the transfer function $F(s)$ is cancelled.

Figure 3 shows the actual and estimated aerodynamic torque. The actual aerodynamic torque represented in Figure 3 by the blue color is obtained for comparison by the following equation:

$$T_a = \text{Rotor}_{\text{Acceleration}} * J_r * \frac{\pi}{180} + \text{Shaft}_{\text{Torque}} * 1000. \quad (7)$$

J_r is the rotor inertia about the shaft of the turbine. The shaft torque (KNm) and the rotor acceleration (deg/sec^2) can be obtained from FAST software as outputs. In the industrial wind turbines, a strain gauge is installed on the mechanical shaft of the wind turbine to measure the shaft torque. The rotor acceleration is measured by an accelerometer.

Note that K_{pTr} is the proportional torque estimator gain; $E_{r\Omega g}$ is the speed tracking error, T_g is the generator torque about the high-speed shaft, N_g is the gearbox ratio, and the term $N_g T_g$ is the generator torque about the low speed shaft; Ω_{gm} is the model generator speed about low speed shaft; Ω_g is the measured generator speed; X is the restoring force of the low speed shaft; Ω_{rm} is the model rotor speed; $\hat{T}_a$ is the estimated aerodynamic torque.

Define the mean convergence error between actual aerodynamic torque T_a and estimated aerodynamic torque $\hat{T}_a$ along N samples of data:

$$\text{error (\%)} = \frac{100}{N} \sum_{i=1}^{N} \left| \frac{T_{a,i} - \hat{T}_{a,i}}{T_{a,i}} \right|. \quad (8)$$

In the present case of simulation, we obtain a relative error of 1.8%. The torque estimation could then be considered sufficiently accurate.

2.4. Estimated Aerodynamic Torque Filtering. In this section, a spectral analysis is performed to identify the frequencies of the wind speed transmitted to the torque and those resulting from the mechanical vibration. The objective is to reconstruct the frequencies specific to the wind speed. Figures 4 and 5 show the power spectral density of the wind profile and aerodynamic torque, respectively.

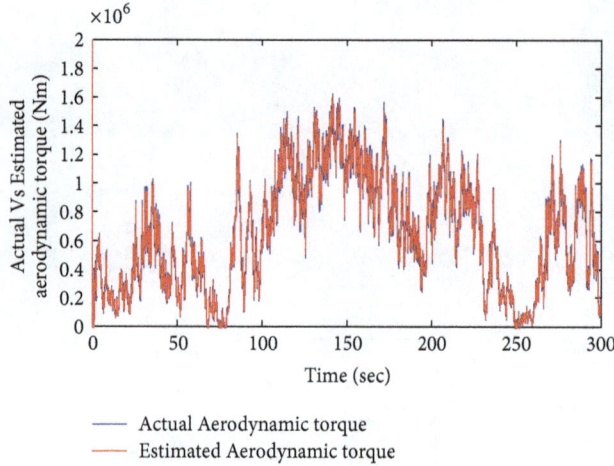

FIGURE 3: The actual and estimated aerodynamic torque.

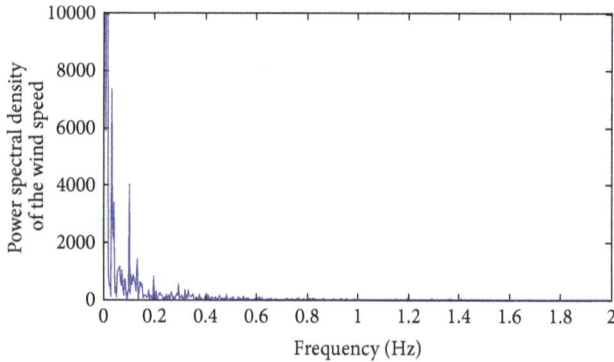

FIGURE 4: Power spectral density of the wind.

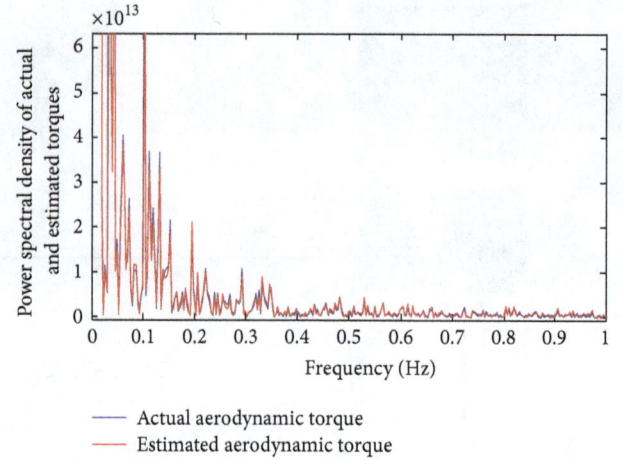

FIGURE 5: Power spectral density of the actual and estimated shaft torque.

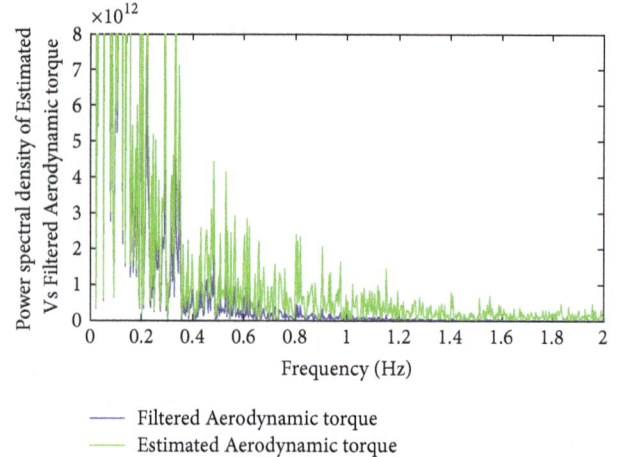

FIGURE 6: The power spectral density of the estimated and filtered aerodynamic torque.

One can notice that the frequencies contained in the wind and having a significant effect on the torque of the rotor resides in the low frequency range ($<0.12\,\text{Hz}$). The peaks that occur beyond this area are the result of vibrations of the mechanical structure as a result of excitations caused by the wind. The estimated aerodynamic torque, which will be used for reconstitution of the wind speed, should therefore be filtered according to the previous remark. In the case of our system, we chose a low pass filter of $0.4\,\text{Hz}$ bandwidth. Figure 6 shows the power spectral density of filtered and the nonfiltered torque.

From Figure 6, one can notice that frequencies above $0.4\,\text{Hz}$ have been attenuated and filtered.

2.5. Gearbox Efficiency Estimation. After estimation of the principal shaft torque, the gearbox efficiency variations η_{gbx} could be estimated through (3) as in [12] by the following manipulations:

$$sJ_g\Omega_g = \eta_{\text{gbx}}T_{\text{lss},e} - N_gT_g$$

$$sJ_g\Omega_g + N_gT_g = T_{\text{lss},e}\eta_{\text{gbx}}$$

$$\underbrace{\Omega_g + \frac{N_gT_g}{sJ_g}}_{Y} = \frac{T_{\text{lss},e}}{sJ_g}\eta_{\text{gbx}}. \tag{9}$$

Figure 7 shows the fault detection scheme of the gearbox efficiency.

Ω_g is the measured generator speed, E_Y is the tracking error loop, η_{gbxm} is the estimated drive train efficiency, N_gT_g is the generator torque about the low speed shaft, and $\hat{T}_{\text{lss},e}$ is the previously estimated shaft torque. Y is the measured variable to be tracked by the model's output. $C_n(s)$ is the proportional action transfer function used for the estimation. $C_n(s)$ is a constant gain M.

In the present case, the proportional gain M is fixed at lower values and progressively increases until we have got good estimation results. For this turbine, we found optimal K at 0.9.

Figure 8 represents different results of estimation for different gearbox efficiencies. The estimate of η_{gbx} constitutes a fault residual, given by (10), used in the activation of the fault tolerant control if a gearbox fault happens. In fact, when $r \neq 1$, it means that the efficiency of the gearbox

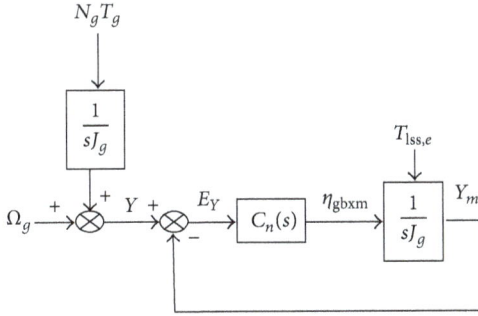

FIGURE 7: Gearbox efficiency detection scheme.

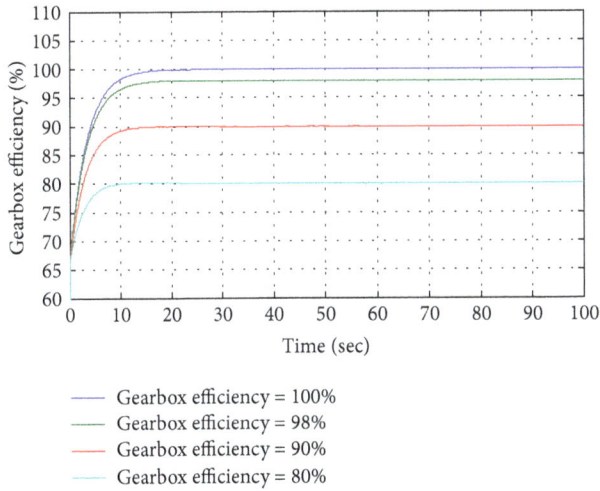

FIGURE 8: Estimation of gearbox efficiency 100%, 98%, 90%, and 80%.

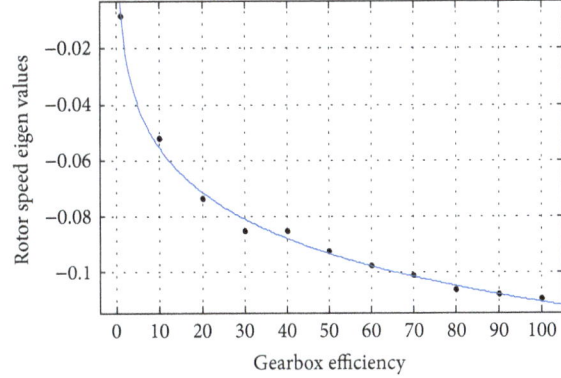

- empirical points
— Eigenvalues vs Gbx Efficiency

FIGURE 9: The variations of the eigenvalues of the dynamics matrix of the rotor speed with respect to the gearbox efficiency.

This paragraph shows that the efficiency fault impacts the stability of the system. Hence, the fault tolerant control becomes necessary.

3. The Fault Tolerant Control Strategy

3.1. Dynamics Equalization Gain. The fault tolerant bloc to be computed should satisfy the following condition:

$$A_f + BKC \xrightarrow{t \to \infty} A, \tag{12}$$

where A, B, and C represent the dynamic, input action, and measurement matrix of the healthy system. A_f represents the dynamic matrix affected by the fault and K is the bloc to be computed, where

$$A_f = \frac{\gamma_f}{J_t};$$

$$A = \frac{\gamma}{J_t};$$

$$B = \frac{\xi}{J_t}; \tag{13}$$

$$C = 1.$$

The resolution of (12) with respect to K while applied on the model of the wind turbine in (2) gives the value of K:

$$K = \frac{\gamma - \gamma_f}{\xi}, \tag{14}$$

where γ and ξ are turbine dependent coefficients defined in (2) of the healthy model of the rotor. γ_f is the coefficient γ but in the faulty situation.

The new control law is then given by

$$\beta_f = \beta + K\Omega_{rf}. \tag{15}$$

decreases and the fault tolerant control bloc should be activated.

$$r = 100\% - \eta_{\text{gbx}}. \tag{10}$$

Since the fault in η_{gbx} affects the generator speed as in (3), and the generator speed is linked to the rotor speed through (4), it can be concluded that this fault affects only the dynamics of the rotor represented by the parameter γ/J_t in (3). This conclusion is verified only under the assumption that no actuator fault neither sensor faults are present at the same time with the gearbox loss of efficiency fault.

2.6. Gearbox Efficiency Impact on the Eigenvalues of the Dynamic Matrix. Figure 9 illustrates the variation of the eigenvalues of the dynamic matrix with respect to the gearbox efficiency. For efficiencies less than 20%, the variations of the eigenvalues are fast with a gradient of 0.62 in absolute values. This gradient becomes slower with 0.1633 for values more than 20% of efficiencies. This means that, for values less than 20%, it is easy to reach eigenvalues near the instability (near 0) than for values more than 20%. The algebraic equation representing the curve in Figure 9 is given by

$$f\left(\eta_{\text{gbx}}\right) = a \times \eta_{\text{gbx}}^b + c, \tag{11}$$

where $a = -0.3203$; $b = 0.0604$; $c = 0.3124$.

3.2. Convergence of the Dynamics Equalization Method.
Let us define the healthy system by (2) and the faulty system by

$$\delta\dot{\Omega}_{rf} = \frac{\gamma_f}{J_t}\delta\Omega_{rf} + \frac{\xi}{J_t}\delta\beta_f + \frac{\alpha}{J_t}\delta\omega$$

$$\delta\beta_f = \delta\beta + K\delta\Omega_{rf}. \tag{16}$$

And the error between the faulty rotor speed and the healthy rotor speed is defined by

$$e = \delta\Omega_{rf} - \delta\Omega_r. \tag{17}$$

The dynamics of the error are given by

$$\dot{e} = \delta\dot{\Omega}_{rf} - \delta\dot{\Omega}_r = \frac{\gamma_f}{J_t}\delta\Omega_{rf} + \frac{\xi}{J_t}K\delta\Omega_{rf} - \frac{\gamma}{J_t}\delta\Omega_r. \tag{18}$$

By taking K as in (14), the error dynamics become

$$\dot{e} = \frac{\gamma_f}{J_t}\delta\Omega_{rf} + \frac{\xi}{J_t}\frac{\gamma - \gamma_f}{\xi}\delta\Omega_{rf} - \frac{\gamma}{J_t}\delta\Omega_r$$

$$= \frac{\gamma_f + \gamma - \gamma_f}{J_t}\delta\Omega_{rf} - \frac{\gamma}{J_t}\delta\Omega_r = \frac{\gamma}{J_t}\delta\Omega_{rf} - \frac{\gamma}{J_t}\delta\Omega_r \tag{19}$$

$$= \frac{\gamma}{J_t}\left(\delta\Omega_{rf} - \delta\Omega_r\right) = \frac{\gamma}{J_t}e.$$

Since γ/J_t is negative (by the nature of the rotor), the error dynamics converges to zero as the time evolves. It can be concluded that the stability of the method depends on the stability of the initial system and any deviation from the healthy speed will be caused only by the imperfections of the model.

3.3. Steady State Reconfiguration by Residual Integration.
The gearbox efficiency impacts also the steady state of the rotor speed. For this, an integral part is needed to cancel the static error between 40 rpm and the real time measured rotor speed. The global fault tolerant control law is given by

$$\delta\beta_f = \delta\beta + K\delta\Omega_{rf} + N\int\left(\delta\Omega_{rf} - \delta\Omega_r\right). \tag{20}$$

In this simulation, a good value of N is -0.03.

3.4. Convergence of the Integrated Strategy.
We take K as in (14), the error dynamics using control law in (20) becomes

$$\dot{e} = \frac{\gamma}{J_t}e + \frac{\xi N}{J_t}\int e. \tag{21}$$

This means that

$$\begin{bmatrix} \int \dot{e} \\ \dot{e} \end{bmatrix} = \begin{bmatrix} 0 & 1 \\ \frac{\xi N}{J_t} & \frac{\gamma}{J_t} \end{bmatrix}\begin{bmatrix} \int e \\ e \end{bmatrix}. \tag{22}$$

The dynamics in (22) are stable, if the eigenvalues of the matrix $\begin{bmatrix} 0 & 1 \\ \xi N/J_t & \gamma/J_t \end{bmatrix}$ are of negative real parts. The eigenvalues of the matrix are given by λ_1 and λ_2 in (23) as

$$\lambda_{1,2} = \frac{\gamma \pm \sqrt{\gamma^2 + 4J_tN\xi}}{2J_t}. \tag{23}$$

To obtain eigenvalues with negative real parts, one should resolve inequality $\lambda_{1,2} < 0$.

$$\gamma < \sqrt{\gamma^2 + 4J_tN\xi} \Longrightarrow$$

$$\gamma^2 < \gamma^2 + 4J_tN\xi \Longrightarrow \tag{24}$$

$$0 < 4J_tN\xi.$$

While ξ is negative and J_t is positive, N should be fixed negative $N < 0$ to have stability of the integrated method.

3.5. Robustness Considerations.
Let define a robustness level θ which should robustify the method (represented by the error e) against the wind disturbance ω. The inequality to be verified is given by

$$|e| < \theta|\omega|, \tag{25}$$

where $e = \delta\Omega_{rf} - \delta\Omega_r$ and $\delta\Omega_r = \delta\Omega_{r\_ref} = 0$ rpm deviation from the operating rotor speed. After applying the FTC method, the closed loop rotor system becomes

$$\delta\dot{\Omega}_{rf} = \left(\frac{\gamma_f}{J_t} + K\right)\delta\Omega_{rf} + \frac{\xi}{J_t}\delta\beta_f + \frac{\alpha}{J_t}\delta\omega + N\int\delta\Omega_{rf}$$

$$- N\int\delta\Omega_r. \tag{26}$$

In the present paper, the desired rotor speed reference $\delta\Omega_{r\_ref}$ is 0 rpm deviation from the operating speed 40 rpm. Equation (26) could be simplified to

$$\delta\dot{\Omega}_{rf} = \left(\frac{\gamma_f}{J_t} + K\right)\delta\Omega_{rf} + \frac{\xi}{J_t}\delta\beta_f + \frac{\alpha}{J_t}\delta\omega$$

$$+ N\int\delta\Omega_{rf}. \tag{27}$$

And the error becomes equivalent to $e = \delta\Omega_{rf}$. So the inequality to be verified becomes

$$\left|\delta\Omega_{rf}\right| < \theta|\omega|. \tag{28}$$

The objective is to extract the transfer function between the turbulence $\delta\omega$ and the rotor speed $\delta\Omega_{rf}$. For this aim, the Laplace transform is applied to

$$s\delta\Omega_{rf}$$

$$= \left(\frac{\gamma_f}{J_t} + K\frac{\xi}{J_t}\right)\delta\Omega_{rf} + \frac{\xi}{J_t}\delta\beta + \frac{\alpha}{J_t}\delta\omega + \frac{N}{s}\frac{\xi}{J_t}\delta\Omega_{rf} \Longrightarrow \tag{29}$$

$$\delta\Omega_{rf}\left(s - \frac{\gamma_f}{J_t} - K\frac{\xi}{J_t} - \frac{N}{s}\frac{\xi}{J_t}\right) = \frac{\xi}{J_t}\delta\beta + \frac{\alpha}{J_t}\delta\omega.$$

If the disturbance effect is considered, the transfer function could be obtained by

$$\frac{\delta\Omega_{rf}}{\delta\omega} = \frac{\alpha/J_t}{\left(s - \gamma_f/J_t - K\left(\xi/J_t\right) - (N/s)\left(\xi/J_t\right)\right)}$$

$$= \frac{\left(\alpha/J_t\right)s}{\left(s^2 - \left(\gamma_f/J_t + K\left(\xi/J_t\right)\right)s - N\left(\xi/J_t\right)\right)}. \tag{30}$$

If the Laplace operator s is replaced by $j\tau$, where j is the complex number and τ is the pulsation $(2\pi f)$. The module of the transfer function is then given by

$$\left|\frac{\delta\Omega_{rf}}{\delta\omega}\right| = \frac{|(\alpha/J_t)\,j\tau|}{\left|(j\tau)^2 - (\gamma_f/J_t + K\,(\xi/J_t))\,j\tau - N\,(\xi/J_t)\right|}$$

$$= \frac{|(\alpha/J_t)\,j\tau|}{\left|-(\tau)^2 - (\gamma_f/J_t + K\,(\xi/J_t))\,j\tau - N\,(\xi/J_t)\right|} \quad (31)$$

$$= \frac{(\alpha/J_t)\,\tau}{\sqrt{\left((\tau)^2 + N\,(\xi/J_t)\right)^2 + \left(\tau\,(\gamma_f/J_t + K\,(\xi/J_t))\right)^2}}.$$

By replacing (31) in (28), inequality (28) becomes equivalent to

$$\frac{(\alpha/J_t)\,\tau}{\sqrt{\left((\tau)^2 + N\,(\xi/J_t)\right)^2 + \left(\tau\,(\gamma_f/J_t + K\,(\xi/J_t))\right)^2}} \prec \theta \implies$$

$$\frac{\alpha}{J_t}\tau \prec \theta\sqrt{\left(\tau^2 + N\frac{\xi}{J_t}\right)^2 + \left(\tau\left(\frac{\gamma_f}{J_t} + K\frac{\xi}{J_t}\right)\right)^2} \implies$$

$$\frac{\alpha}{J_t}\theta \succ \sqrt{\left(1 + N\frac{\xi}{J_t\tau^2}\right)^2 + \left(\frac{\gamma_f}{J_t} + K\frac{\xi}{J_t}\right)^2} \implies$$

$$\left(\frac{\alpha}{J_t}\theta\right)^2 \succ \left(1 + N\frac{\xi}{J_t\tau^2}\right)^2 + \left(\frac{\gamma_f}{J_t} + K\frac{\xi}{J_t}\right)^2 \implies$$

$$\left(1 + N\frac{\xi}{J_t\tau^2}\right)^2 \prec \left(\frac{\alpha}{J_t}\theta\right)^2 - \left(\frac{\gamma_f}{J_t} + K\frac{\xi}{J_t}\right)^2 \implies$$

$$1 + N\frac{\xi}{J_t\tau^2} \prec \sqrt{\left(\frac{\alpha}{J_t}\theta\right)^2 - \left(\frac{\gamma_f}{J_t} + K\frac{\xi}{J_t}\right)^2} \implies$$

$$N\frac{\xi}{J_t\tau^2} \prec \sqrt{\left(\frac{\alpha}{J_t}\theta\right)^2 - \left(\frac{\gamma_f}{J_t} + K\frac{\xi}{J_t}\right)^2} - 1 \implies$$

$$N \succ \frac{J_t\tau^2}{\xi}\sqrt{\left(\frac{\alpha}{J_t}\theta\right)^2 - \left(\frac{\gamma_f}{J_t} + K\frac{\xi}{J_t}\right)^2} - 1. \quad (32)$$

The inequality symbol change between the two last lines of (32) comes because the term $\xi/J_t\tau^2$ is inferior to 0.
   Finally,

$$\frac{J_t\tau^2}{\xi}\sqrt{\left(\frac{\alpha}{J_t}\theta\right)^2 - \left(\frac{\gamma_f}{J_t} + K\frac{\xi}{J_t}\right)^2} - 1 \prec N \prec 0. \quad (33)$$

Then, to ensure robustness level θ, and given a pulsation $\tau = 2\pi f$ (rad/s) the design parameter N should verify inequality (33). It is recommended to take the maximal frequency contained in the wind equal to 1 Hz as in Section 2, Figure 4.

4. Results and Discussions

Figure 10 illustrates the rotor speed in the nominal, faulty, and fault tolerant cases. In order to evaluate the effectiveness

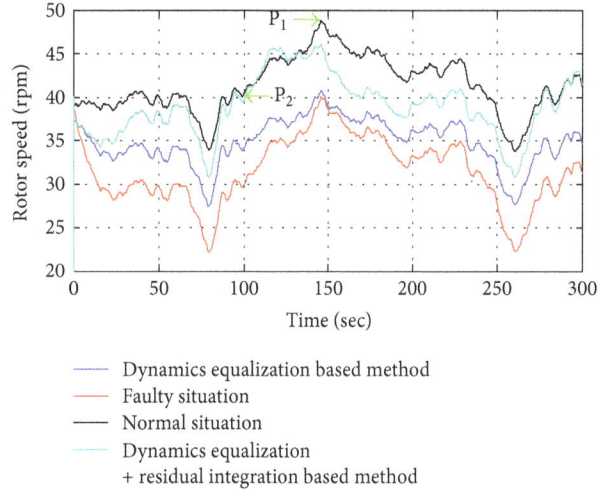

— Dynamics equalization based method
— Faulty situation
— Normal situation
— Dynamics equalization
   + residual integration based method

FIGURE 10: The rotor speed in the normal, faulty, and fault tolerant situation.

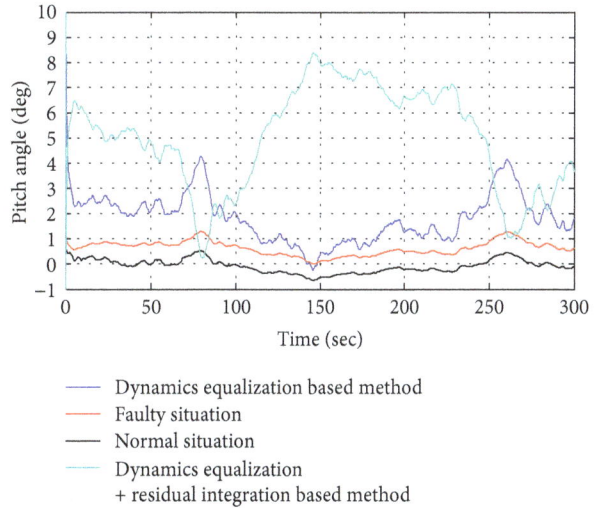

— Dynamics equalization based method
— Faulty situation
— Normal situation
— Dynamics equalization
   + residual integration based method

FIGURE 11: The pitch angles (control effort) in the normal, faulty, and fault tolerant situation.

of the method, the distance (according to the ordinates axis) between two points on the curve is considered. The first point is P1 with abscissa 143 seconds and the second point P2 with 90.34 seconds of abscissa. In the nominal situation, the difference between the ordinates of P1 and P2 is of 8.1 rpm; this value becomes 11.5 rpm in the faulty situation. By using the fault tolerant control strategy, this value is reduced to 9.2 rpm. As shown in the same Figure 10, not only the dynamics are impacted but also the steady state value of the rotor speed which become closer to 40 rpm operating point due to the integral term $N\int(\delta\Omega_{rf} - \delta\Omega_r)$ in the control law.
   Figure 11 illustrates the pitch angle in the three situations. It can be stated that, in the faulty situation (red), the nominal regulator "tries" to beat the deviation in the speed by generating some pitch angles but without satisfactory results. By adding the term $K\delta\Omega_{rf}$ to the old control signal, the efforts become bigger and the difference between the points P1 and

P2 is reduced. In the integrated method, the efforts become bigger and the integral part is evidently impacting the pitch angles to cancel the steady state error.

Compared to other methods, the proposed strategy reconstructs the rotor speed while keeping the generated power constant as before the fault occurrence. In fact, another way to do is to decrease the operating point generator torque in (3) to maintain the same generator speed variations and then the same rotor speed variations $\delta\dot{\Omega}_r$. This could be done by multiplying the generator torque input by the estimated gearbox efficiency. This method helps keeping the rotor speed as it was but impacts the generator torque and then the generated power whose expression is given by

$$\delta P_g = \delta\Omega_r \times \delta T_g. \tag{34}$$

The method of this paper uses pitch actuators instead of generator torque and this helps keeping the generated power constant and then satisfies continuously the customer demand of electricity.

5. Conclusion

In this paper, a fault tolerant control method composed of several steps is proposed to deal with the loss of gearbox efficiency. This fault occurs due to the dusty environment of the modern large scale wind turbines. The first step is to estimate the shaft torque through a suitable transfer function between the rotor speed and the shaft torque. The last torque is used to estimate in real time the gearbox efficiency through a suitable loop. The estimated efficiency is used to select a corresponding eigenvalue from Figure 9. A design coefficient K is computed so the dynamics of the prefault and postfault cases are equalized. The rotor speed steady state is reconstructed by adding an integral part of the residual between the measured and reference rotor speed (zero deviation from operating point is desired).

On the other hand, a performance based conditions were given to help choosing the design integration parameter N. Based on the fact that the residual should not be impacted by the wind disturbance; a performance level θ has been proposed to verify $|e| \prec \theta|\omega|$. The last inequality helps to give a negative (because ξ is by construction negative) inferior limit to N function of θ as $(J_t\tau^2/\xi)\sqrt{((\alpha/J_t)\theta)^2 - (\gamma_f/J_t + K(\xi/J_t))^2} - 1$.

The method is characterized by its ease of elaboration and parameters design and especially helps keeping a continuous production of the power in contrast to the generator torque based method.

Conflicts of Interest

The authors declare that there are no conflicts of interest regarding the publication of this paper.

References

[1] H. P. Wang, A. Pintea, N. Christov, P. Borne, and D. Popescu, "Modelling and recursive power control of horizontal variable speed wind turbines," *Control Engineering and Applied Informatics*, vol. 14, no. 4, pp. 33–41, 2012.

[2] K. T. Magar, M. J. Balas, and S. A. Frost, "Smooth transitioning of wind turbine operation between region II and region III with adaptive disturbance tracking control," *Wind Engineering*, vol. 38, no. 3, pp. 337–348, 2014.

[3] Y.-D. Song, P. Li, W. Uu, and M. Qin, "An overview of renewable wind energy conversion system modeling and control," *Measurement + Control*, vol. 43, no. 7, pp. 203–208, 2010.

[4] J. Ribrant and L. M. Bertling, "Survey of failures in wind power systems with focus on Swedish wind power plants during 1997–2005," *IEEE Transactions on Energy Conversion*, vol. 22, no. 1, pp. 167–173, 2007.

[5] V. Rezaei and F. R. Salmasi, "Robust adaptive fault tolerant pitch control of wind turbines," *Wind Engineering*, vol. 38, no. 6, pp. 601–612, 2014.

[6] P. F. Odgaard, J. Stoustrup, and M. Kinnaert, "Fault-tolerant control of wind turbines: A benchmark model," *IEEE Transactions on Control Systems Technology*, vol. 21, no. 4, pp. 1168–1182, 2013.

[7] P. Li, Y. Song, W. Liu, and M. Qin, "Monitoring of Wind Turbines: A Bio-Inspired Fault Tolerant Approach," *Measurement + Control*, vol. 44, no. 4, pp. 111–115, 2011.

[8] J. M. Jonkman and M. L. Buhl, "FAST User's Guide - Updated August 2005," Tech. Rep. NREL/TP-500-38230, 2005.

[9] Y. Ait Elmaati, L. El Bahir, and K. Faitah, "An integrator based wind speed estimator for wind turbine control," *Wind and Structures, An International Journal*, vol. 21, no. 4, pp. 443–460, 2015.

[10] K. Z. Østergaard, P. Brath, and J. Stoustrup, "Estimation of effective wind speed," *Journal of Physics: Conference Series*, vol. 75, no. 1, Article ID 012082, 2007.

[11] G. Hafidi and J. Chauvin, "Wind speed estimation for wind turbine control," in *Proceedings of the 2012 IEEE International Conference on Control Applications, CCA 2012*, pp. 1111–1117, Croatia, October 2012.

[12] Y. Ait Elmaati, L. El Bahir, and K. Faitah, "Residual generation for the gearbox efficiency drop fault detection in the NREL 1.5WindPact turbine," in *Proceedings of the 1st International Conference on Electrical and Information Technologies, ICEIT 2015*, pp. 77–81, Morocco, March 2015.

15

Oscillatory Tip Leakage Flows and Stability Enhancement in Axial Compressors

Feng Lin ⓘ **and Jingyi Chen**

Institute of Engineering Thermophysics, Chinese Academy of Sciences, Beijing, China

Correspondence should be addressed to Feng Lin; fenglinfenglin@hotmail.com

Academic Editor: Paolo Pennacchi

Rotating stall axial compressor is a difficult research field full of controversy. Over the recent decades, the unsteady tip leakage flows had been discovered and confirmed by several research groups independently. This paper summarizes the research experience on unsteady tip leakage flows and stability enhancement in axial flow compressors. The goal is to provide theoretical bases to design casing treatments and tip air injection for stall margin extension of axial compressor. The research efforts cover (1) the tip flow structure at near stall that can explain why the tip leakage flows go unsteady and (2) the computational and experimental evidences that demonstrate the axial momentum playing an important role in unsteady tip leakage flow. It was found that one of the necessary conditions for tip leakage flow to become unsteady is that a portion of the leakage flow impinges onto the pressure side of the neighboring blade near the leading edge. The impediment of the tip leakage flow against the main incoming flow can be measured by the axial momentum balance within the tip range. With the help of the theoretical progress, the applications are extended to various casing treatments and tip air recirculation.

1. Background and Motivation

Modern high-performance axial compressors in gas turbines, especially in aeroengines, are mostly unshrouded due to the high rotating speed of the shafts, which means that the tip clearance is necessary. The tip clearance brings additional complication into the corner of rotor blade and the casing. In 1996, Lakshminarayana [1] published a comprehensive illustration on the flow structures within rotor blade passages. The shock, the tip leakage flow (TLF), and the tip leakage vortex (TLV) are the three flow structures that are unique to the blade tip region. Two boundary layers, the end-wall boundary layer at casing and the blade surface boundary layer, provide a battle ground for all these flows interacting with each other. Please note that TLF is different from TLV in that in some cases only a portion of the TLF would be revolved into TLV and the rest of the TLF would either mix with the incoming main flow or leak again through the tip gap of the neighboring blade.

Lakshminarayana's illustration is meant to present the situation at compressor's design condition. It would become even more complicated when the compressor is throttled to close-to and near stall conditions, which will be the main theme of this paper. The tip clearance not only changes the flows structures at the design condition but also alters the Stall Inception and stalling behaviors. We noticed the debate on whether or not the stalling mechanism requires a nonzero tip clearance. Vo et al. [2] hypothesized that spikes (one of the two basic types of stall precursors) emerged when the interface between the TLF and the main flow spilled out of the leading edge of rotor blade. This hypothesis thus can only be true when the tip clearance is not zero. Pullan et al. [3] argued that spikes can be recognized as a vortex tube spanning from blade's suction surface near the tip to the casing, a result of blade leading edge separation due to high incident angle of the incoming main flow, which can happen even with zero tip clearance.

As a matter of fact, most modern axial compressors are tip critical; that is, their rotating stall is initiated at the tip region. In other words, as long as the tip clearance is not zero, the tip leakage flow takes part in Stall Inception process. Thus, the hypothesis from either side of the debate is applicable to the compressors we are dealing with daily. Pullan's argument, the flow separation at rotor blade's suction surface near leading edge as the cause of the rotating stall, is a continuation of the traditional Emmons' model [4] that is widely accepted in engineering practices.

More than ten years ago, we started to study the tip flow structures at near stall and estimate the location of the interface between the incoming main flow (MF) and TLF. At first, we were able to repeat the finding of others [4–7] that TLF would become oscillatory at operating points close to stall limit, while the whole compressor is completely stable. This brings an unexpected benefit, making the tip air injection become closed-loop without sensing spikes. We then realized that this had been done several years before us without knowing the existence of the oscillatory tip leakage flow [8–10]. In order to estimate the MF/TLF interface location, axial momentum equation within the tip region was then integrated over a sequence of control volumes, which ended up with a curve of cumulative axial momentum distribution along blade's axial chord. This curve resembles a bell shape, whose peak point marks the "boundary" between main flow dominated region and the TLF dominated region. This makes it possible to compare the ability to extend the stall margin for various casing treatments, which is very helpful in screening out of bad designs from a large pool of casing treatment candidates in early design stage.

Note that the purpose of this paper is not to review the entire research field of compressor instability. Day [11] presented an excellent review recently as a Scholarly Lecture in the annual conference of International Gas Turbine Institute in 2017. In this paper, the tip leakage flow structure at close-to and near stall is reviewed in Section 2, followed by the features of oscillatory TLF in both low- and high-speed compressors (Section 3). The role of axial momentum and the "bell-shaped" curve are explained in Section 4. The applications of both the "bell-shaped" curve and the oscillatory TLF are reviewed in Section 5. Conclusions are given in the last section for convenience, together with discussions on future work.

2. The Tip Leakage Flow Structure in Close-to and Near Stall

It is necessary to clarify a few terms before reading further. As a compressor is throttled from large flow rate to stall, it would experience several steps. Figure 1 illustrates them along a typical compressor characteristic. Several points are identified in the figure. The point of peak efficiency (PE) is where the efficiency reaches the maximum, while the design point (DP) is the point that the compressor is designated to work. While these two often referred to the same point, there are occasions where these two are different. For instance, when a compressor has to work on a designated flow rate to match the other components in the same system (e.g., a gas turbine), its DP would be different from PE. For the cases studied in this paper, DP and PE are the same because all the cases are single rotors. This ensures that the flow angles around the blade are well organized, which forms a fair ground for comparison against the rest of the off-design points.

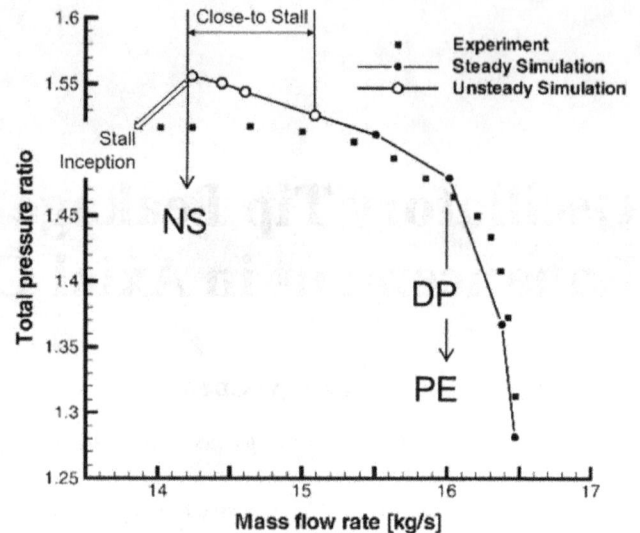

FIGURE 1: The comparison of experimental and computational characteristics of Darmstadt Rotor 1 ([12]). Several technical terms are also illustrated. Figure reproduced from LIN et al. (2010).

As a compressor is throttled to the stall limit along its characteristic, it goes through a segment on the curve with totally steady flow, followed by a segment with only the TLF becoming unsteady (close-to stall) before it reaches the point of near stall (NS). The NS point is a point right before stall happens. In experiments, a compressor at NS may ramp down into stall by itself without further closing the throttle. In CFD computation, NS is tricky to identify because it depends on how the state of NS is approached. When the compressor is throttled by gradually increasing back pressure step by step, NS is identified as the last point at which the time-accurate unsteady simulation can "converge" into a periodic solution, because no more solutions could be found when increasing the back pressure any further. This is a common practice in the compressor CFD community. The segment with only the tip leakage flow becoming unsteady is called "close-to stall" in this paper. In 1996, Graf [5] first observed periodicity and motion of the tip leakage vortex in his study based on CFD simulations of the flow through a single-stage compressor utilizing a 3D time-accurate Reynolds-averaged Navier-Stokes solver. In 2001, Mailach et al. [4] published test results of periodic fluctuation of tip leakage vortex in a large-scale low-speed compressor, which was named "Rotating Instability" thereafter. Bae et al. [6] studied the phenomenon in a cascade wind tunnel and explained the vortex oscillation with Crow Instability, a model that was used to explain the unstable vortex pairs behind the aircraft wings. Deng et al. [7] found that this vortex oscillation strongly depended on the size of the tip clearance, and one of its necessary conditions was that the trajectory of the vortex core impinged the pressure side of the neighboring blade.

The process through which a compressor ramped down into stall is called "Stall Inception." There are two basic stall precursors in Stall Inception, *modal waves* and *spikes*. The distinguishing among "close-to stall," "near stall," and

"Stall Inception" is very important, as stall precursors only happen in Stall Inception. Near stall is a point that belongs to the segment of close-to Stall, in which no spike would be formed yet. Therefore, according to the opinion of this paper's authors, the oscillation of tip leakage vortex or the Rotating Instability is NOT part of Stall Inception, although it may trigger spikes or modal waves that later eventually initiate rotating stall.

The flow structure in the tip region is the key to understand the mechanism behind the flow phenomena. The early efforts included Adamczyk et al. (1993) [20] and Suder and Celestina (1996) [21]. At off design conditions, a large "blockage," that is, a low-energy region within the blade passage, was found due to vortex breakdown caused by the shock. Their excellent work was so much ahead of their time that no much more knowledge in this regard had been gained since then until unsteady oscillatory TLF was found and confirmed.

The modern powerful tools of unsteady Navier-Stokes solvers empower us to revisit this flow structure at a level much closer to stall and from a viewpoint of unsteady flow. One of the rotors was Darmstadt Rotor 1, which is a transonic rotor of a single-stage transonic compressor rig at Technische Universität Darmstadt [22, 23]. The comparison of experimental and computational characteristics is depicted in Figure 1 [12]. Note that the steady simulation fits well with the test curve. While overpredicting the total pressure rise up to 2.5%, the unsteady simulation was able to make the computed stall limit very close to that of the test result (less than 1.5% of the test flow rate at stall). Figure 2 depicts the regions of influences by tip leakage flows. At PE, the loading is distributed over about 80% of the blade chord. The shock is attached to the leading edge and the tip leakage vortex flows out of the blade passage. At NS, the loading fluctuates within the first half of the blade chord. The shock is detached from the blade leading edge and interacts with the tip leakage vortex, causing a large "blockage" as seen in Adamczyk's paper [20]. Note that the pictures at NS are only one instant of the oscillatory movement.

Particles were released within the tip gap to trace the trajectories of TLF in numerical results. Figure 3 compares them at PE and at one of the instants at NS. They are the 3D views of those in Figure 2. For the one at NS, a schematic is plotted in Figure 4 to summarize the flows at NS for clarity. It can be seen that the TLF are divided into two parts along the blade chord based on the location of shock on the suction surface. The first portion of TLF forms the vortex core and flows through the blade passage, while the second portion goes across the blade passage and hits the pressure side of the neighboring blade. It then splits there, a part of which leaks over the blade tip one more time.

3. The Unsteady Features of Oscillatory Tip Leakage Flow

The fact that the tip leakage vortex impinges the pressure side of the neighboring blade is believed to be one of the necessary conditions for TLF oscillation. Deng et al. [7] were the first to notice this in our research group when simulating a low-speed rotor. This had later been observed in serval other compressor rotors, including Darmstadt Rotor 1 as shown in Figures 2 and 3 and NASA Rotor 67 in Figure 5. The top row of Figure 5 lists six instantaneous contours of static pressure coefficients with the last one almost exactly repeating the first one. The bottom row is the corresponding pressure coefficient distribution in the pressure side of the blade. The low-pressure spot on the pressure side is marked as A1, which is clearly casted by the low pressure core of the tip leakage vortex. At time 0/30T, where "T" represents the time interval to store data set by the time-accurate CFD solver, A1 is located at the leading edge. Because of it, the pressure difference across the tip clearance is low, which weakens the tip leakage vortex and pushes the TLF/MF interface downstream. As time goes on (e.g., at 20/30T and 30/30T), the low-pressure spot A1 moves towards the trailing edge, while the high pressure regains its control on the leading edge causing the first half of the TLF/MF interface to swing back. Close to the end of one period, the low-pressure spot A1 moves out of the blade chord, the TLF/MF interface returns to its starting position, and a new A1 emerges at the leading edge of the pressure side. The entire process repeats. The frequency of this oscillation is calculated as 0.586 BPF (Blade Passing Frequency) for this particular rotor in rotor relative frame.

The features of the aforementioned unsteady process were observed in compressor experiments. The first feature was that the location of the highest amplitude should be at the leading edge of the pressure side, neither at the spot where the shock/vortex interact nor at the shock itself. It is not even at the starting place of the tip leakage vortex. Figure 6 depicts the comparison of the root mean square (RMS) of static pressure distribution at casing. The left was the result of phase-locked RMS obtained from experiments of Darmstadt Rotor 1, while the right was the numerical simulation [12]. Both match well with each other.

The second feature is signature frequency bands of the casing static pressure measurements. Figure 7 demonstrates the relation between rotating relative frame and the casing stationary frame using numerical results. The top row shows the time series and its frequency spectrum for a probe fixed on the rotor relative frame. The first peak at 3056.68 Hz or 0.57 BPF is the main frequency and the second peak is its harmonic. The second row is the data taken from a probe at the same axial location but fixed on the casing. The frequency components are much richer due to the modulation of frequencies between the relative frame and the rotor rotating frame. Figure 8 compares the frequency spectra between the experiments and CFD simulations [12]. Note that the test results contain noises from turbulent, the variations of the blade geometry and/or assembly, and so forth; thus the frequency spectra appear as a few signature bands, not as pronounceable individual peaks.

The comparison of the oscillatory TLF introduced here and Rotating Instability as described in [8] is of interest. Since there is no chance to study the compressors that were reported to possess Rotating Instability such as those in [8, 24], we are not able to justify whether or not these two are the same. However, there are at least two features that are common to both phenomena: (1) both embark a

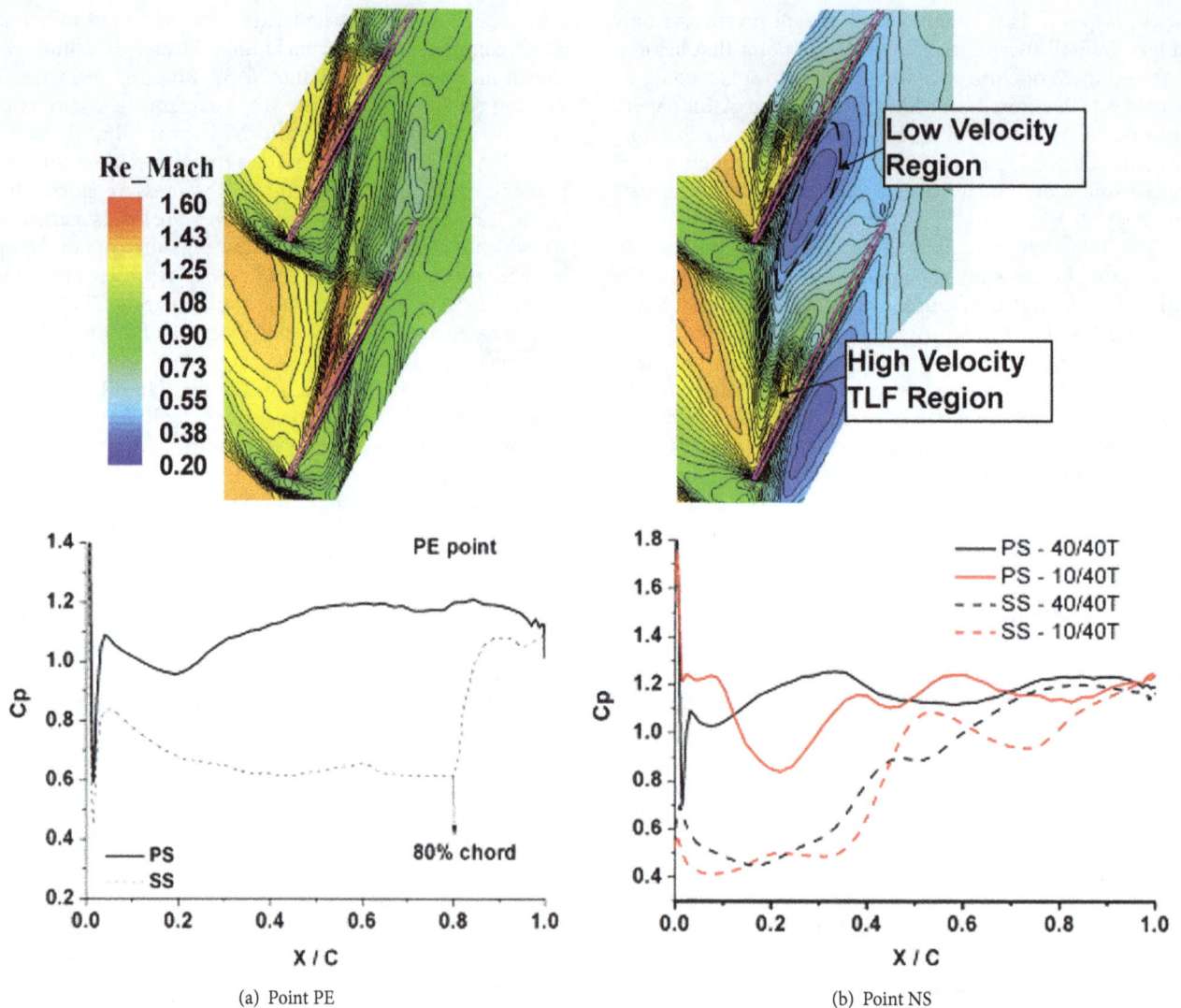

FIGURE 2: The top two pictures show the relative Mach number contours at 99% span, a plane at the middle of the tip clearance, at points PE and NS. The bottom two pictures depict the blade loadings using the surface pressure coefficient distributions along the blade's axial chord. ([12]). Figure reproduced from LIN et al. (2010).

signature frequency band; (2) the frequency band is about 0.5 BPF. Regardless of whether this frequency is measured in the rotor rotating frame or in the casing stationary frame, as long as it is about 0.5 BPF, it would appear as the tip leakage vortices (TLV) alternate their trajectories between neighboring blade passages. That is, when one TLV is within the blade passage, its neighboring one would impinge the blade's pressure surface. Both features were observed in our low-speed compressor.

4. The Role of Axial Momentum and the "Bell-Shaped" Curve

According to Vo's hypothesis [2], it is necessary to be able to estimate the location of the interface between the incoming main flow (MF) and the tip leakage flow (TLF). Since such an MF/TLF interface exists in the rotor's rotating reference frame, how it looks like on the casing stationary reference

frame becomes crucial to experimentists. Cameron et al. [14] demonstrated that on casing the complex 3D curvy oscillatory surface of MF/TLF in the rotor rotating frame can be observed as a straight line on the casing stationary frame, as seen in Figure 9 [14]. This is because, when observed at casing, the spatial and temporal variations within the rotor frame are all naturally averaged. Figure 9(a) depicts the experimental result of the casing streaklines, which was taken from a transparent window when the compressor was operating at Point B. In Figure 9(b), on the left, it is the computational result of one instant showing the streaklines on casing, while on the right the resultant streaklines after spatial and temporal averaged display flow patterns qualitatively the same as those of experiments. The rotor under investigation was the transonic rotor in University of Notre Dame in USA, abbreviated as ND-TAC.

Cameron et al. [14] then proposed a simple model to estimate the location of the MF/TLF interface on casing, as

(a) Point PE   (b) Point NS

FIGURE 3: Streaklines that depict the TLF structures ([12]). Figure reproduced from LIN et al. (2010).

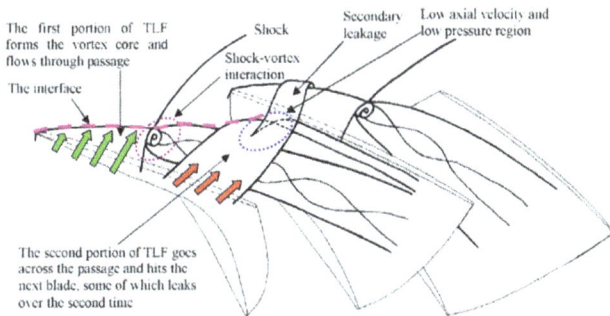

FIGURE 4: The schematic of TLF structure at NS ([12]). Figure reproduced from LIN et al. (2010).

illustrated in Figure 10. The axial momentum equation near the tip clearance was simplified as the balance between the pressure due to main flow and the axial momentum due to tip leakage jet. The details of derivation, which can be found in [14, 15], are omitted here due to the space limitation of this paper. Despite its simplicity, the trend of Xzs versus the compressor's incoming flow coefficient fitted well with that of the throttling process, which suggests that this simple model did capture some physics embedded in the stalling mechanism.

Since the MF/TLF interface is a 3D surface, such as the one in Figure 11, it is not surprising that using Xzs alone is not sufficient to correlate it to stall. A control volume approach is thus proposed to include the 3D effect into a new model [16, 19]. Unfortunately, it is impossible to establish an analytical equation like the one for Xzs due to the complexity of the 3D unsteady flow. The new model is indeed a method of postdata processing based on 3D unsteady Reynolds-Averaged Navier-Stokes solutions. It starts with the same strategy as the Xzs equation: observing the flows within the rotor while sitting on the casing stationary reference frame. The details can be described below.

Consider a series of discrete control volumes installed at the tip region between the casing and the rotor passage tip as illustrated in Figures 12 and 13. If we integrate all the linear momentum in axial direction, according to the Newton's second law, such an integral would be equal to the total axial force acting on the control volume. Because this is done on a fixed and stationary control volume, it is equivalent to pitch-wise smear the spatial variation. For unsteady cases, the total axial linear momentum at every time instant should first be averaged before spatial integration. The end result is a number representing the net axial force on the control volume, as measured by an observer on casing stationary frame. If this number is positive, it means that the net axial force on this control volume is pushing the fluid downstream. If it is negative, it means that the net force is pushing the fluid upstream. Therefore, the control volume approach proposed here makes the justification of the flow stability become simple. The key is where the net force is zero. The closer this location is to the leading edge, the easier the compressor would run into stall.

The rotor in Figure 12 is NASA Rotor 67. The smooth casing is placed on the left, while the casing treated with 6 grooves is on the right. The bottom row depicts the entropy contours on the suction surface of the blade, indicating the radial depth of influence by the tip leakage flow. This is how the depth of the control volumes is decided. For instance, in the case of Figure 12, the control volumes are taken as deep as up to 90% span. A volumetric illustration of a typical control volume is given in Figure 13. Considering only the axial direction, the momentum equation in a finite control volume form can be written as

$$\int_{z_-} P_{z_-} dA_- + \int_{z_+} P_{z_+} dA_+ + \int_{CS} P_{CS} dA_{CS}\Big|_z$$

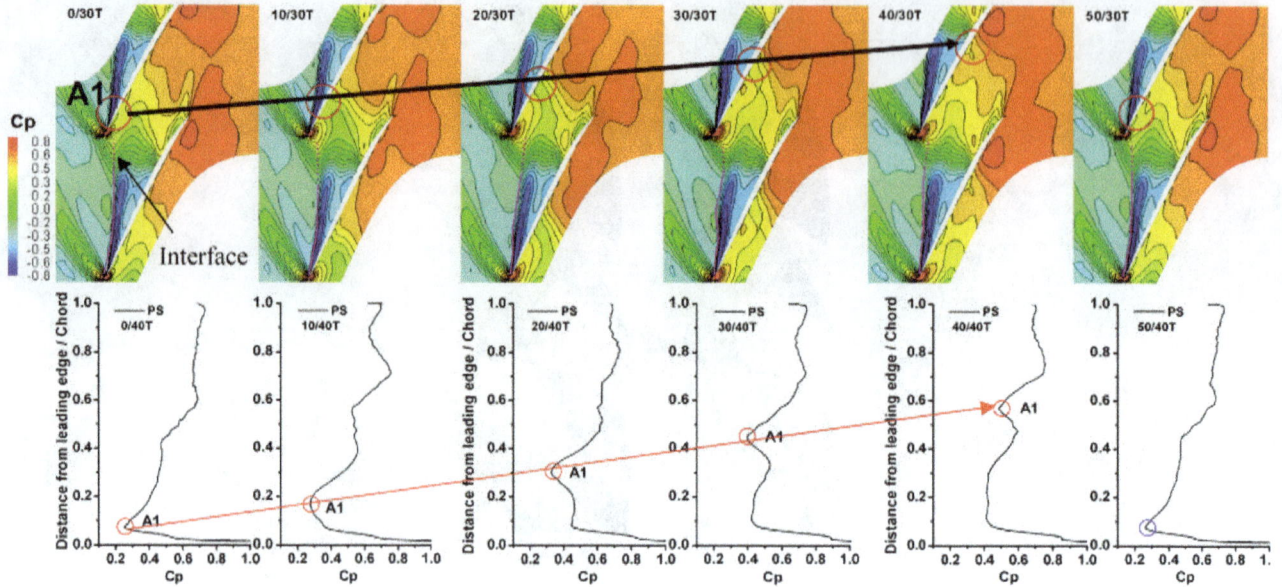

FIGURE 5: Snapshots of 6 instantaneous moments during one period of TLD oscillation for Rotor 67 ([13]). "T" represents the time interval to store the data set by the time-accurate CFD solver. Figure reproduced from DU et al. (2010).

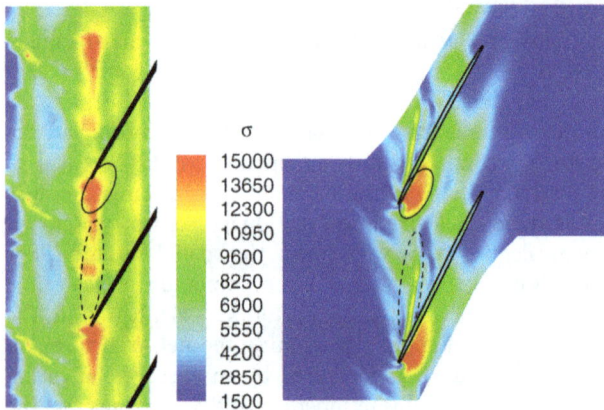

FIGURE 6: Experimental verification of ([12]). Figure reproduced from LIN et al. (2010).

$$
\begin{aligned}
+ &\left. \int_{BT} P_{BT} dA_{BT} \right|_z + \left. \int_{CS} \tau_{CS} dA_{CS} \right|_z \\
+ &\left. \int_{BT} \tau_{BT} dA_{BT} \right|_z + F_{blade-Z} \\
= &\int_{Z-} \rho W_z \left(\vec{W} \cdot \vec{n} \right) dA_{z-} \\
+ &\int_{Z+} \rho W_z \left(\vec{W} \cdot \vec{n} \right) dA_{z+} \\
+ &\int_{BT} \rho W_z \left(\vec{W} \cdot \vec{n} \right) dA_{BT} \\
+ &\int_{CS} \rho W_z \left(\vec{W} \cdot \vec{n} \right) dA_{CS}
\end{aligned}
\tag{1}
$$

Here, z refers to axial direction. Each control volume covers one pitch. The two periodic surfaces of each control volume are ignored in the equation due to the periodic boundary conditions for the single passage simulation.

By adding the right-hand side of the above equation together and averaging it over the oscillation period, we will obtain the net axial momentum (or force) on the control volume. Plotting all the net forces for all the control volumes, we obtain a distribution curve of local net axial momentum (or force) (Figure 14(a)). Two curves in the plot are for peak efficiency (PE) and near stall (NS) points, respectively. The location where the net momentum is zero for PE is at 38% axial chord, while the one for the NS is at 14%, much closer to leading edge as expected. In Figure 14(b), the curves of cumulative axial momentum for both the smooth casing (SC) and the 6-groove casing treatment (CG6 as in Figure 12) at the NS point of SC are given (the comparison between them will be explained later). The cumulative axial momentum is equal to the cumulative distribution of the net force acting on the enlarged control volume from the inlet of the first control volume gradually to the outlet of the last control volume on the curve. This cumulative momentum curve reaches its peak at which the local momentum curve crosses the zero. After that, the local momentum becomes negative and the net momentum starts to decrease. Therefore, the curve forms a bell shape. It is thus called "bell-shaped curve" or simply "bell curve" for breviate.

The peak of the bell curve divides the tip region into two. The one in the front (near the leading edge) is dominated by the incoming main flow, while the other one on the back is dominated by the tip leakage flow. The location of the peak is more valuable than the peak value itself because the peak value involves the blade force that is so complicated that it is hard to explain the meaning of the peak value in general. If needed, it has to be carefully examined case by case.

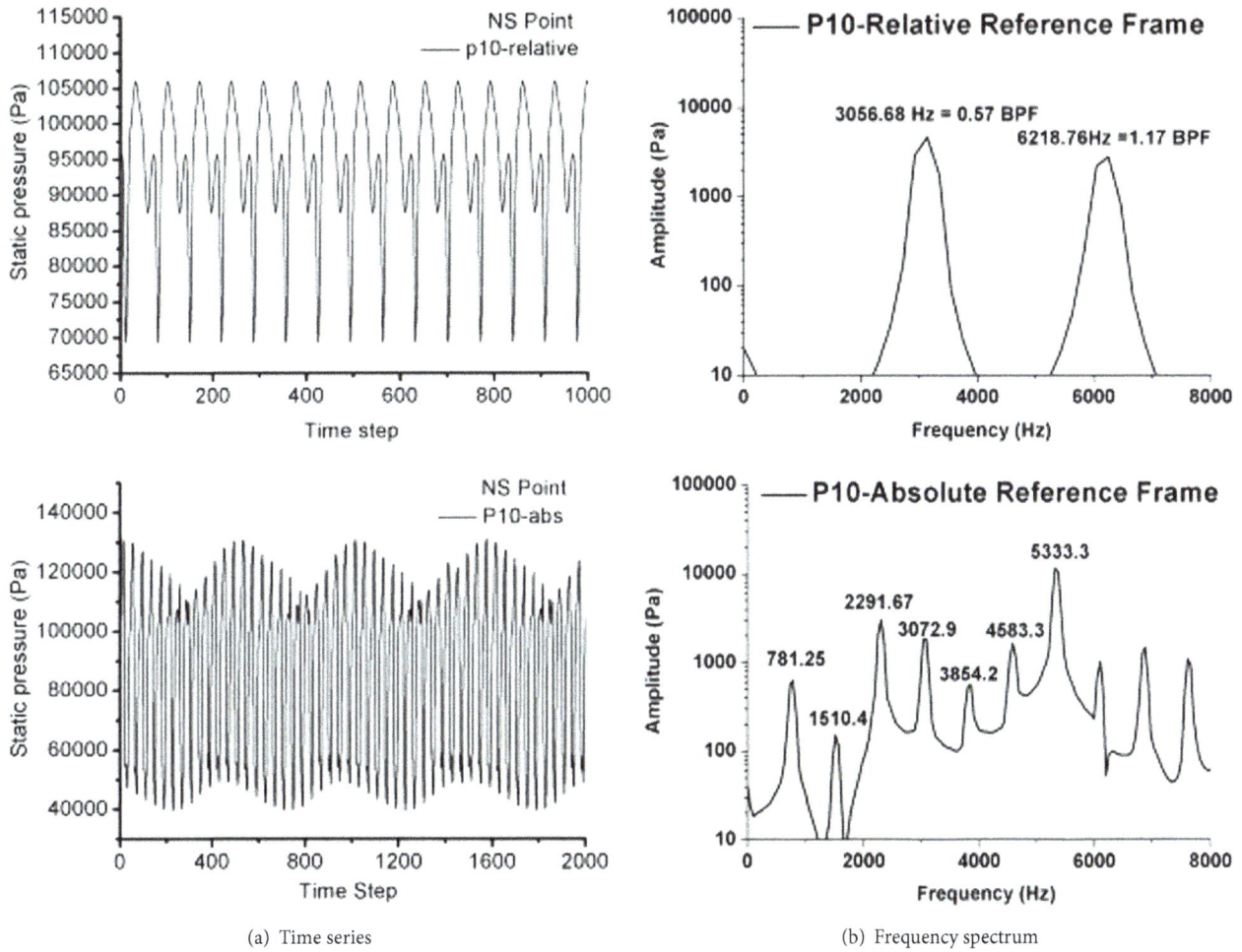

(a) Time series   (b) Frequency spectrum

FIGURE 7: The relation between relative and absolute frames at NS ([12]). Figure reproduced from LIN et al. (2010).

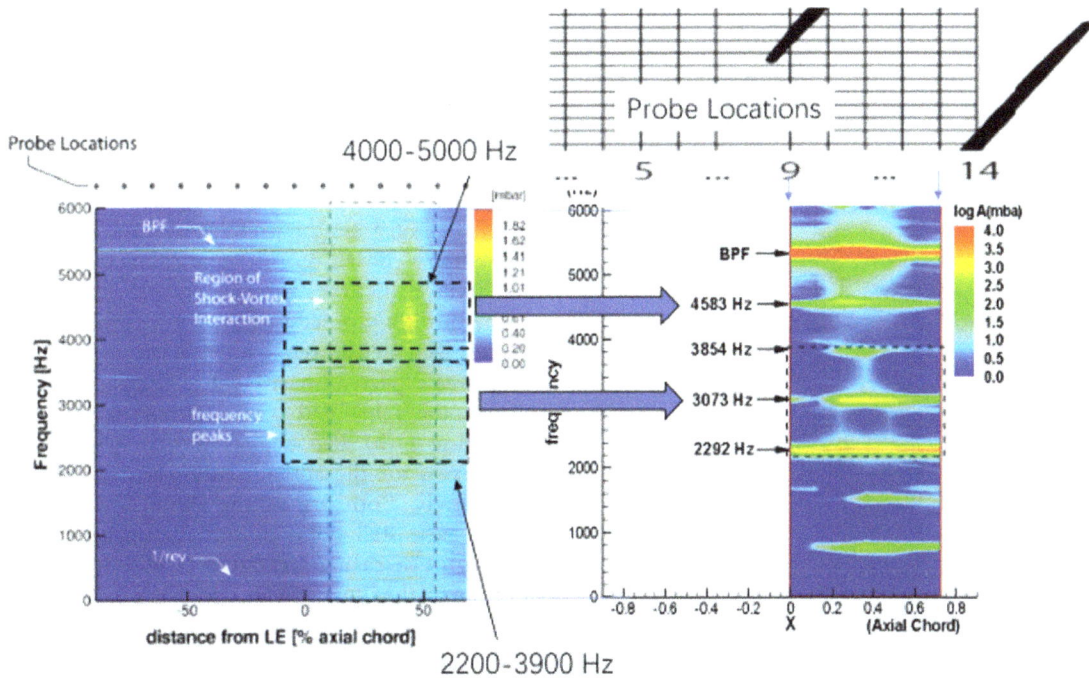

FIGURE 8: The comparison of frequency spectra between experiments (left) and simulations (right) at NS ([12]). Figure reproduced from LIN et al. (2010).

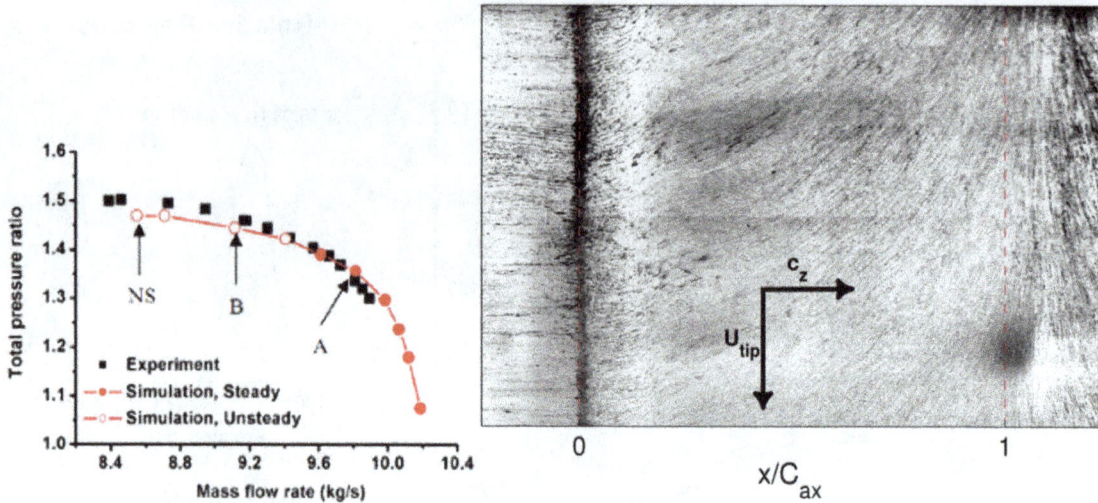

(a) Experimental result

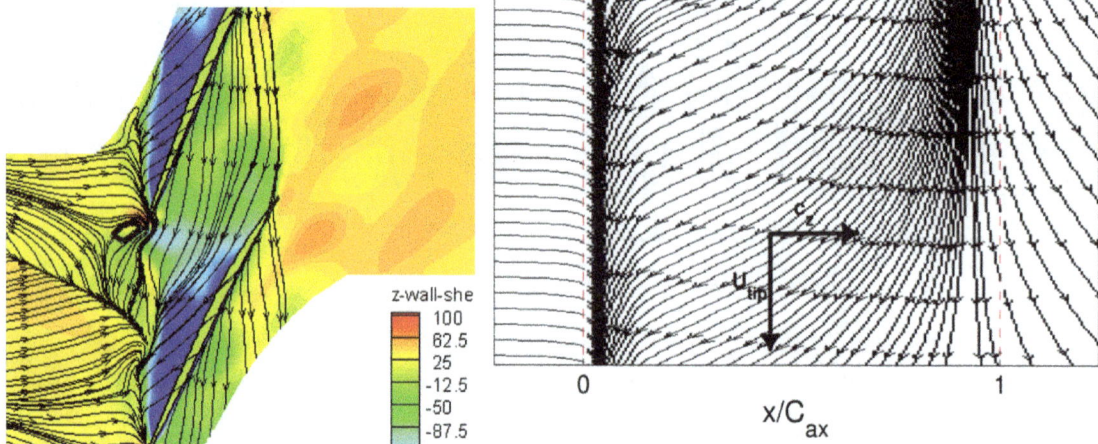

(b) Numerical result

FIGURE 9: The MF/TLF interface as observed in the rotor rotating frame and at casing stationary frame ([14]). Figure reproduced from CAMERON et al. (2013).

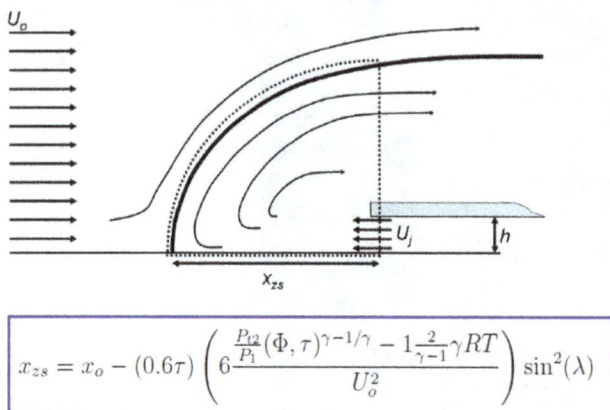

$$x_{zs} = x_o - (0.6\tau) \left(6 \frac{\frac{P_{t2}}{P_1}(\Phi, \tau)^{\gamma - 1/\gamma} - 1\frac{2}{\gamma - 1}\gamma RT}{U_o^2} \right) \sin^2(\lambda)$$

FIGURE 10: The simplified model of axial momentum balance on casing and its vicinity (illustrated based on the information in [15]). Figure reproduced from BENNINGTON et al. (2008).

To demonstrate the usefulness of the bell curves, seven points were chosen on the same characteristic of Rotor 67 and seven bell curves were generated, which corresponded to the throttling process of the rotor [16]. Figure 15 depicts the results. One can see that the locations of the peak move exactly as expected towards the leading edge monotonically as the rotor is throttled to near stall.

Figure 14(b) also depicts the comparison of the two bell curves for the smooth casing and the treated casing, respectively. Both are at the same flow coefficient, that is, the near stall point of smooth casing. It is clear that the 6-groove casing treatment is able to move the peak location further downstream compared to the smooth casing. Therefore, one can predict that the CG6 would extend the stall margin of the rotor by comparing the bell curves at the near stall point with smooth casing, without even being numerically calculated to CG6's own stall limit. The

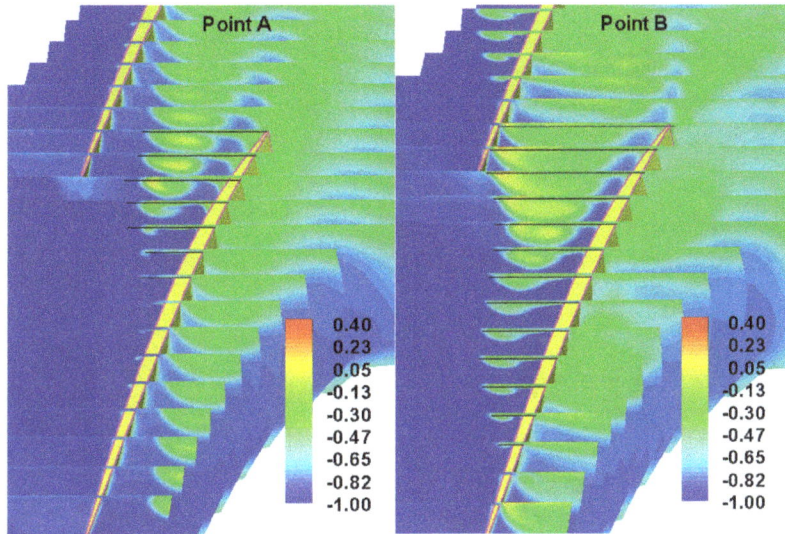

FIGURE 11: The axial-radial slices of entropy contours that show the 3D complex surface of MF/TLF interface ([14]). Points A and B are the ones marked on the left of Figure 9(a). Figure reproduced from CAMERON et al. (2013).

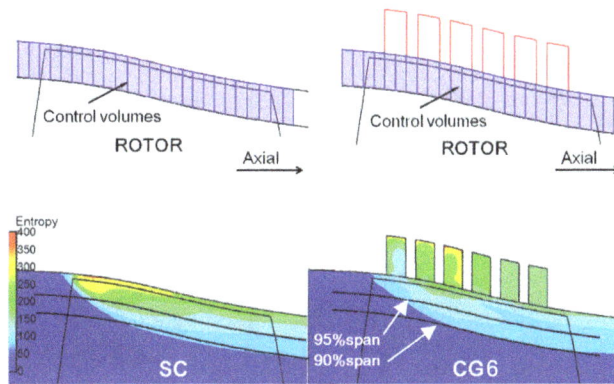

FIGURE 12: The installment of control volumes ([16]). Figure reproduced from NAN et al. (2014).

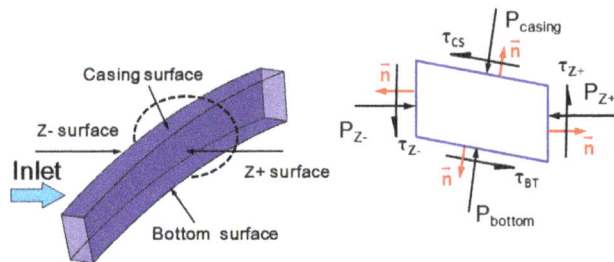

FIGURE 13: The volumetric illustration of typical control volumes ([16]). Figure reproduced from NAN et al. (2014).

feature could greatly reduce the need of unsteady CFD simulations when it comes to design or optimize casing treatments.

Both Figures 14 and 15 are CFD results. In the next section, we will provide examples that have experimental results to validate the bell curves and demonstrate how they can be used in design or optimization of casing treatments and tip air injections. However, it must be noted that the idea of bell curves relies on the hypothesis that rotating stall would be triggered by the MF/TLF interface spillage out of the rotor leading edge. There is at least one counter example. Houghton et al. [25] showed that when testing the effect of single groove's axial location on the stall margin improvement (SMI), there are two peaks on the curve of SMI versus axial chord. The hypothesis of MF/TLF spillage cannot explain what happens in this case. However, other than this special case, the method of bell curves works well with many stability enhancement techniques.

(a) Local axial momentum

(b) Cumulative axial momentum

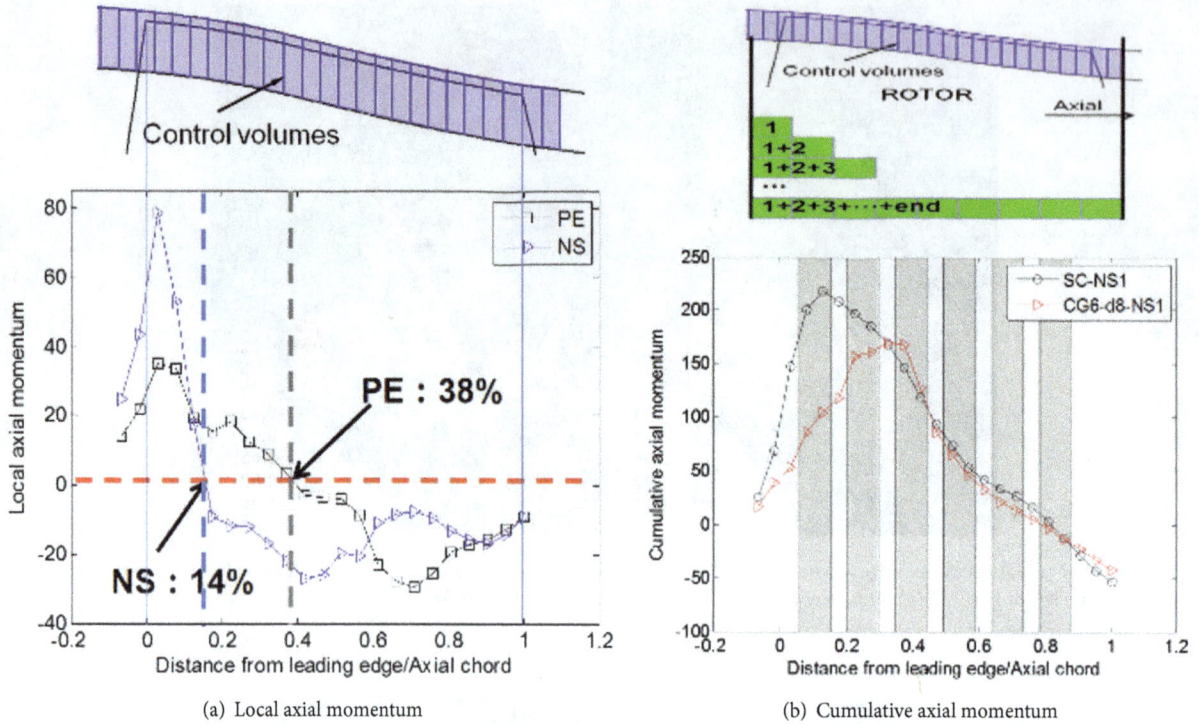

FIGURE 14: The net axial momentum distributions ([16]). Figure reproduced from NAN et al. (2014).

5. The Examples of Application

In this section, we will demonstrate how to apply the bell curves to design or optimize stability enhancement techniques. Every example had been experimentally validated.

5.1. Casing Grooves. The experimental data of the first example is taken from Ross [26], in which a set of casing grooves were tested. There are three grooves available on the especially decided casing (called variable casing treatment, or VCT). One can open or block any of the grooves to make 7 combinations of casing grooves, which are marked as 001, 010, and so forth on Figure 16. Number 1 means that the groove is open and number 0 means that the groove is closed. The seven SMIs together with that of the smooth casing were listed in Figure 16. The total eight bell curves were calculated and plotted together in one chart. The SMIs can be grouped into four groups, A, B, C, and D, each of which corresponds to two cases of casing grooves that exhibit similar SMIs. The order of SMI groups is A>B>C>D. The eight bell curves can also be grouped into four groups, each of which contains two bell curves whose peaks located almost at the same axial location. Furthermore, the order of the four groups of peaks along the axial chord corresponded EXACTLY to the order of SMI groups. The bell curves were plotted based on the CFD results at the same incoming flow coefficient as the NS point of smooth casing. At this point, the flow fields in most of the cases with grooves were still steady, which saved lots of computation time and expenses.

5.2. Skewed Axial Slots. Four skewed axial-slot casing treatments with different geometries were chosen to check the applicability of the bell curve (Table 1), which were named as CT-a, CT-b, CT-c, and CT-d for short. The data were taken from a large-scale low-speed compressor test rig [18]. In this study, each individual slot component covers exactly the tip chord of the blade and has a skew angle of 60° along the same direction of blade rotation. The depth of all slots is 10 mm. The comparison of the bell-shaped curves for these four CTs is shown in Figure 17. The experimental results are listed in Table 1. The trend of those peaks of the bell curves matches very well with the trend of SMIs.

5.3. Tip Air Injection with Self-Recirculation. Up to this point, the experiments were done prior to CFD and the bell curves, so the results were used for validation purposes. Hereby we are giving an example where the bell curves were done before the experiments and predicted the test results of SMI. A type of casing treatment (SELF-INJ), which recirculates the air from the trailing edge into the leading edge, was designed as shown in Figure 18. A high-fidelity time-resolved CFD simulation was done to predict its SMI by gradually rising the back pressure. The stall limit was recognized as the last point before the CFD collapsed. At the same time, the same CFD result at the near stall point of the smooth casing was used to construct the bell curve and compare it with those of the smooth casing, a double-grooved casing (generated separately for another paper [16]), a five-grooved casing, and a SAS casing. The resultant bell curves are depicted in

TABLE 1: The effectiveness for different axial slot schemes.

Configurations		Slots /passage	Slot width (% of C_{ax})	*SMI (EXP)*
	CT-a	7	15%	**8.2%**
	CT-b	5	15%	**6.6%**
	CT-c	7	7.5%	**1.8%**
	CT-d	17	5%	**5.4%**

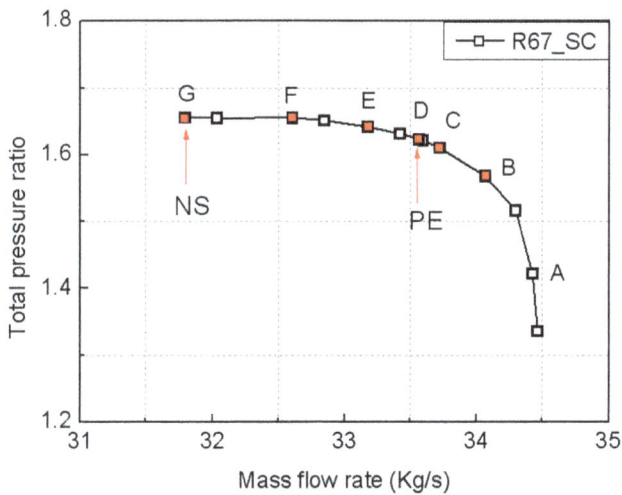

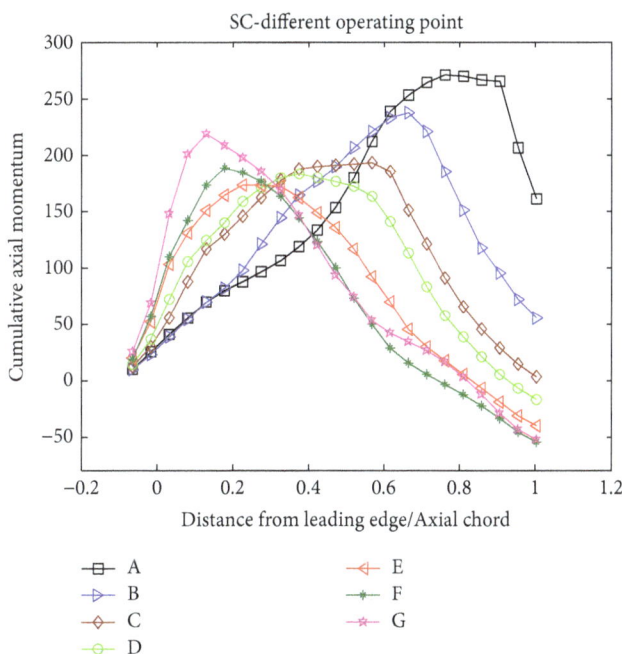

TABLE 2: Z_{bp} in bell-shaped curves and the experimental SMI [19].

End-wall treatment	Z_{bp}	ΔZ_{bp}	Experimental SMI
SC	0.273	–	–
SELF-INJ	0.341	0.068	5.26%
Double-groove CG13	0.274	0.064	5.28%
CG5	0.409	0.136	10.8%
SAS	0.544	0.271	27.4%

Figure 19 [19], which predicted that the SMI of SELF-INJ should be similar to the double-grooved CG13 (not shown in the figure), much less than the other two casing treatments. The test results are given in Table 2 [19], which confirmed the prediction in Figure 19.

6. Summary and Future Work

Research on oscillatory TLF and its applications are reviewed in this paper. The oscillatory TLF is a phenomenon that has been observed and studied by many researchers. As one of the research teams worldwide, we offered both CFD and test results and explained them from a view angle of flow structure at near stall. Since the oscillation of TLF was considered as a consequence of the axial momentum imbalance, a novel control-volume-based method, the bell curve, is proposed to estimate the axial momentum balance between the main flow and the tip leakage flow. The bell curve was then applied to make comparison of SMIs for various casing treatment methods and other stability enhancement methods that involve tip leakage flows.

These research efforts have been centered on one single purpose, that is, enlarging the stall margin of axial compressors. However, towards this goal, there is still a long way to go. Four ideas are proposed here for future research:

(1) Momentum transport in radial direction from PE to NS

(2) Identification of critical stage in multistage environment

FIGURE 15: The bell curves during the throttling process of Rotor 67 ([16]). Figure reproduced from NAN et al. (2014).

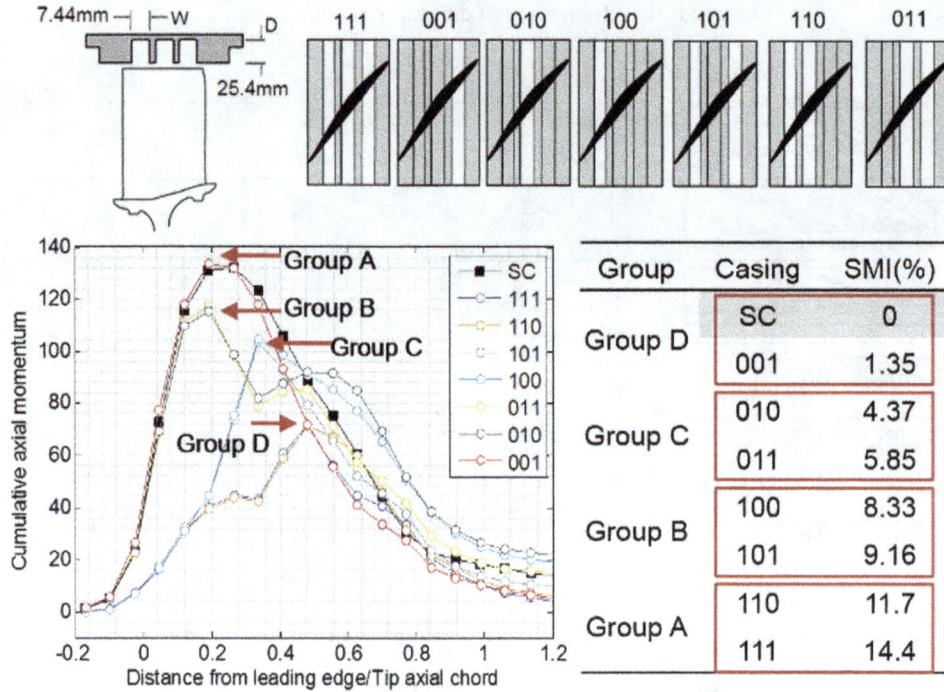

FIGURE 16: The bell curves for prediction of VCT ([17]). Figure reproduced from NAN (2014).

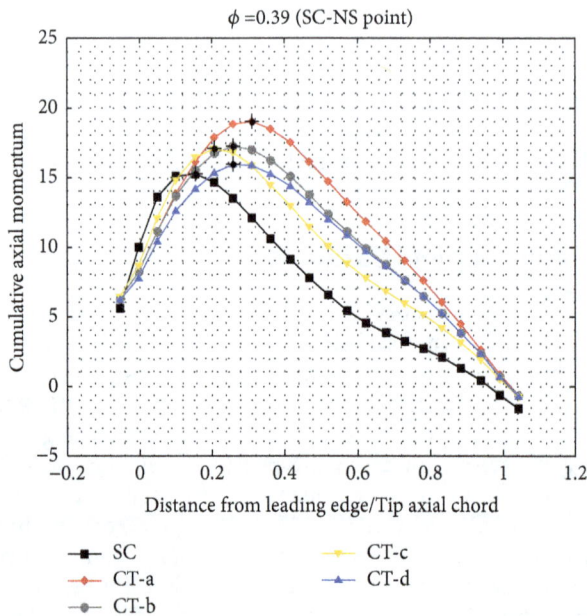

FIGURE 17: The bell-shaped curves for the four different axial-slots configurations together with that of the smooth casing ([18]). Figure reproduced from MA (2016).

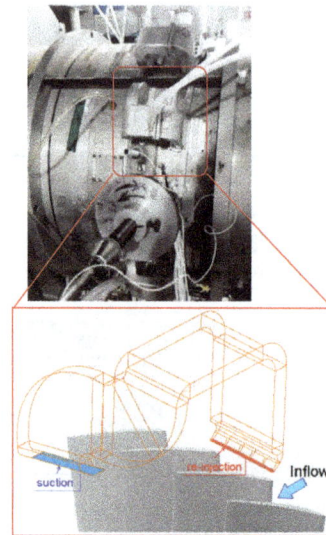

FIGURE 18: Tip air recirculation as a means of casing treatment ([19]). Figure reproduced from NAN et al. (2015).

(3) Integrated optimization of casing treatment and blade design for high-pressure compressors

(4) Active control of heavily loaded low-pressure compressors

The first idea is to extend the momentum analysis from tip region to all radial section of the blade span through the throttling process from PE to NS. It will help to understand the blade loading transition as a compressor approaches to stall. The second is to extend the current research from single rotors to multistages. These two ideas are basic research. The third and fourth ideas are for high-pressure compressor and low-pressure compressor, respectively. They are the applications of the first two ideas, with a hope to design safer modern compressors.

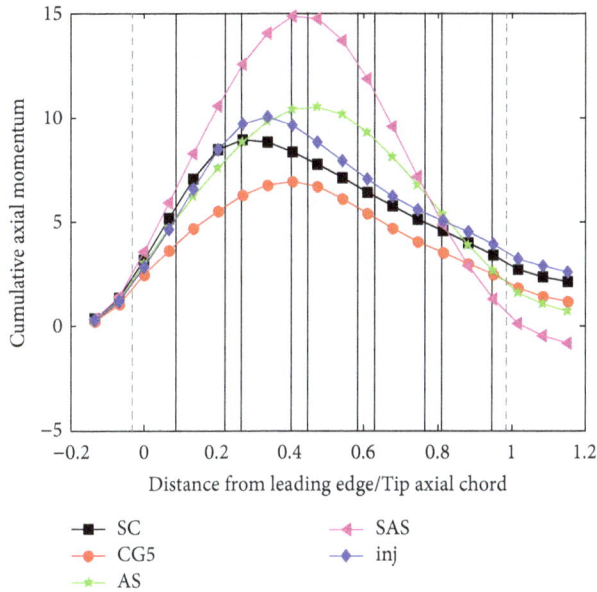

FIGURE 19: The bell-shaped curves for the studied casing treatments ([19]). Figure reproduced from NAN et al. (2015).

Nomenclature

ρ: Density
σ: Root mean square of static pressure
Φ: Mass flow coefficient of compressor
τ: Shear stress on solid surfaces
A: Area
C_{ax}:   Tip axial chord
h: Tip clearance
Mz: Axial momentum
F: Force
n: Unit vector normal to given surface
p: Static pressure
T: Time interval to store data in unsteady CFD
U: Absolute velocity
W: Relative velocity
Xzs: Axial location of zero shear region
Z: Axial direction
Z_{pb}:   Axial location of bell curve peak.

Superscripts and Subscripts

BT : Blade tip
CS: Casing surface
j: Leakage jet through the tip clearance
$Z+$:   Positive axial direction
$Z-$:   Negative axial direction.

Abbreviations

BPF: Blade passing frequency
CFD: Computational fluid dynamics
CT: Casing treatment
DP: Design point
$IGTI$:   International Gas Turbine Institute

MF: Main flow
NS: Near stall point
PE: Peak efficiency point
RMS: Root mean square
SAS: Skewed axial slots
SC: Smooth casing
SFB: Signature frequency band
SMI: Stall margin improvement
TLF: Tip leakage flow
TLV: Tip leakage vortex
$URANS$:   Unsteady Reynolds-averaged Navier-Stokes
$UTLF$:   Unsteady tip leakage flow
VCT: Variable casing treatment.

Conflicts of Interest

The authors declare that they have no conflicts of interest.

Acknowledgments

The support of the National Natural Science Foundation of China through Grants no. 51676184 and no. 51606189 is acknowledged. The authors thank many colleagues and former students at Chinese Academy of Sciences who have contributed and are still contributing their efforts to enrich this ongoing research experience.

References

[1] B. Lakshminarayana, *Fluid Dynamics and Heat Transfer of Turbomachinery*, John Wiley & Sons, 1995.

[2] H. D. Vo, C. S. Tan, and E. M. Greitzer, "Criteria for spike initiated stall," *ASME Journal of Turbomachinery*, vol. 130, no. 1, Article ID 011023, 2008.

[3] G. Pullan, A. M. Young, I. J. Day, E. M. Greitzer, and Z. S. Spakovszky, "Origins and structure of spike-type rotating stall," in *Proceedings of the ASME urbine Technical Conference and Exposition (GT '12)*, pp. 2567–2579, June 2012.

[4] R. Mailach, I. Lehmann, and K. Vogeler, "Rotating instabilities in an axial compressor originating from the fluctuating blade tip vortex," *Journal of Turbomachinery*, vol. 123, no. 3, pp. 453–460, 2001.

[5] M. B. Graf, *Effects of Stator Pressure Field on Upstream Rotor Performance, [Doctoral Thesis]*, Massachusetts Institute of Technology, 1996.

[6] J. Bae, K. S. Breuer, and C. S. Tan, "Periodic Unsteadiness of Compressor Tip Clearance Vortex," in *Proceedings of the ASME Turbo Expo Power for Land, Sea, and Air*, pp. 457–465, Vienna, Austria, 2004.

[7] X. Deng, H. Zhang, J. Chen, and W. Huang, "Unsteady Tip Clearance Flow in a Low-Speed Axial Compressor Rotor With Upstream and Downstream Stators," in *Proceedings of the ASME Turbo Expo Power for Land, Sea, and Air*, pp. 1371–1381, Reno, Nevada, USA, 2005.

[8] N. Tahara, T. Nakajima, M. Kurosaki, Y. Ohta, E. Outa, and T. Nisikawa, "Active stall control with practicable stall prediction system using auto-correlation coefficient," in *Proceedings of the 37th Joint Propulsion Conference and Exhibit*, Salt Lake City,UT,U.S.A..

[9] M. Dhingra, Y. Neumeier, J. V. Prasad, A. Breeze-Stringfellow, H. Shin, and P. N. Szucs, "A Stochastic Model for a Compressor

Stability Measure," in *Proceedings of the ASME Turbo Expo Power for Land, Sea, and Air*, pp. 833–843, Barcelona, Spain, 2006.

[10] D. Christensen, P. Cantin, D. Gutz et al., "Development and Demonstration of a Stability Management System for Gas Turbine Engines," in *Proceedings of the ASME Turbo Expo: Power for Land, Sea, and Air*, pp. 165–174, Barcelona, Spain, 2006.

[11] I. J. Day, "Stall, surge, and 75 years of research," *Journal of Turbomachinery*, vol. 138, no. 1, Article ID 2478223, 2015.

[12] F. Lin, J. Du, J. Chen, C. Nie, and C. Biela, "Flow Structures in the Tip Region for a Transonic Compressor Rotor," in *Proceedings of the ASME Turbo Expo 2010: Power for Land, Sea, and Air*, pp. 2561–2572, Glasgow, UK.

[13] J. Du, F. Lin, H. Zhang, and J. Chen, "Numerical investigation on the self-induced unsteadiness in tip leakage flow for a transonic fan rotor," *Journal of Turbomachinery*, vol. 132, no. 2, 2010.

[14] M. A. Bennington, M. H. Ross, J. D. Cameron et al., "An Experimental and Computational Investigation of Tip Clearance Flow and Its Impact on Stall Inception," in *Proceedings of the ASME Turbo Expo 2010: Power for Land, Sea, and Air*, pp. 501–512, Glasgow, UK.

[15] M. A. Bennington, J. D. Cameron, S. C. Morris et al., "Investigation of Tip-Flow Based Stall Criteria Using Rotor Casing Visualization," in *Proceedings of the ASME Paper*, pp. 641–651, Berlin, Germany.

[16] X. Nan, F. Lin, S. Wang, L. Liu, N. Ma, and J. Chen, "The Analysis of Axial Momentum of the Rotor Tip Flows for Axial Compressors With Circumferential Grooves," in *Proceedings of the ASME Turbo Expo Turbine Technical Conference and Exposition*, Düsseldorf, Germany, 2014.

[17] X. Nan, *The compressor rotor tip control volume method and its application on circumferential groove casing treatment , Doctoral Thesis [Doctoral, thesis]*, Chinese Academy of Sciences, 2014.

[18] N. Ma, *Research on the Low-speed Modeling Similarity Criteria for High-speed compressors and the flow mechanisms of Axial-slot Casing Treatments , Doctoral Thesis [Doctoral, thesis]*, Chinese Academy of Sciences, 2016.

[19] X. Nan, N. Ma, Q. Lu, and F. Lin, "The Control Volume Analysis of the Effectiveness of Casing Treatments for a Low-speed Compressor," in *Proceedings of the International Gas Turbine Congress IGTC-2015-0260*, Tokyo, 2015.

[20] J. J. Adamczyk, M. L. Celestina, and E. M. Greitzer, "The role of tip clearance in high-speed fan stall," *Journal of Turbomachinery*, vol. 115, no. 1, pp. 28–38, 1993.

[21] K. L. Suder and M. L. Celestina, "Experimental and computational investigation of the tip clearance flow in a transonic axial compressor rotor," *Journal of Turbomachinery*, vol. 118, no. 2, pp. 218–229, 1996.

[22] J. Bergner, M. Kinzel, H. Schiffer, and C. Hah, "Short Length-Scale Rotating Stall Inception in a Transonic Axial Compressor: Experimental Investigation," in *Proceedings of the ASME Turbo Expo 2006: Power for Land, Sea, and Air*, pp. 131–140, Barcelona, Spain.

[23] C. Hah, J. Bergner, and H.-P. Schiffer, "Tip clearance vortex oscillation, vortex shedding and rotating instabilities in an axial transonic compressor rotor," in *Proceedings of the 2008 ASME Turbo Expo*, pp. 57–65, deu, June 2008.

[24] J. März, C. Hah, and W. Neise, "An Experimental and Numerical Investigation Into the Mechanisms of Rotating Instability," in *Proceedings of the ASME Turbo Expo 2001: Power for Land, Sea, and Air*, p. V001T03A084, New Orleans, Louisiana, USA.

[25] T. Houghton and I. Day, "Enhancing the stability of subsonic compressors using casing grooves," in *Proceedings of the ASME Paper*, pp. 39–48, Orlando, Fla, USA, June 2009.

[26] M. Ross, *Tip clearance flow interaction with circumferential groove casing treatment in a transonic axial compressor , [Doctoral, thesis]*, University of Notre Dame, 2013.

16

Global Sensitivity Analysis of High Speed Shaft Subsystem of a Wind Turbine Drive Train

Saeed Asadi ⓘ, Viktor Berbyuk, and Håkan Johansson

Department of Mechanics and Maritime Sciences, Chalmers University of Technology, 412 96 Göteborg, Sweden

Correspondence should be addressed to Saeed Asadi; saeed.asadi@chalmers.se

Academic Editor: Ryoichi Samuel Amano

The wind turbine dynamics are complex and critical area of study for the wind industry. Quantification of the effective factors to wind turbine performance is valuable for making improvements to both power performance and turbine health. In this paper, the global sensitivity analysis of validated mathematical model for high speed shaft drive train test rig has been developed in order to evaluate the contribution of systems input parameters to the specified objective functions. The drive train in this study consists of a 3-phase induction motor, flexible shafts, shafts' coupling, bearing housing, and disk with an eccentric mass. The governing equations were derived by using the Lagrangian formalism and were solved numerically by Newmark method. The variance based global sensitivity indices are introduced to evaluate the contribution of input structural parameters correlated to the objective functions. The conclusion from the current research provides informative beneficial data in terms of design and optimization of a drive train setup and also can provide better understanding of wind turbine drive train system dynamics with respect to different structural parameters, ultimately designing more efficient drive trains. Finally, the proposed global sensitivity analysis (GSA) methodology demonstrates the detectability of faults in different components.

1. Introduction

Nowadays wind turbine industry is growing very fast and wind power has become one of the most cost effective of all renewable energy sources because of environmental concerns and finite resource of fossil fuels [1]. Therefore, it has attracted political and business interest and extensive research is carried out at different universities and institutes [2].

It is very important to reduce the costs of design, manufacturing, and maintenance and optimize the power performance of wind turbines in order to be cost competitive in the current market. A better understanding and consequently improvement of wind turbine drive train dynamics performance could lead to overall cost efficiency in wind turbine operations. This is done by advanced system modelling, identifying the reasons for premature failures and fault diagnoses [3] in different components in drive train of wind turbines. Examples of multibody dynamic analysis to predict the components' loads in wind turbine drive trains are in

[4–6]. Generally, the dynamic performance of wind turbine drive train can be investigated based on different points of view, such as power performance, operational speeds, loads in specific components like coupling and bearings in both healthy and faulty situations, and misalignment parameters [7–14]. Different structural dynamic parameters can directly influence a wind turbine ultimate performance and in terms of loading, high speed shaft components are the most important since the components experience major extreme loads during different regimes, especially in transient one.

The current research deals with mathematical modelling of high speed shaft drive train test rig. A typical wind turbine drive train is composed of a generator as an electrical part with a control set, flexible couplings, a speed-up gearbox, a main shaft, and a wind rotor as an air-operated system [15]. The power from the rotor is transferred through main shaft, gearbox, and high speed shaft to the generator.

We propose the global sensitivity analysis (GSA) as a tool to better investigate wind turbine drive train dynamics, more specifically, high speed shaft drive train models. The aim is to quantify the input structural parameters' effect to objective functions of the system dynamics (interaction quantification set [16]). From GSA the most influential design parameters on objective functions of wind turbines can be determined, which could ease the optimization process by narrowing down the number of input design parameters.

Methodology of sensitivity analysis is various and it has been used for different applications. Reviews on different GSA methods can be found in [16–20]. Some examples of sensitivity analysis are derivative-based Morris one-at-a-time (OAT) local sensitivity method [21] and Sobol/Saltelli method providing the variance based sensitivity [22, 23], whereas the Sobol/Saltelli variance based sensitivity method provides indices that quantify the relative contribution of each input structural parameter to set of objective functions [22, 23]. In the variance based GSA, input variables are chosen randomly in a proper range. Then, the contribution of input parameter to objective functions is investigated via the sensitivity indices. These sensitivity indices are quantifying the contribution of a random variable to the total variance of the response, which is basis of variance based global sensitivity analysis in [23].

Computational effort when design analysis requires a large set of loading conditions to be evaluated, such as wind turbine drive trains, is very important. As one of the most common methods for GSA, Monte Carlo simulation for these systems is not efficient since it is computationally expensive. On the other hand, GSA within analysis of variance decomposition (ANOVA) or high dimensional model representation (HDMR) theory is suitable method in terms of being cost effective. It decomposes the response function of a high dimensional system into combination of low dimensional system set. This reduces the computational effort. A multiplicative dimensional reduction method (M-DRM) for GSA was proposed by Zhang and Pandey [24] entailing simple representation of GSA indices. The method has proved to be as accurate as Monte Carlo with much less computational effort [22].

Several works have been done within sensitivity analysis of rigid and flexible multibody systems with respect to structural parameters such as mass, moment of inertia, stiffness, and damping values [25–28]. In [29] extended Fourier amplitude sensitivity test method was used and the results of a single turbine and a set of large group of turbines was compared by GSA. Sensitivity analysis to perform a large set of data analysis for a wind farm was reported in [30].

Considering that the faults occur in high speed shaft drive trains, this could be a subject of study [31, 32]. The test rig for high speed shaft drive train as subsystem of wind turbine has been constructed in order to do experimental study and acquire data for validation of developed mathematical model. Another purpose of the test rig is to investigate strategies to detect faults. Early fault detection is important since it can reduce the cost of maintenance drastically.

Ultimately, by introducing faults and defects and different components of high speed shaft drive train, we investigate the detectability of faults and failures within GSA. Such an analysis could lead to fault diagnose in components of wind turbine drive train and prevent failures in early stages.

In the following sections, the proposed mathematical model of high speed shaft drive train test rig with the experimental setup at the lab is presented. Then, input variables and different operational scenarios and objective functions are introduced. Different sources of faults and failures such as defect in the bearing, backlash in the coupling, and torque ripple in the motor are introduced and modelled. Ultimately, the GSA indices for the validated model have been investigated. Results of GSA allow one to understand the priorities in contribution of system parameters for any specific objective function and lead one to focus on the areas of operation that show greatest opportunities for optimization. The outcome of the current research leads to useful information regarding wind turbine drive train dynamics behaviour on different operational scenarios which not only provide possibility to reduce the number of input parameters for optimization and speed up the process, but also give insight to design functional components in proper ranges and in a smarter way. In particular, the main aim of the paper is to apply the GSA for the proposed mathematical model of high speed shaft drive train including aforementioned fault sources. Quantifying the faults effects on the drive train objective functions could be applied as a methodology to contribute to virtual condition monitoring in order to detect and predict faults and prevent failures in early stages in different components of drive trains.

2. High Speed Shaft Drive Train System Test Rig

The most common wind turbine is of lateral (horizontal) axis type, which entails main supporting tower upon which sits a nacelle and rotor. Inside the nacelle sits a main shaft supported by bearings on which the rotor hangs. The main shaft is connected to a gearbox, which, in turn, is connected to the generator with control system (Figure 1). The high speed shaft subsystem is one part where faults in bearing are often observed.

To support experimental and numerical studies of wind turbine drive train dynamics, a scaled down test rig has been built and instrumented as reported in [33, 34]. It entails a motor, rotor disk with shaft, where the rotor shaft is passing through the bearing housing and connected to the motor by a coupling. A small eccentric mass can be added on the rotor disk to give a varying load. The setup of the drive train test rig is represented in Figure 2.

The test rig is driven by ABB motor model M3BP160 MLA 6 (6 pole, 75 kW) with a variable frequency controller (ACS355 from ABB providing speed up to 1000 rpm). In addition, the test rig entails motor shaft adapter, shaft's coupling, bearing housing with flexible support, a disk on which an eccentric mass can be added, and bedplate on which all the components are mounted.

The data acquisition system is SKF @ptitude Observer; SKF offers under its Windcon 3.0 condition monitoring system. The set of sensors comprise accelerometers and Eddy

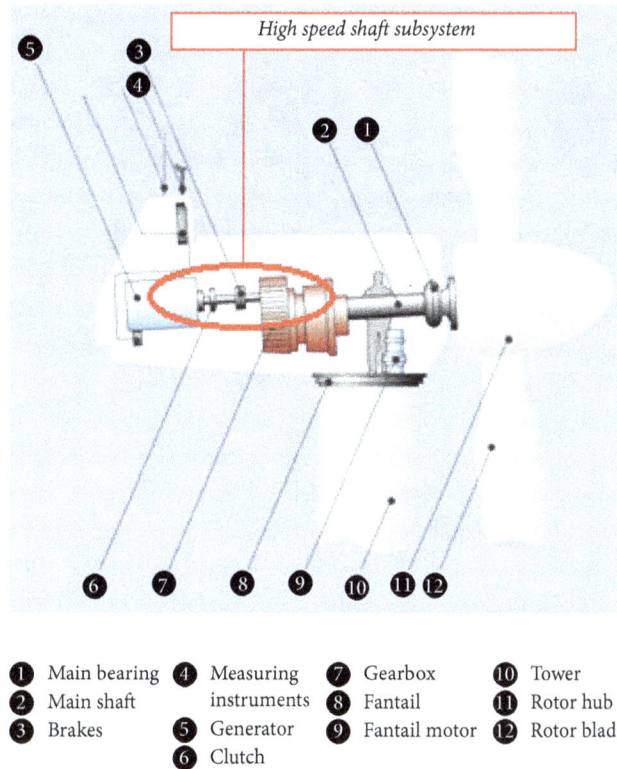

FIGURE 1: Overview of a high speed shaft subsystem as part of wind turbine drive train [52].

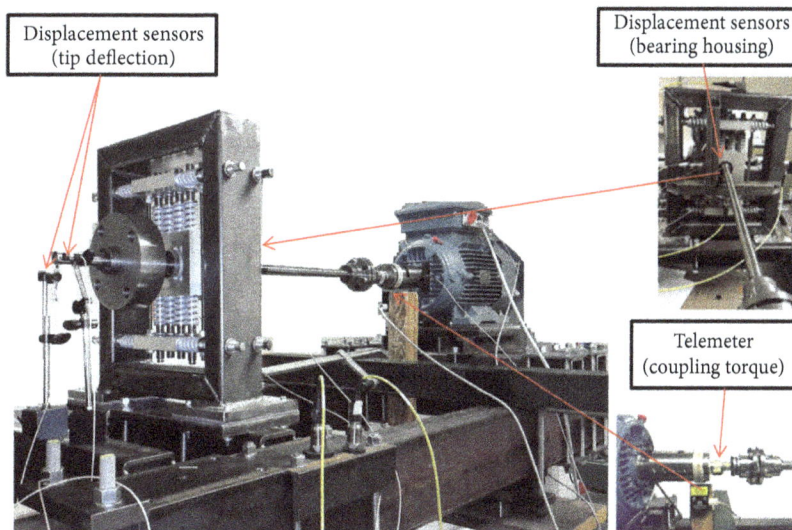

FIGURE 2: The test rig with sensors location.

probes displacement sensors, which detect vibrations in the test rig and deflection at both tip and bearing housing. Additionally, in order to measure the torque in coupling, full bridge strain gauges are put on the motor shaft, and the signal is acquired using a telemeter system (TEL 1-PCM).

In previous works [33, 34], the coupling torque, deflection fields in both lateral and vertical directions in tip, and bearing housing have been measured and used for model validation. To support future studies of defect detection, the objective functions derived in this paper are based on these sensors.

2.1. Mathematical Model of Drive Train System Test Rig. An engineering abstract of the test rig is shown in Figure 3. The mathematical model has been developed based

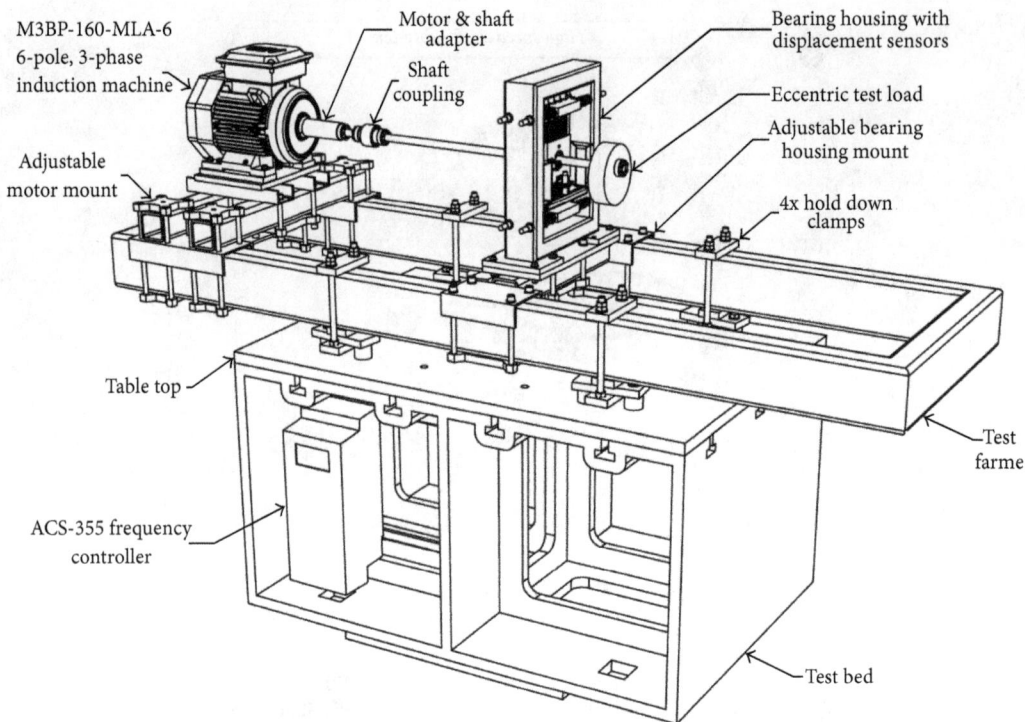

(a)

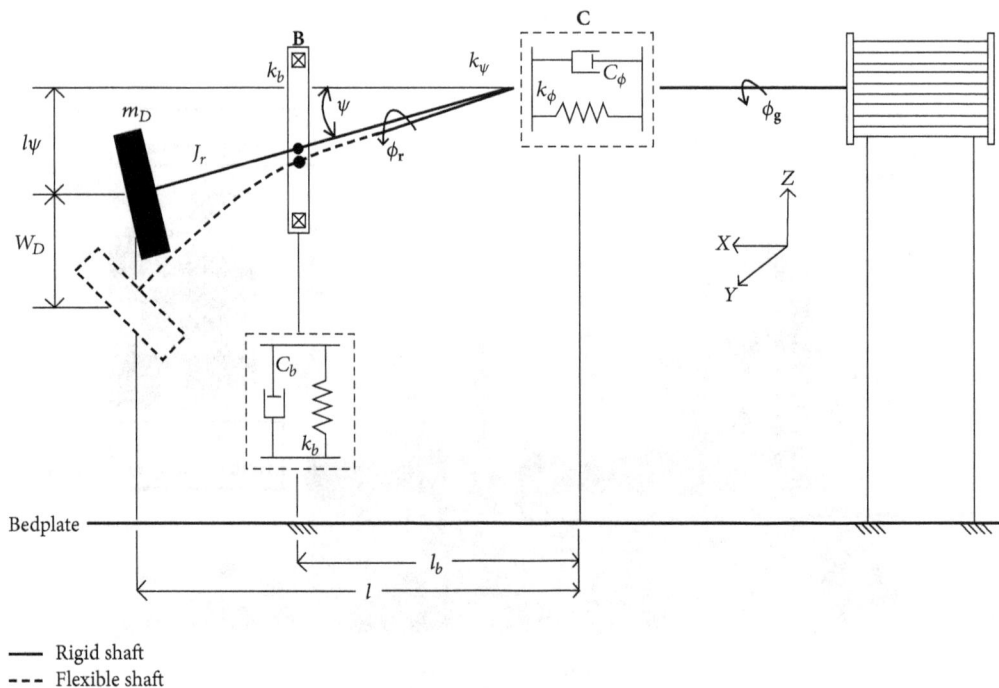

—— Rigid shaft
- - - Flexible shaft

(b)

FIGURE 3: A sketch of the test rig (a) and an engineering abstract (b).

on both rigid and flexible shaft assumption in bending modes, considering that the torsional flexibility is concentrated at coupling. The motor is rigidly attached to the bedplate.

2.1.1. Governing Equations. The kinetic, potential, and dissipative energies and nonconservative forces are derived considering that the shaft is rotating, and its bending deflection $w(x)$ is assumed to occur in the radial direction only. The disk is seen as a rigid body, and the shaft torsional flexibility is concentrated in coupling as well as linear along the shaft length. A linear spring model is adopted to simplify the flexibility of structure of the bearings [35].

$$T_D \doteq \frac{1}{2} m_D \left[\left[\dot{w}_D \varphi'(l) \right]^2 + \left[w_D \varphi(l) \right]^2 \dot{\phi}_s^2 \right]$$
$$+ \frac{1}{2} m_e \left[\dot{w}'^2(l) + \left(w(l) + r_e \right)^2 \dot{\phi}_s^2 \right] + \frac{1}{2} J_D \dot{\phi}_D^2$$
$$+ \frac{1}{2} I_D \left[\dot{w}_D \varphi'(l) \right]^2 ,$$

$$T = T_D + \frac{1}{2} J_g \dot{\phi}_g^2 + \frac{1}{2} \left[J_c + \frac{J_{sh}}{2} \right] \dot{\phi}_s^2 + \frac{1}{2} \left[\frac{J_{sh}}{2} \right] \dot{\phi}_D^2 , \tag{1}$$

$$V = \left(m_D + m_e \right) g w(l) \cos \phi_s + \frac{1}{2} k_\phi^{sh} \left(\phi_D - \phi_s \right)^2$$
$$+ \frac{1}{2} k_\phi^C \left(\phi_s - \phi_g \right)^2 + \frac{1}{2} k_b^\psi \left[w_D \varphi(l_b) \right]^2 + V_{sh},$$

$$R = \frac{1}{2} C_\phi^{sh} \left(\dot{\phi}_D - \dot{\phi}_s \right)^2 + \frac{1}{2} C_\phi^C \left(\dot{\phi}_s - \dot{\phi}_g \right)^2 + \frac{1}{2} C_{bf} \left| \dot{\phi}_s \right|$$
$$+ \frac{1}{2} C_b \left[\dot{w}_D \varphi(l_b) \right]^2 .$$

Here $V_{sh} = \int_0^l EI(w''(x))^2 dx$ is the contribution of the shaft bending to potential energy and the indices sh and D denotes for shaft and disk, respectively. The complete list of structural parameters is given in Table 1. The shaft torsion is parametrized by ϕ_D, ϕ_s, and ϕ_g, as an angle of rotation in the disk (rotor tip), shaft, and motor, respectively.

Applying Lagrange formalism, the nonlinear set of equations of motion can be written as matrix equation with time variant coefficients as follows:

$$M(q) \ddot{q} + B(q, \dot{q}, t) = F(q, \dot{q}, t), \tag{2}$$

where M is the inertia matrix, B is the vector related to internal forces acting on the test rig, and q is the vector of generalized coordinate to be defined in Section 2.1.2.

F is an external load vector and it is a combination of $M_{BH}^\bullet(q(t))$, $T_C^{BL}(q(t))$, and $M_g^{rip}(q(t))$ which applies to specific degree of freedom.

$M_{BH}^\bullet$ corresponds to external forces due to the disturbance of the bearing defects imposed to the system (described in Section 2.2.3). The defect may impose the disturbance torque into bending and torsional degrees of freedom ($\bullet = \{\psi, \phi_r\}$).

T_C^{BL} refers to the backlash of the coupling which appears as an internal load in the governing equations. The backlash has a contribution in the potential energy since it is modelled as a spring with gap.

M_g^{rip} is torque ripples imposed by the motor, a source of faults in drive train dynamics representing electromechanical interaction between components.

2.1.2. Parametrization of the Model. Since the coupling is assumed to be very stiff in torsion (apart from its backlash),

the shaft torsional flexibility and coupling play are here added and included in k_ϕ, and T_C^{BL}. We introduce ϕ_r as the rotor (disk plus shaft) angle, such that the shaft twist is represented by $\Delta \overset{\text{def}}{=} \phi_g - \phi_r$ where ϕ_g is the motor shaft angle. Hence, ϕ_s and ϕ_D are both replaced by ϕ_r in the governing equations in Section 2.1.1. We assume that the shaft inertia and coupling are abstracted as $J_r = J_{sh} + J_C$. Coupling torsional stiffness is represented in a way that includes both shaft and coupling torsional stiffness's. Similarly, the shaft and coupling damping, C_ϕ^{sh} and C_ϕ^C, are represented by C_ϕ.

The shaft deflection is parametrized in terms of rigid and flexible contribution to total deflection as $w(x) = \psi x + w_D \varphi(x)$, where ψ is the angular deflection at the coupling, w_D is the rotor shaft tip deflection due to shaft flexibility, and $\varphi(x)$ is taken as the shape of the fundamental bending eigenmode as described in [33]. In [33], we developed and validated the mathematical model of high speed shaft drive train test rig considering the shaft flexibility in terms of bending. To obtain a suitable shape function $\varphi(x)$, the Euler-Bernoulli beam theory is applied [36]. Only radial vibration amplitudes of the beam were considered. The fundamental bending eigenmode based on structural parameters in Table 1 was found at $\beta = 0.8734$ and its shape function φ for a shaft rotating at speed $\Omega = 1000\,\text{rpm}$ gives the eigenfrequency at $\omega^* = 109.4608\,\text{rad/s}$. The next eigenmode was found at $\omega^* = 6151\,\text{rad/s}$, which is much higher than relevant excitation frequency. The deflection fields in two locations, namely, tip ($x = l$) and bearing hub ($x = l_b$), are assessed within the aforementioned assumptions. Thus, the tip radial deflection would be $\psi l + w_D \varphi(l)$, and the bearing housing radial deflection becomes $\psi l_b + w_D \varphi(l_b)$. The displacements in vertical and lateral directions become mirrored by rotational angle ϕ_r. Note that we simplify the deflection field in the bearing hub. For more realistic model, shaft prebending and relative motion of inner ring with respect to the bearing hub frame must be considered.

To conclude, the vector of generalized coordinate is defined as $q = [\psi \;\; \phi_r \;\; \phi_g \;\; w_D]$.

2.2. Faults Modelling in the Drive Train

2.2.1. Electromechanical Modelling of the Motor: Motor Torque with Ripple. Considering that wind turbine drive train contains both mechanical and electrical components, it is of important concern to have a model which studies electromechanical interactions. In induction motors, many kinds of excitation occur such as electromagnetic forces exist, that is, unbalanced magnetic pull. Torsional vibration of drive train system could lead to significant fluctuation of motor speed. Such oscillation of the angular speed superimposed on the average motor angular speed causes severe oscillation of electric currents in motor windings [37]. In this section, we model torque ripple as source of this fluctuations in motor speed.

In the current mathematical model, we assume the motor as a 6-pole induction machine with rated power 7.5 kW at speed 975 rpm. The speed is set from a frequency converter, such that the torque is given by the difference from set speed

TABLE 1: Test rig nominal and randomized input parameters.

Parameter	Symbol	Nominal value $(X^\star)$	$X_{\min}/X^\star$	$X_{\max}/X^\star$
Motor inertia with respect to x axis	J_g	0.087 kgm^2	0.5	2
Rotor inertia with respect to x axis	J_r	0.095 kgm^2	0.5	2
Rotor shaft inertia with respect to x axis	J_{sh}	0.0003 kgm^2		
Rotor inertia with respect to ψ (minor axis)	I_r	19.083 kgm^2	0.5	2
Disk inertia with respect to y axis	I_D	18 kgm^2		
Shaft inertia with respect to y axis	I_{sh}^Y	0.5 kgm^2		
Disk mass	m_D	14.52 kg	0.75	1.5
Rotor shaft mass	m_{sh}	3.85 kg		
Eccentric mass value mounted in the rotor disk	m_e	74.4 gr	0	4
Distance of eccentric mass from center	r_e	8 cm		
Rotor shaft and disk mass	m_r	18.35 kg		
Length from coupling to the tip	l	0.85 m	0.8	1.2
Length from coupling to bearing housing	l_b	0.72 m	0.8	1.2
Length from coupling to CG of disk and rotor shaft	l_0	0.75 m		
Damping factor	ζ_d	0.01 s		
Coupling torsional stiffness	k_ϕ	2600 Nm/rad	0.75	5
Coupling bending stiffness	k_ψ	56 kNm/rad	0.75	5
Bearing mount lateral and vertical stiffness coefficients	k_b^0	4308 kN/m	0.75	5
Bearing hub bending stiffness	k_b^ψ	$0.1 k_\psi$	0.75	5
Coupling torsional damping	C_ϕ	26 kNm/rad	0.5	10
Coupling bending damping	C_ψ	56 kNm s/rad	0.5	10
Bearing mount lateral and vertical damping coefficients	C_b	4308 kN s/m	0.5	10
Bearing friction damping coefficient	C_{bf}	0.0023 Ns/m	0.8	3
Rotor torsional damping coefficient plus bearing friction	C_{br}^f	0.006 Ns/m	0.8	3
Motor torsional damping coefficient	C_{bg}	0.006 Ns/m	0.5	10
Shaft bending stiffness	EI	7.6699 kNm^2	0.5	10
Angle dependency factor of k_b	α_b	1	0.8	1.5
Torque ripples parameters	$M_{rip}^{(i)}$	10 Nm	0	3
Backlash gap	BL	0.002 m	0	3
Backlash regularization parameter	σ	40		
Bearing torque disturbance zone for defect	ε	0.1		
Maximum value of rotational bearing defect torque	$M_{BH}^{\phi_r}$	30 Nm	0	5
Maximum value of radial bearing defect torque	M_{BH}^ψ	30 Nm	0	5

ω_g^{set} and actual speed $\dot{\phi}_g$. A cycle of ramp startup-steady state-shutdown can thus be described by

$$M_{mech} = \frac{M_{mark}}{\Delta\omega_{mark}}\left(\omega_g^{set} - \dot{\phi}_g\right), \tag{3}$$

where

$$\omega_g^{set} = \frac{t}{t_0}\omega_g^{max} \quad t < t_0$$

$$\omega_g^{set} = \omega_g^{max} \quad t_0 \le t < t_1 \tag{4}$$

$$M_{mark} = 0 \quad t_1 \le t,$$

where $\Delta\omega_{mark} = 25/60 \times 2\pi$ rad/s, $M_{mark} = 73.4561$ Nm, t_0 is the startup-time, and t_1 is the instance of time when motor

is turned off (without braking). From electrical measurement of the voltage from the frequency converter, it was found that, in addition to M_{mech}, there is a small periodic torque (ripple) acting on the generator rotor at frequencies ω_g^{set}, $3\omega_g^{set}$, and $6\omega_g^{set}$. Hence, the following electrical torque is considered:

$$M_g(t) = \frac{M_{mark}}{\Delta\omega_{mark}}\left(\omega_g^{set} - \dot{\phi}_g\right) + M_{rip}^{(1)}\sin\left(\phi_g\right)$$
$$+ M_{rip}^{(2)}\sin\left(3\phi_g\right) + M_{rip}^{(3)}\sin\left(6\phi_g\right), \tag{5}$$

where $M_{rip}^{(i)}$ ($i = \{1, 2, 3\}$) are coefficients that contribute to torque ripple with corresponding frequencies and have been obtained based on comparison of the torque ripple range in steady state in simulation and experiments.

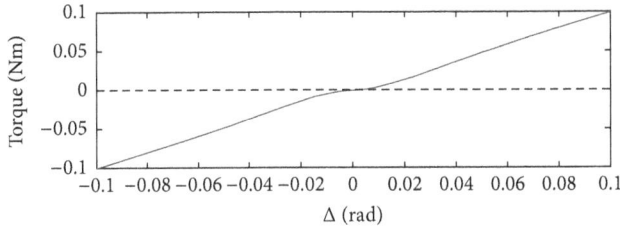

FIGURE 4: Backlash effect as a torque to the coupling.

2.2.2. Backlash Modelling of the Coupling. Here, we propose backlash as nonlinear stiffness properties to the system structure, which is common source of faults in transmission system dynamics (Figure 4). This phenomenon affects the governing equations. Properly functioning mechanical systems need to have a certain clearance (gap, play) between the components transmitting motion under load. As part of internal forces, T_{int} is defined as follows [38]:

$$T_{int}(\Delta) = k_\phi (\Delta - BL) + (k_\phi - BL)$$
$$\times \frac{[g(\Delta - BL) - g(\Delta + BL)]}{2}, \quad (6)$$

$$g(x) = |x| \tanh(\sigma x).$$

2.2.3. Modelling of Defects in Rolling Elements (Bearing): Disturbance Torque from Bearing Inner Ring. Bearing fault diagnosis is important in condition monitoring of any rotating machine. Early fault detection is one of the concerns within the wind turbine system modelling. To this end, we need to model source of faults in the bearing. Dynamic modelling and vibration response simulation of rolling bearings with surface defects is a common assumption within fault modelling. When bearing balls roll over defects, shock pulses with high frequency are generated with excessive vibration leading the whole system to failure. This characteristic impulse-response signal is detected by the vibration monitor. An inner raceway rotates with shaft's rotation and goes through the loaded zone of bearing every rotation. This mechanism shows that the vibration of inner raceway with defect is modulated by the shaft rotation [39–41].

Here we propose a simple modelling of surface defect for bearing which imposes a disturbance torque in certain angles in the inner raceway connected to the rotor shaft (similar to when a ball passes from the defects and causes a excitation to the system). Figure 5 shows the external torque imposed in radial and rotational directions. As part of external forces, $M_{BH}^\bullet$ is defined based on impulse factor ε and maximum torque M_{BH} as shown in Figure 5(a).

Proper value for maximum disturbance is $M_{BH}^{max} = 30$ Nm. $\varepsilon(=0.1)$ is the angular length of active contact zone, where the defect is assumed (Figure 5).

Depending on the rolling elements number included in the bearing structure, the geometric relationship among its components, and shaft rotational speed, characteristic rotational frequencies initiating from respective bearing elements

could be obtained. In this regard, the characteristic passing frequency in the inner ring is defined as follows:

$$f_i = f_r \frac{N_b}{2} \left(1 + \frac{D_b}{D_c} \cos\varphi\right), \quad (7)$$

where f_r is the rotor rotational frequency ($=\dot{\phi}_r/2\pi$), N_b is number of rolling balls ($N_b = 9$), $D_b(=72\,\text{mm})$ and $D_c(=53.5\,\text{mm})$ are the balls diameter and bearing pitch (($D_o + D_i$)/2) diameters, respectively. φ is the contact angle ($\varphi = 0$ rad). The bearing defect location (here, the inner race) will determine that $f_i(\approx 10.55 f_r)$ will be exhibited in the machine vibration signal.

2.2.4. Uneven Mounting Stiffness. For subsequent study of excitation of uneven mounting stiffness, the bearing mount stiffness has been defined as function of rotor shaft angle $\phi_r(t)$ as follows:

$$k_b \doteq k_b^0 \sqrt{\cos^2\phi_r + \frac{1}{\alpha_b}\sin^2\phi_r}, \quad (8)$$

where α_b is a parameter corresponding to angle dependency of k_b and defines different loads in vertical and lateral directions imposed to the bearing springs.

2.3. Numerical Values of Parameters in the High Speed Shaft Drive Train Test Rig. Table 1 represents the parameters and their numerical values assumed in the mathematical model:

The damping coefficients are chosen as proportional to respective stiffness coefficient: $C_\bullet \doteq \zeta_d k_\bullet$, where $\bullet = \{\phi, \psi, b\}$.

The value of C_{bf} which represents the bearing friction damping has been defined from SKF manual handbook [42].

2.4. Operational Scenarios. In order to study the GSA in different conditions, we introduce the following set of operational scenarios in wind turbine drive train dynamic analysis such as transient, steady state, and shutdown:

(i) Transient: a torque is applied to reach the preset speed.

(ii) Steady state: it is where the structure is under almost constant speed control.

(iii) Shutdown: in this regime there is no torque imposed to the system, and it is left to slow down to stop.

Note that the corresponding motor speeds ([1000 800 600] rpm) correspond to the specific frequency in the frequency controller ([50 40 30] Hz, resp.).

2.5. System Model Response for the Nominal Values. Here, we present the simulation response of the mathematical model based on structural parameters nominal values represented in Table 1. The displacement fields in both lateral and vertical direction with their FFT have been presented.

The deflection fields include three different operational scenarios based on Figure 6 with steady state of $t = [1\ 6]$ sec and shutdown starting at $t = 6$ sec. The tip deflection fields in both vertical and lateral directions are presented with zoom

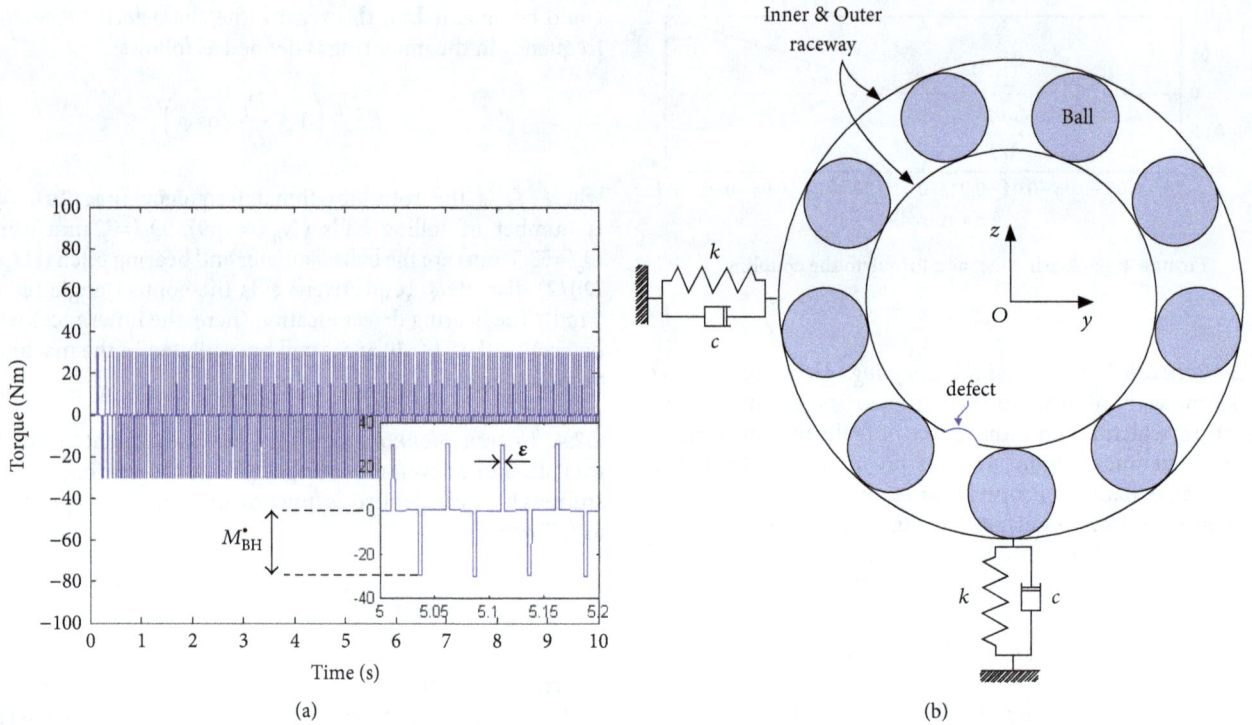

FIGURE 5: Typical disturbance torque ($M_{BH}^{\bullet}$) with zoom (a) and inner race defect modelling sketch with rolling bearing elements (b).

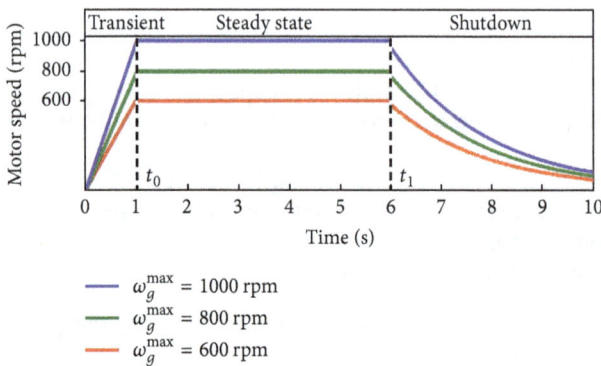

FIGURE 6: Motor speed for operational scenarios transient, steady state, and shutdown for 3 different maximum speeds.

in steady state ($t = [5 \ 6]$ sec) and shutdown ($t = [7 \ 8]$ sec) in Figure 7.

The bearing housing deflection fields in both vertical and lateral directions are presented with zoom in steady state ($t = [5 \ 6]$ sec) and shutdown ($t = [7 \ 8]$ sec) are shown in Figure 8.

The frequency response functions of the tip deflection in both vertical and lateral directions are presented with zoom on the dominant frequency in Figure 9.

The coupling torque with its frequency response is presented here in Figure 10.

The equations of motion have been solved using Newmark numerical procedure method implemented in Matlab

software. In order to satisfy the convergence criteria, the proper time step has been chosen as $\Delta t = 10^{-4}$.

3. Objective Functions

In this section we introduce the set of objective functions. The introduced objective functions are assessed in order to evaluate the dynamic performance of the developed high speed shaft models.

(i) Tip Deflection in Both Vertical and Lateral Directions in Steady State and Shutdown Regimes (OF_B^{1-4}). The RMS values of the tip deflection in both lateral and vertical in two operational scenarios (steady state and shutdown) (RMS(Y^{std}), RMS(Z^{std}) and RMS(Y^{SD}), RMS(Z^{SD}), resp.) are obtained within the mathematical model and the sensitivity indices based on the structural inputs have been defined for these objective functions.

(ii) Bearing Hub Deflection in Both Vertical and Lateral Directions in Steady State and Shutdown Regimes (OF_B^{5-8}). RMS value of the bearing housing bending in both lateral and vertical directions in two operational scenarios (steady state and shutdown) (RMS(BH_Y^{std}), RMS(BH_Z^{std}) and RMS(BH_Y^{SD}), RMS(BH_Z^{SD}), resp.) is obtained within the mathematical model and the sensitivity indices based on the structural inputs have been defined for these objective functions. A model of gearbox is required to consist of shaft and bearing housing stiffness, which allows interaction between housing and the internal components to be investigated.

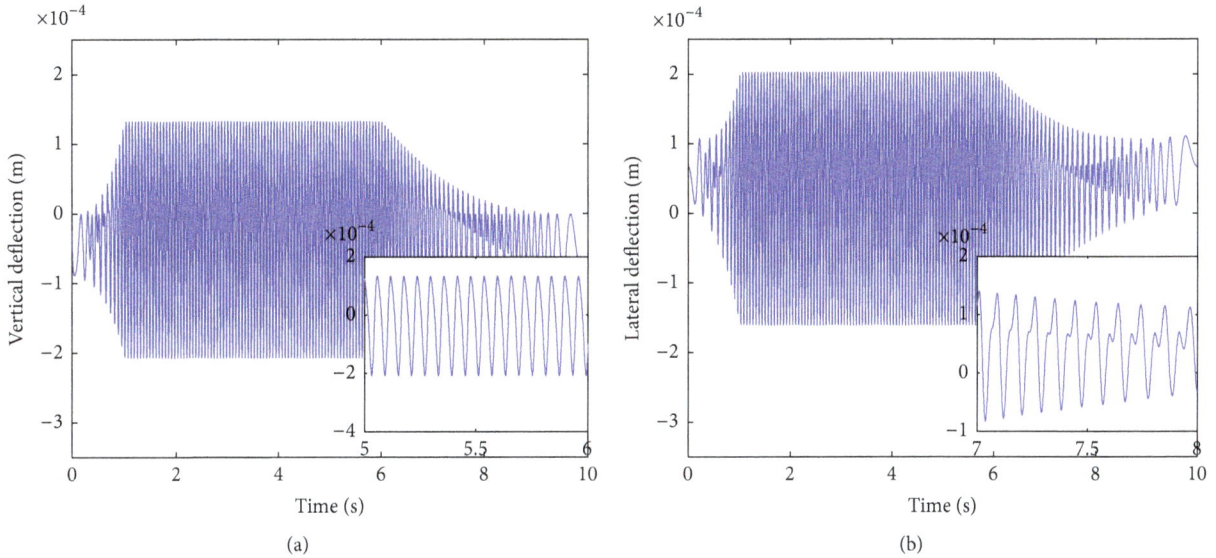

FIGURE 7: Tip deflection with zoom in steady state in vertical (a) and with zoom in shutdown regime in lateral (b) directions for the total model.

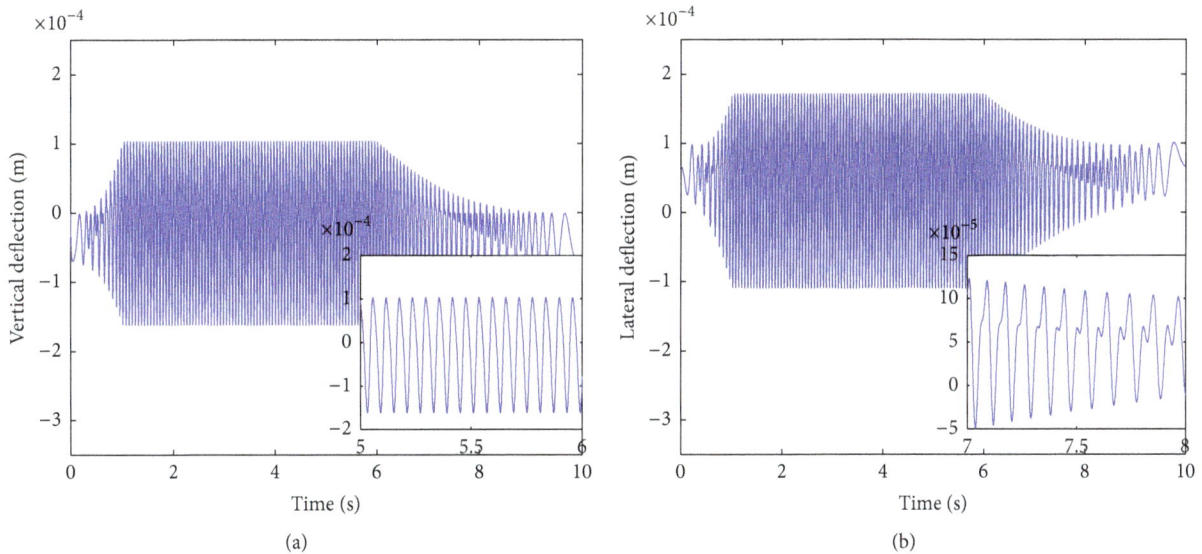

FIGURE 8: Bearing housing deflection with zoom in steady state in vertical (a) and with zoom in shutdown regime in lateral (b) directions for the total model.

(iii) Frequency Response of the Tip Deflection in Lateral and Vertical Directions in Steady State and Shutdown (OF_B^{9-12}). These are the desired objective function representations: FFT(Y^{std}), FFT(Z^{std}) and FFT(Y^{SD}), FFT(Z^{SD}), respectively. The modal sensitivity analysis allows one to recognize the eigenfrequency response to design parameters changes. It can also help for the better identification of important dynamic properties.

(iv) Coupling Torque in Different Operational Scenarios Transient, Steady State, and Shutdown Cases (OF_T^{1-3}). In order to determine the influence of the design parameters in

torsional vibration model, RMS values of the coupling torque in different operational scenarios transient, steady state, and shutdown cases are investigated as objective functions (RMS(CT^{tr}), RMS(CT^{std}), and RMS(CT^{SD}), resp.). The coupling torque is defines as $M_\phi = k_\phi \Delta$.

Coupling torque in transient regime is very important to measure and quantify the parameters which have the most effect, since the component experiences the most severe loadings due to this regime nature. By shutting down the motor, there is not any load imposed by motor side and the range of the coupling torque is decreased from steady state value to zero in short period of time.

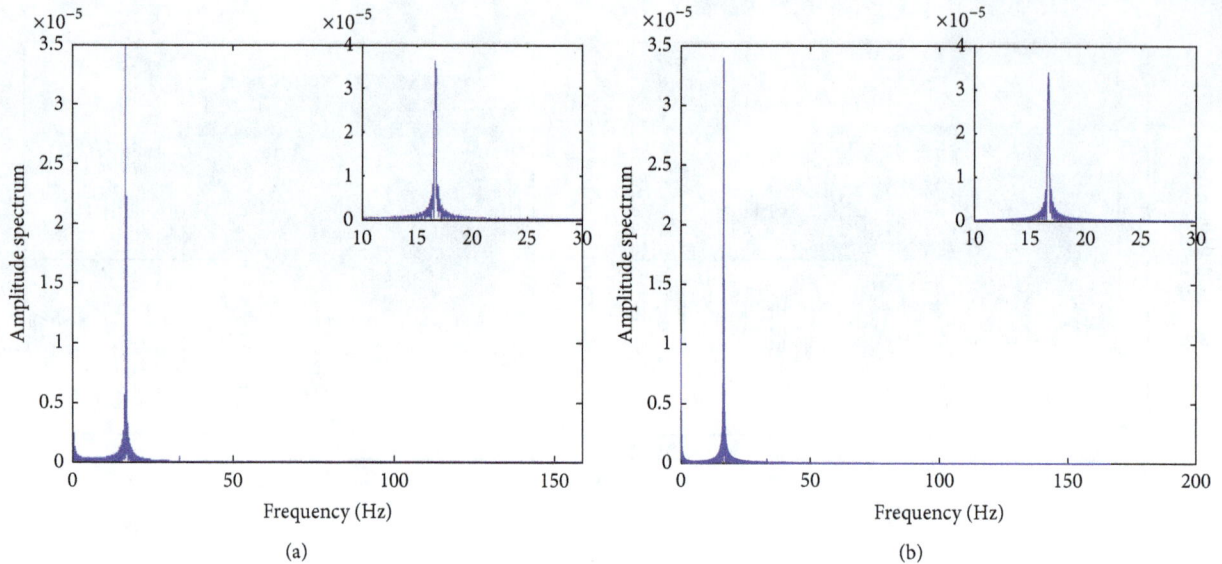

FIGURE 9: FFT of tip deflection in steady state regime, in vertical (a) and lateral (b) directions for the total model.

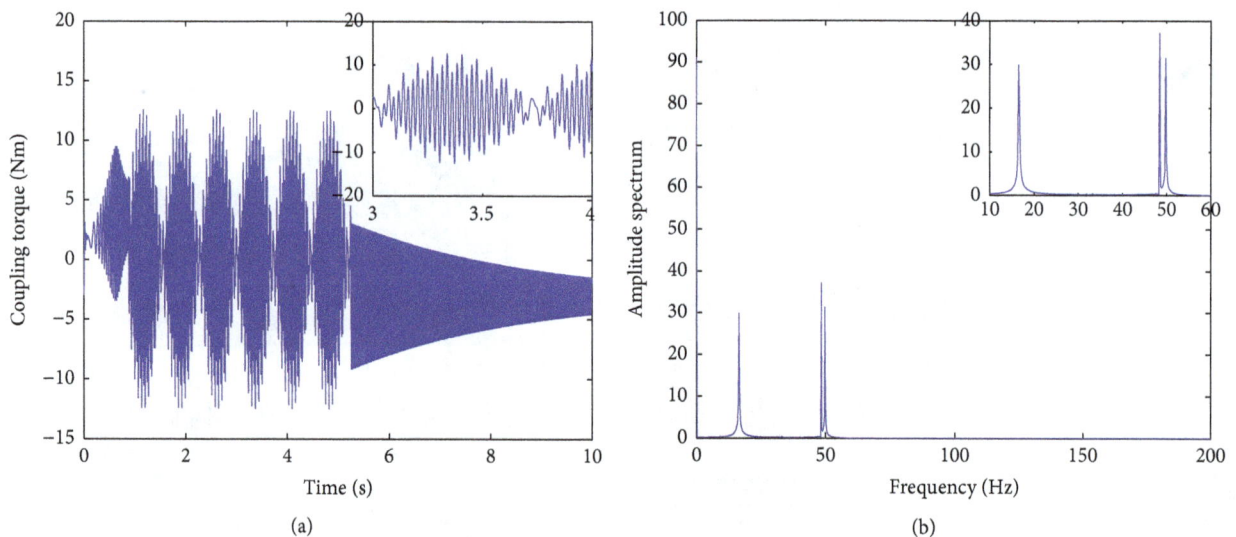

FIGURE 10: Coupling torque with zoom in the steady state (a) and its FFT with zoom in the peaks (b) for the total model.

(v) The First and Second Peaks in FFT Diagram of Coupling Torque in Steady State (OF_T^{4-5}). There are several picks in the coupling torque frequency response, of which the most dominant one is of interest to study, and investigating the most influential parameter would give valuable conclusion form the vibrational point of view. Studying the torsional model, we came up with a conclusion that some peaks in FFT are related to the ripple due to motor. Quantifying those peaks would give the contribution of the fault source to the objective function, which is very valuable information in terms of optimization.

(vi) The Peak in FFT Diagram of Coupling Torque in Shutdown Regime (OF_T^6). During shutdown we expect that only torsional eignefrequency of the structure plays the role in FFT.

4. Global Sensitivity Analysis of High Speed Shaft Test Rig

The efficiency and applicability of global sensitivity analysis is already proven by some mathematical and mechanical examples [24]. The aim here is to apply this method to the high speed shaft drive train dynamics with respect to the structural parameters.

4.1. Theoretical Concept of Variance Based Sensitivity Analysis. Each of the objective functions in Section 3 can be seen as a function of the test rig structural parameters, represented as vector of n variables, $\mathbf{X} = [X_1, X_2, \ldots, X_n]^T$. We have adopted the variance based sensitivity indices developed by Saltelli et al. [43] and mainly founded on the previous work of the Russian mathematician Zhang and Pandey [22]. These methods are defined as global sensitivity analysis (GSA), since the set of sensitivity indices are calculated considering the entire range of variations in parameters, compared to OAT (once at a time) or differential approaches [44].

Upon considering the structural parameters $\mathbf{X}$ as random, the objectives OF $= h(\mathbf{X})$ are also random. The expectation operation is denoted by $E[\bullet]$, such that the mean and variance of h are calculated as follows:

$$\mu_{\mathrm{OF}} = E[h(\mathbf{X})],$$
$$V_{\mathrm{OF}} = E\left[(h(\mathbf{X}) - \mu_{\mathrm{OF}})^2\right]. \tag{9}$$

Under certain assumptions of independence of X_i s, the ANOVA decomposition of $h(\mathbf{X})$ can be expressed as follows [45, 46]:

$$h(\mathbf{X}) = h_0 + \sum_{i=1}^{n} h_i(X_i) + \sum_{i<j} h_{ij}(X_i, X_j) + \cdots$$
$$+ h_{12\cdots n}(X_1, X_2, \ldots, X_n). \tag{10}$$

Upon taking the variance of (10), the total variance of OF can be decomposed in the same manner [47]:

$$V_{\mathrm{OF}} = \sum_{i=1}^{n} V_i + \sum_{i<j} V_{ij} + \cdots + V_{12\cdots n}, \tag{11}$$

where it was assumed that the components of the function (11) are orthogonal and can be expressed in terms of $h(\mathbf{X})$ integrals as

$$V_{i_1 \cdots i_s} = \int \left(h_{i_1 \cdots i_s}\right)^2 dx_{i_1} \cdots dx_{i_s}, \tag{12}$$

$$V_{\mathrm{OF}} = \int (h)^2 d\mathbf{X} - (h_0)^2. \tag{13}$$

In case X_i's are not uniformly distributed, it is possible to define a probability function for random variables $X_1, X_2, \ldots, X_s$ and adjust (12) and (13) accordingly [22].

The constants V_{OF} and $V_{i_1 \cdots i_s}$ are called variances of h and $h_{i_1 \cdots i_s}$, respectively. The first-order terms V_i represent the partial variance in the response due to the individual effect of a random variable X_i, while the higher order terms show the interaction effects between two or more random variables. This decomposition puts forward two concepts, namely, the primary effect referring to the effect of a term associated with only one random variable [48] and the total effect referring to both the individual effect of a random variable as well as its interaction with other random variables. The primary effect index of a random variable X_i is obtained by the normalization of the main-effect variance over the total variance in OF:

$$S_i = \frac{V_i}{V_{\mathrm{OF}}} \tag{14}$$

and the total sensitivity index is defined as

$$S_{T_i} = \frac{V_{T_i}}{V_{\mathrm{OF}}}, \tag{15}$$

where $V_{T_i} = V_{\mathrm{OF}} - V_{-i}[E_i(\mathrm{OF} \mid \sim X_i)]$. Note that $V_{-i}[\bullet]$ refers to the variance of quantity with randomized set of $\mathbf{X}$, except X_i taken as nominal value. More elaborate derivation of equations is presented in [22].

4.1.1. Multiplicative Dimensional Reduction Method (M-DRM): To Compute the Simplified Sensitivity Indices. In order to reduce the computational effort associated with the numerical evaluation of (13), [22], we propose to utilize a multiplicative form of dimensional reduction method (M-DRM) for sensitivity analysis, of which the corresponding deterministic model, OF $= h(X)$, is written with reference to a fixed input point, that is, $\mathbf{X} = \mathbf{c}$, referred to as the cut point with coordinates $\mathbf{c}$. To this end, the ith invariance functions defined by fixing all input variables, but X_i, to their respective cut point coordinates are derived as

$$h_i(X_i) = h(c_1, \ldots, c_{i-1}, X_i, c_{i+1}, \ldots, c_n). \tag{16}$$

Based on this concept, a multiplicative dimensional reduction method approximates the deterministic function $h(\mathbf{X})$ as follows [24]:

$$h(\mathbf{X}) \approx [h(\mathbf{c})]^{1-n} \cdot \prod_{i=1}^{n} h_i(X_i, \mathbf{c}_{-i}), \tag{17}$$

where $h(\mathbf{c}) = h(c_1, c_2, \ldots, c_n)$ is a constant value and $h(X_i, \mathbf{c}_{-i})$ represents the function value for the case that all inputs except X_i are fixed at their respective cut point coordinates. M-DRM is particularly useful for approximating the integrals required for evaluation of the sensitivity indices described in the previous section.

In order to write compact mathematical expression, ρ_k and θ_k denoted as the k mean and mean square values, respectively, associate with input parameters X_k. The Gaussian quadrature is computationally efficient for one-dimensional integrals calculation. The definitions are as follows:

$$\rho_k = E[h_k(X_k)]$$
$$\approx \sum_{l=1}^{N} w_{kl} h(c_1, \ldots, c_{k-1}, X_{kl}, c_{k+1}, \ldots, c_n),$$
$$\theta_k = E\left[[h_k(X_k)]^2\right]$$
$$\approx \sum_{l=1}^{N} w_{kl} [h(c_1, \ldots, c_{k-1}, X_{kl}, c_{k+1}, \ldots, c_n)]^2, \tag{18}$$

where N is the total number of integration points and X_{kl} and w_{kl} are the lth Gaussian integration abscissa and weight function, respectively. The total number of function evaluations required to calculate the sensitivity indices using this approach is $N_{\mathbf{X}} \times N$ with $N_{\mathbf{X}}$ as number of design parameters.

For computed θ_i and ρ_i ($i = \{1, \dots, n\}$), the primary and total sensitivity indices are approximated as shown as follows:

$$S_i = \frac{E_i\left(V\left(\text{OF} \mid X_i\right)\right)}{V\left(\text{OF}\left(\mathbf{X}\right)\right)} \approx \frac{\theta_i/\rho_i^2 - 1}{\prod_{k=1}^{n}\theta_k/\rho_k^2 - 1},$$

$$S_{T_i} = 1 - \frac{E_i\left(V\left(\text{OF} \mid \sim X_i\right)\right)}{V\left(\text{OF}\left(\mathbf{X}\right)\right)} \approx \frac{1 - \rho_i^2/\theta_i}{1 - \prod_{k=1}^{n}\rho_k^2/\theta_k}.$$

(19)

The primary sensitivity index, S_i, gives an indication of how strong the direct influence of X_i is on OF without any interactions with other variables. The total sensitivity index, S_{T_i}, gives an indication of the total influence of X_i on OF from its own direct effects along with its interaction with other variables [49, 50].

The differences between the primary effect and the total effect then give an indication of how important the interactions of X_i are with other variables in influencing OF. If the main effect is small, whereas the total effect is large, then X_i does influence OF but only through interactive effects with other system inputs.

The introduced sensitivity indices accuracy depends on number of integration points used that can be determined via a convergence study [24, 51].

4.2. System Input Parameters Used in GSA. Assuming the bending model with torsional parameters imposed to the system, we will have a model which contains bending and torsional response of the system. In order to apply GSA for this model, we propose the following set of input parameters as follows:

$$\mathbf{X}_{\text{TOT}} = \begin{bmatrix} J_r & J_g & I_r & m_D & m_e & l & l_b & k_\phi & k_\psi & k_b & k_b^\psi & C_\phi & C_\psi & C_b & C_{bf} & EI & \alpha_b & M_{\text{rip}}^{(1)} & M_{\text{rip}}^{(2)} & M_{\text{rip}}^{(3)} & \text{BL} & M_{\text{BH}}^{\phi_r} \end{bmatrix}.$$

(20)

The nominal values of the input parameters as well as the range of parameters variation are given in Table 1. X^* is chosen as cut point (see Section 4.1.1).

5. Results from GSA of High Speed Shaft (HSS) Model

In this section, the global sensitivity analysis of the developed mathematical models corresponding to high speed shaft drive train system has been presented. Both torsional and bending models are examined under three different operational scenarios with different motor speeds in steady state (OS$_i$ = ω_g^{max} = [50 40 30] Hz).

The analysis presented here focuses on the response characteristics particular to chosen objective functions. It is expected that the operational scenarios will influence the dynamics and structural response to different input variables.

5.0.1 Convergence Study. The number of function evaluations using M-DRM method is $N_{\mathbf{X}} \times N$. For specific set of input variables and on each operational scenario, all objective functions are evaluated at the same time by a single time integration of the model. In order to achieve an acceptable approximation of the sensitivity analysis indices using the M-DRM method, it is important to assess the convergence of the results by increasing number of Gaussian quadrature integration abscissas denoted by N in (18).

The convergence study shows that for a nonlinear system such as high speed shaft drive train with system input parameters prescribed in Section 5, sufficient level of accuracy for the sensitivity analysis can be obtained with $21 \times 20 = 420$ ($N_{\mathbf{X}}$ = 21 design parameters and N = 20 integration abscissas) function evaluations. This is one of the most prominent aspects of the global sensitivity analysis using

this approach, which could reduce the computational effort drastically.

5.1. GSA of Drive Train High Speed (HSS) Shaft Subsystem Model. In this section we apply GSA to the total model which comprises bending and torsional parameters all together. The input parameters are defined in (20).

For the model containing all bending and torsional parameters, the primary and total sensitivity indices are as shown in Figure 11.

It can be concluded from this that the torsional input parameters (coupling torsional stiffness k_ϕ and bearing damping due to friction C_{bf} and motor torque ripples $M_{\text{rip}}^{(i)}$, backlash parameter BL, and bearing impact disturbance $M_{\text{BH}}^{\phi_r}$) do not have significant influence to the bending objective functions (OF$_B^{1-11}$). The same conclusion could be made based on the bending input parameters (coupling bending stiffness k_ψ and bearing mounting stiffness k_b and shaft bending stiffness EI) which do not seem to have major effect on torsional objective function (OF$_T^2$).

Based on this analysis, we conclude that the torsional and bending response of the model can be analysed separately, which allows for substantially simpler analysis, which will discussed next.

5.2. HSS Bending Model. After separating the two models, we analyse the sensitivity of HSS within the frame of bending model based on bending objective functions such as deflection field in bearing hub and tip in both lateral and vertical directions and the corresponding frequency domains.

Considering the separation of model, the simulation results of the bending model have been obtained based on semi-inverse dynamics approach ("perfect" speed control), by prescribing the motor angular speed, where it constitutes

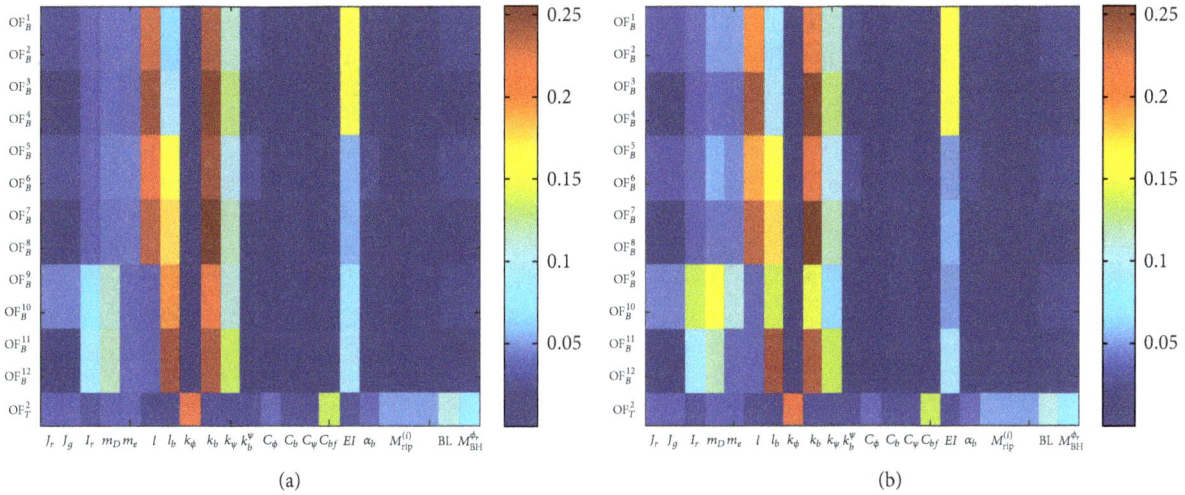

FIGURE 11: Primary sensitivity S_i (a) and total sensitivity ST_i (b) indices for combined bending and torsional model for OS_1.

the input to the system and is prescribed as a ramp loading expressed by Heaviside function:

$$\omega_g(t) = \frac{\omega_g^{\max}}{t_0} t H(t_0 - t) + \omega_g^{\max} H(t - t_0) H(t_1 - t) + \omega_g^{\max} \left(\frac{t_{\text{end}} - t}{t_{\text{end}} - t_1} \right) H(t - t_1), \tag{21}$$

where t_0 and t_1 denote the instant of time of the end of transient and instant of time of the start of the shutdown operations, respectively. t_{end} is duration of the operational scenarios (see Figure 6).

In order to investigate the bending model sensitivity, the input parameter appearing in the governing equations corresponding to the model is chosen as follows:

$$\mathbf{X}_B = \begin{bmatrix} J_r & J_g & I_r & m_D & m_e & l & l_b & k_\phi & k_\psi & k_b & k_b^\psi & C_\phi & C_\psi & C_b & C_{bf} & EI & \alpha_b & M_{\text{BH}}^\psi \end{bmatrix}. \tag{22}$$

Here, the sensitivity analysis has been performed for bending model for 3 different operational scenarios with different motor speeds in steady state regime.

It was shown that within these 3 different motor frequencies, the contribution of the input parameters is not changing drastically. The obtained results of GSA are presented in Figure 12. The results indicates that the most influential structural parameters which have the major effect to the objective functions are coupling bending stiffness k_ψ and bearing mounting stiffness k_b, shaft bending stiffness EI, and bearing hub and tip locations l_b, l. Also, the damping parameters do not have significant effect on both primary and total sensitivity indices.

More detailed analysis of the results based on each objective functions and influence of the input parameters is presented as follows.

(i) Tip Deflection in Vertical and Lateral Directions in Steady State (OF_B^{1-2}) and Shutdown Scenarios (OF_B^{3-4}). It is illustrated that the coupling bending stiffness k_ψ and bearing hub mounting stiffness k_b are the most dominant influential parameters, respectively.

It could be seen from Figure 12 that the shaft bending stiffness EI has more influence in tip deflection field compared to

the other objective function. This is somehow logical since the first bending mode shape is higher in tip ($x = l$) compared to bearing hub ($x = l_b$) [33].

Moreover, the geometric locations of bearing hub and tip (l_b, l) are influential parameters in terms of deflection field in both bearing hub and tip.

(ii) Bearing Hub Deflection in Vertical and Lateral Directions in Steady State (OF_B^{5-6}) and Shutdown Scenarios (OF_B^{7-8}). In bearing deflection field the influence of l_b is more than l. In tip deflection field the influence of l is more than l_b.

The bearing bending stiffness k_b^ψ does not have a major effect to the system objective functions.

It could be seen that the angle dependency of the bearing stiffness α_b is not having a major effect to all of objective functions, since the realistic range for α_b is small.

(iii) Frequency Response of the Tip Deflection in Vertical and Lateral Directions in Steady State (OF_B^{9-10}) and Shutdown Scenarios (OF_B^{11-12}). The influences of stiffness parameters k_ψ and k_b and EI and l_b and also inertia parameters I_r, m_D, and m_e are the most dominant parameters, respectively. This tells that aforementioned parameters are the most important ones in terms of frequency response of the tip deflection.

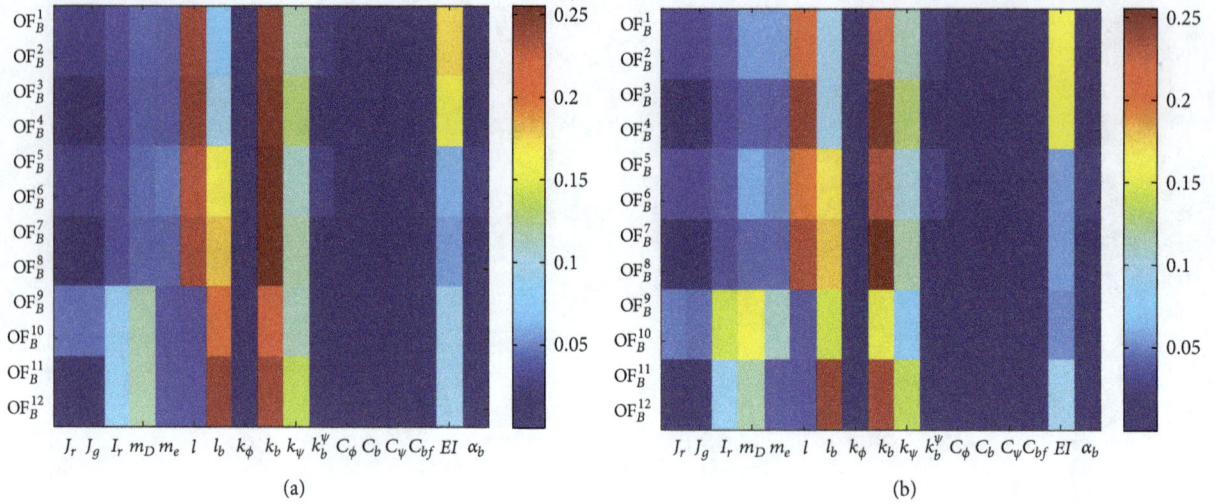

FIGURE 12: Primary sensitivity S_i (a) and total sensitivity ST_i (b) indices for bending model for $OS_{1,2,3}$.

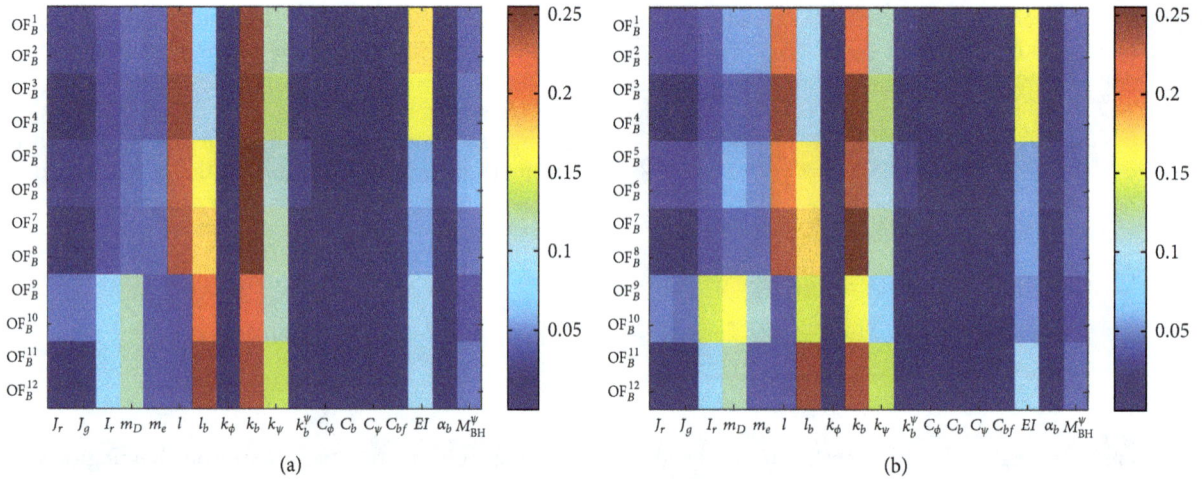

FIGURE 13: Primary sensitivity S_i (a) and total sensitivity ST_i (b) indices for bending model with bearing defect assumption for OS_1.

Among the mass inertia parameters, I_r, m_e, and m_D affect most of the bending objective functions.

It could be also seen form Figure 12 that the damping input parameters have the least influences on both primary and total sensitivity indices.

To sum up the conclusion related to the GSA of HSS subsystem within the frame of bending model, k_b and k_ψ and EI and I_r and m_D and m_e have the major effects on both primary and total sensitivity indices.

However, according to the sensitivity analysis results, rest of parameters stated in the system modelling do not have a remarkable effect on the mentioned objective functions. Such information would be particularly useful in attenuating the number of design parameters for optimization applications of wind turbine drive trains setups which leads to cost efficient simulations.

5.3. Bending Model with Bearing Defect: Detection of Faults in Inner Ring. Here, we study the bearing defect influence to the

objective functions. The aim is to investigate the sensitivity indices with respect to the bearing defect parameter M_{BH}^ψ imposing a radial load to the bearing housing and seek its effect to the bearing deflection and loads.

It is hypothesized that if the sensitivity index with respect to parameter M_{BH}^ψ is comparable to other sensitivities in certain OF, we can expect that a sensor measuring this OF is able to detect defects of size related to the magnitude of M_{BH}^ψ. The obtained results on GSA with bearing defect are presented in Figure 13 and summarized below.

The minimum value of M_{BH}^ψ which makes the sensitivity indices to be detectable has been investigated.

(i) Deflection fields in vertical and lateral directions in steady state and shutdown scenarios for the tip (OF_B^{1-4}) and bearing hub (OF_B^{5-8}), respectively: in Figure 13, comparing M_{BH}^ψ sensitivity indices for objective functions tip and bearing hub deflections (OF_B^{1-4} versus OF_B^{5-8}), the influence of the bearing defect M_{BH}^{ψ} in bearing hub is more than the tip. This is due to direct influence of the defect in the bearing by imposing the defect torque, which means that the bearing may experience more damage by introducing a defect in the inner ring.

(ii) Frequency response of the tip deflection in steady state ($\mathrm{OF}_B^{9\text{-}10}$) and shutdown ($\mathrm{OF}_B^{11\text{-}12}$) scenarios: in FFT of tip deflection, bearing defect has shown different influence compared to the healthy case (no defect). This means that the frequency response of structure in both healthy and faulty conditions is different and by studying FFT of system response and investigating their sensitivity indices, we could detect some faults sources in specific components.

5.4. HSS Torsional Model. The torsional model has been studied based on the faults sources introduced in the previous sections. Backlash effects on coupling creating nonlinearities to the system structure, motor torque ripple with specific peaks in the frequency domain, and bearing defect disturbance torque imposing on the rotor angular direction are applied within system modelling. In order to optimize the design and upgrade the structural performance, the system response must be studied based on these assumptions and compared with the healthy conditions. This gives a better insight to understand the faults sources and could facilitate the way to prevent the failures in drive train by creating structures which are less sensitive with respect to these faults sources. This could be studied within GSA by investigating and optimizing the sensitivity indices.

The sensitivity analysis for torsional model has been performed based on 3 different operational scenarios (see Figure 6) for motor speeds in steady state regime. In order to investigate the torsional model sensitivity analysis, the input parameters appearing in the governing equations corresponding to the model are chosen as follows:

$$\mathbf{X}_T = \left[J_r \ \ J_g \ \ m_e \ \ k_\phi \ \ C_{br}^f \ \ C_{bg} \ \ M_{\mathrm{rip}}^{(1)} \ \ M_{\mathrm{rip}}^{(2)} \ \ M_{\mathrm{rip}}^{(3)} \ \ \mathrm{BL} \ \ M_{\mathrm{BH}}^{\phi_r} \right]. \tag{23}$$

5.4.1. Torsion Vibration Model of Drive Train Test Rig with Backlash. Upon considering only the torsional response of the test rig, the complete model in Section 2.1 is reduced, a lumped torsion vibration model. An engineering abstract of the model is depicted in Figure 14 and entails two inertias, disk J_r and generator rotor J_g, respectively. Each of these inertias is supported by bearings. The shaft and coupling flexibility is lumped together and represented by a torsion spring. Due to gravity and eccentric mass, there will be an external torque $M_r (=m_e r_e g \cos \phi_r)$ acting on the disk, and the induced electrical torque from the motor M_g acts on the generator rotor.

Hence, we have the following set of equations to describe the test rig torsional vibration response:

$$\left[J_r + m_e r_e^2 \right] \ddot{\phi}_r + C_{br}^f \dot{\phi}_r + k_\phi \left(\phi_r - \phi_g \right) = M_r (t),$$

$$J_g \ddot{\phi}_g + C_{bg} \dot{\phi}_g + k_\phi \left(\phi_g - \phi_r \right) = M_g (t). \tag{24}$$

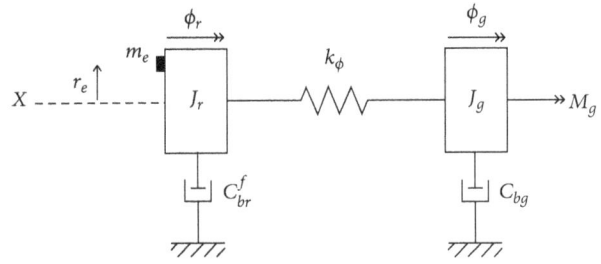

FIGURE 14: Engineering abstract of the test rig in torsional vibration.

Here, we define $C_{br}^f (=C_{br} + C_{bf})$ as combination structural damping in rotor and the bearing damping due to the friction.

Based on the results (see Figure 15), it could be demonstrated that the most influential input parameters to the torsional objective functions are coupling torsional stiffness k_ϕ and backlash in coupling BL, motor torque ripples $M_{\mathrm{rip}}^{(i)}$, and bearing defect parameter $M_{\mathrm{BH}}^{\phi_r}$. Also, the structural damping of the rotor and bearing friction coefficient C_{br}^f has some effects to both primary and total sensitivity indices. The effects of damping coefficient (except the bearing damping) and rotor and motor mass inertia values (J_r and J_g) are not significant compared to the other aforementioned parameters. In total sensitivity S_{T_i}, also eccentric mass has shown some impact which is expectable since it imposes an external torque to the structure.

More detailed analysis of the results based on each objective functions and influence of the input parameters is stated as follows.

(i) Coupling Torque in Transient and Steady State Regimes ($\mathrm{OF}_T^{1\text{-}2}$). The coupling torque has strong sensitivity in terms of torque ripple parameters $M_{\mathrm{rip}}^{(i)}$ ($i = \{1, 2, 3\}$) arising from the motor. This tells the significance of motor ripple assumption in drive trains as one type of fault source.

(ii) Coupling Torque and Its FFT in Shutdown Regime ($\mathrm{OF}_T^{3,6}$). It is clear that in shutdown regime there is no any torque ripple imposed by the motor, of which objective functions OF_T^3 and OF_T^6 sensitivity indices for torque ripples are identically zero.

In FFT, coupling torque influence is more than steady state case, since there is no ripple imposed by the motor. Also eccentric mass has some contribution especially in total sensitivity index.

(iii) Frequency Response of the Coupling Torque in Steady State Regime ($\mathrm{OF}_T^{4\text{-}5}$). In FFT of the coupling torque in steady state regime, the most dominant parameter is k_ϕ. Then the backlash and motor torque ripples have significant contribution. This tells that in vibration analysis it is very important to identify the vibration faults sources namely here backlash and ripple, which may affect the system structure in a severe way.

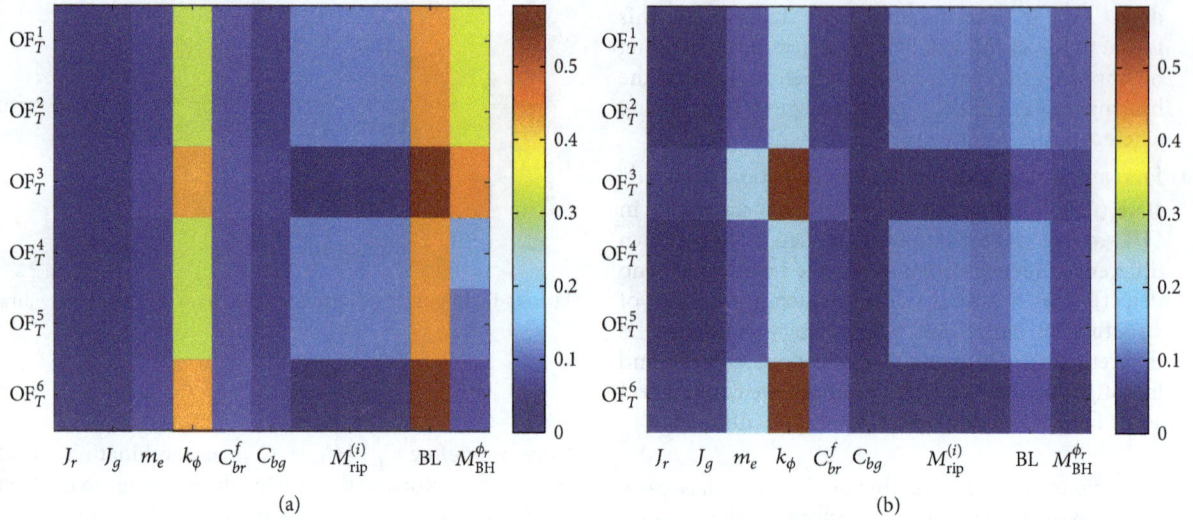

FIGURE 15: Primary sensitivity S_i (a) and total sensitivity ST_i (b) indices for torsional model for OS_1 with backlash.

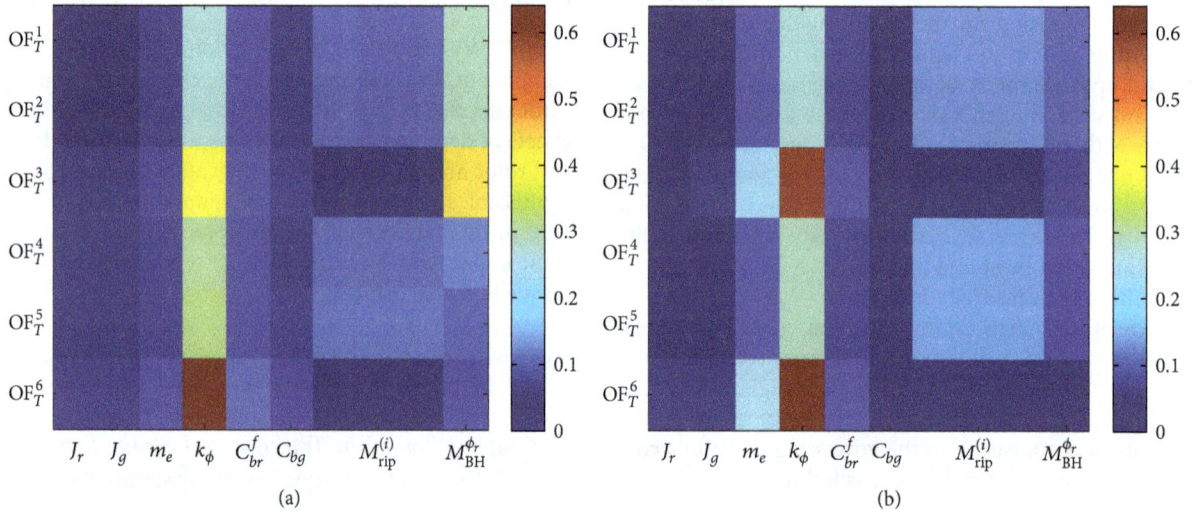

FIGURE 16: Primary sensitivity S_i (a) and total sensitivity ST_i (b) indices for torsional model for OS_1 with no backlash.

(iv) Bearing Defect Maximum Value $M_{BH}^{\phi_r}$. By increasing the bearing defect maximum torque (Figure 5), we could investigate its contribution to the sensitivity indices. The aim is to define the minimum value of defect parameter M_{BH}^{max} such that the sensitivity indices related to this parameter would be larger than a specific value ($M_{BH}^{max} > 100$ Nm). This methodology contributes to detect, predict, and prevent the faults and failures in bearings.

Summing up the results of GSA for torsional model, we could state that the most influential parameters are k_ϕ and faults sources BL and $M_{rip}^{(i)}$ and also C_{bf}, respectively.

For disturbance torque coming from bearing housing $M_{BH}^{\phi_r}$, we notice that its influence is more in total sensitivity index rather than primary one. This means that $M_{rip}^{(i)}$ influence is more when the other parameters are changed. The same conclusion could be made for ripple parameters $M_{rip}^{(i)}$.

5.4.2. Torsional Model with No Backlash. In order to emphasize into the bearing fault detection, we study the system structure response based on no backlash (Figure 16). Then, we investigate the influence of the input parameters to the objective functions in case when the shaft coupling has a backlash.

The obtained results have led to the following conclusions:

(i) The bearing defect (quantified in terms of $M_{BH}^{\phi_r}$) could be detected easier in case where no backlash compared to with backlash (more contribution sensitivity indices especially the primary one). This is very important conclusion and means that, in the systems with a couple of sources of faults, the detectability of each fault may not be easy and it get more complex by introducing more complex system and assuming more faults.

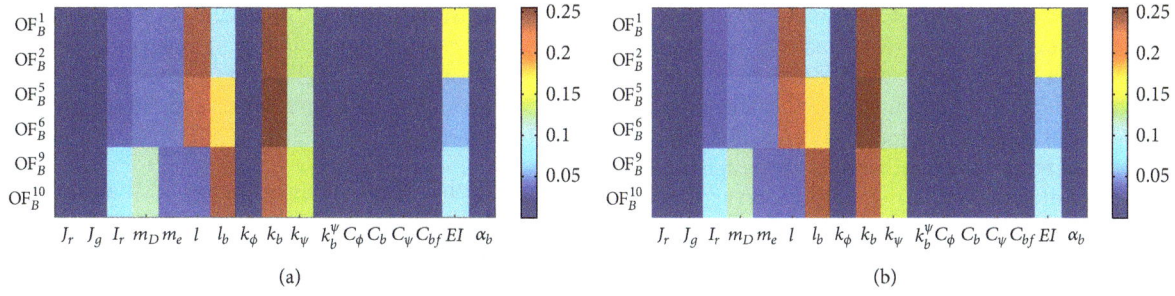

FIGURE 17: Primary sensitivity S_i (a) and total sensitivity ST_i (b) indices for bending model for h_{res}.

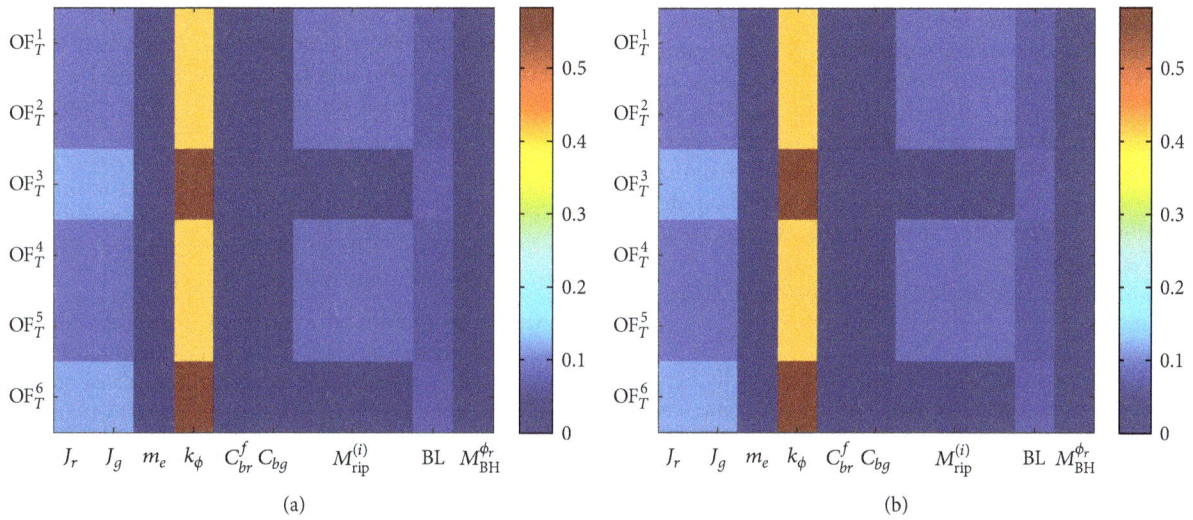

FIGURE 18: Primary sensitivity S_i (a) and total sensitivity ST_i (b) indices for torsional model for h_{res}.

(ii) The influence of bearing defect in the FFT for first peak and the second peaks is small, though recognizable. This draws the same conclusion for the bending model with defect, which tells that some specific peaks in the FFT correspond to the system structural faults.

5.5. Special Cases: Resonance Investigation. The operating speed ranges in wind turbines and system excitation frequency have a wide range. The system may encounter resonances leading to severe vibration and affect the normal operation of the machine. Thus, the drive train dynamics must be studied with more focus in excitation ranges.

To focus on the excitation ranges, we investigate the primary and secondary sensitivity indices within the case in which the input parameters leads to resonance. It means that we only take the randomized input parameters which are near to the resonance frequency of the system in terms of bending and torsion:

$$X_{res} = \{X \text{ such that norm} (eig (SYS (X))$$
$$- eig (SYS (c))) \leq TOL\}, \quad (25)$$

where $SYS(X)$ and $SYS(c)$ denote for the system eigenvalue based on randomized set of X and nominal structural

parameters presented in Table 1, respectively. Also TOL = 1 Hz is considered as a tolerance for frequency fluctuation.

Some conclusions could be made based on aforementioned cases, and the results are presented in Figures 17 and 18.

(i) By selecting those randomized variables which based on them the eigenfrequency of the structure is near to specific value (50 Hz), the sensitivity of eigenfrequency variables takes the most contribution. This is expectable since in the resonance points the system influential parameters are only the ones which the eigenfrequency of the system is based on them.

(ii) In optimum case, where the randomized values are identical to the bending and torsional eigenfrequencies, we expect that only the input parameters have influence that they are in the eigenfrequency definition.

6. Conclusion and Outlook

The global sensitivity analysis for both torsional and bending models of high speed shaft drive train test rig with respect to its structural parameters has been carried out using M-DRM method.

The current study showed the feasibility and efficiency of the M-DRM for global sensitivity analysis of nonlinear dynamical system (high speed shaft drive train), which provided the results in a computationally efficient framework. The M-DRM applied here revealed practical significant results which can reduce the number of input design parameters for optimization of wind turbine drive train systems.

By comparing the sensitivity of objective functions with respect to different input structural parameters, a better understanding of drive train system dynamics is obtained. Vertical and lateral deflections in bearing housing and tip, coupling torque in 3 different operational scenarios, FFT of these functions, have been introduced as the objective functions to represent the test rig dynamic performance. To reflect the effect of operational scenarios, the aforementioned outputs have been evaluated within transient, steady state, and shutdown cases in 3 different motor speeds. The effect of the fundamental structural parameters on the dynamic response of the HSS drive train test rig has been recognized. By comparing the sensitivity indices of corresponding objective functions, a better understanding of drive train system dynamics is obtained.

Based on the present analysis, the following conclusions could be stated:

(i) GSA for complex systems is computationally expensive by using ordinary algorithms like Monte Carlo. Nevertheless, M-DRM method has been successfully applied to high speed shaft drive train model and provided physically reasonable results in an efficient computational framework.

(ii) For considered HSS drive train system model, the estimated primary and total sensitivity indices coincide, revealing no variance interaction effects between the input parameters within different operational scenarios. Also there is a direct correspondence between S_i and S_{T_i}, which means that the higher order effects are negligible.

(iii) Deflection fields in both tip and bearing hub are mostly sensitive with respect to bearing mounting stiffness k_b and coupling bending stiffness k_ψ and geometric parameters l, l_b and shaft bending stiffness EI.

(iv) Coupling torque mostly in all regimes is influenced mainly by coupling torsional stiffness k_ϕ, coupling backlash BL, and motor torque ripples $M_{rip}^{(i)}$, respectively. This means that the proposed faults sources may lead to drastic contributions in components loads. In this research we have quantified these contributions.

(v) Dynamic modelling and vibration responses simulation are important for fault mechanism studies to provide proofs for defect detection and fault diagnose. The most significant demonstration in the presented research is the applicability of GSA towards defect detection and faults diagnose in functional components of drive train. The results suggest that, from sensitivity indices, detectability of some faults such as bearing defect could be analysed. This analysis illustrates the applicability of the GSA method for advancing modern condition monitoring systems for bearings.

The global sensitivity analysis provided a useful method to deeply understand some aspects of the initial structural parameters to the dynamic outputs. At this stage, a preliminary assessment of the key parameters has been performed. Moreover, fault modelling in functional components such as motor ripples, bearing defects, coupling backlash, and their detectability within GSA has been investigated. This facilitates understanding the faults sources and their effect to other functional components, which leads to better design and optimize drive trains more efficiently with less failure.

Finally, the items for future work could be outlined as follows:

(i) Apply GSA for real wind turbine structural parameters such that we first propose to apply the same analysis for NREL 5 MW wind turbine and investigate GSA sensitivity indices.

(ii) Carry out dimensional analysis and upscaling of the test rig and apply GSA for Pi Buckingham theory: the aim is to prescribe system structure based on minimal number of dimensionless parameters, which could lead to upscale the structure and predict the real wind turbine behaviours.

(iii) Apply wind turbine aerodynamic loads as an external load to the rotor (M_r): in order to get more realistic response of the system, considering the stochastic nature of wind loads, it would be interesting to study the current model based on realistic calculated aerodynamic loads extracted from FAST software.

(iv) Study the bearing modelling in more advanced and detailed approach: the bearing defects here have been modelled by introducing an radial excitation to certain DOFs imposed on inner ring of the bearing. For future work, it is important to consider also ball mass inertia and outer ring dynamics [40, 41].

(v) Model the electromechanical interaction of wind turbine drive train: since wind turbine contains both mechanical and electrical components, to have better understanding of drive train dynamics, the study must be done within this interaction.

Conflicts of Interest

The authors declare that they have no conflicts of interest.

Acknowledgments

This project is financed through the Swedish Wind Power Technology Centre (SWPTC). SWPTC is a research centre for design of wind turbines. The purpose of the centre is to support Swedish industry with knowledge of design techniques as well as maintenance in the field of wind power.

The centre is funded by the Swedish Energy Agency and Chalmers University of Technology as well as academic and industrial partners. In particular, a donation of motors and frequency converters by ABB CR, WindCon3.0 condition monitoring system, and set of sensors by SKF is gratefully acknowledged. Besides, the authors would like to thank Jan Möller for his contribution in the laboratory.

References

[1] International Renewable Energy Agency, *Wind Power*, vol. 1, Power sector, Issue 5/5 of *Renewable energy technologies: Cost analysis series*.

[2] F. Dincer, "The analysis on wind energy electricity generation status, potential and policies in the world," *Renewable & Sustainable Energy Reviews*, vol. 15, no. 9, pp. 5135–5142, 2011.

[3] J. Ribrant and L. Bertling, "Survey of failures in wind power systems with focus on Swedish wind power plants during 1997-2005," in *Proceedings of the 2007 IEEE Power Engineering Society General Meeting, PES*, USA, June 2007.

[4] J. L. M. Peeters, D. Vandepitte, and P. Sas, "Analysis of internal drive train dynamics in a wind turbine," *Wind Energy*, vol. 9, no. 1-2, pp. 141–161, 2006.

[5] F. Oyague, "Gearbox Modeling and Load Simulation of a Baseline 750-kW Wind Turbine Using State-of-the-Art Simulation Codes," Tech. Rep. NREL/41160, 2009.

[6] W. Shi, H.-C. Park, S. Na, J. Song, S. Ma, and C.-W. Kim, "Dynamic analysis of three-dimensional drivetrain system of wind turbine," *International Journal of Precision Engineering and Manufacturing*, vol. 15, no. 7, pp. 1351–1357, 2014.

[7] S. Asadi, *Drive train system dynamic analysis: application to wind turbines, Licentiate Thesis*, Chalmers University of Technology, 2016.

[8] F. Oyague, *Fault Identification in Drive Train Components Using Vibration Signature Analysis*, TP-500-41160, National Renewable Energy Laboratory, 2009.

[9] K. G. Scott, D. Infield, N. Barltrop, J. Coultate, and A. Shahaj, "Effects of extreme and transient loads on wind turbine drive trains," in *Proceedings of the 50th AIAA Aerospace Sciences Meeting Including the New Horizons Forum and Aerospace Exposition*, January 2012.

[10] G. Semrau, S. Rimkus, and T. Das, "Nonlinear Systems Analysis and Control of Variable Speed Wind Turbines for Multiregime Operation," *Journal of Dynamic Systems, Measurement, and Control*, vol. 137, no. 4, Article ID 041007, 2015.

[11] T. M. Ericson and R. G. Parker, "Natural frequency clusters in planetary gear vibration," *Journal of Vibration and Acoustics*, vol. 135, no. 6, Article ID 061002, 2013.

[12] M. Singh and S. Santoso, "Dynamic Models for Wind Turbines and Wind Power Plants," Tech. Rep. NREL/SR-5500-52780, 2011.

[13] J. Wang, D. Qin, and Y. Ding, "Dynamic behavior of wind turbine by a mixed flexible-rigid multi-body model," *Journal of System Design and Dynamics*, vol. 3, no. 3, pp. 403–419, 2009.

[14] S. Struggl, V. Berbyuk, and H. Johansson, "Review on wind turbines with focus on drive train system dynamics," *Wind Energy*, vol. 18, no. 4, pp. 567–590, 2015.

[15] C. Zhu, Z. Huang, Q. Tang, and Y. Tan, "Analysis of nonlinear coupling dynamic characteristics of gearbox system about wind-driven generator," *Chinese Journal of Mechanical Engineering*, vol. 41, no. 8, pp. 203–207, 2005.

[16] B. Iooss and L. Lemaitre, "A review on global sensitivity analysis methods," in *Uncertainty management in Simulationoptimization of Complex Systems: Algorithms and Applications*, C. Meloni and G. Dellino, Eds., pp. 101–122, Springer US, 2015.

[17] C. I. Reedijk, *Sensitivity Analysis of Model Output: Performance of various local and global sensitivity measures on reliability problems, [Master, thesis]*, Delft University of Technology, 2000.

[18] F. Pianosi, K. Beven, J. Freer et al., "Sensitivity analysis of environmental models: A systematic review with practical workflow," *Environmental Modeling and Software*, vol. 79, pp. 214–232, 2016.

[19] E. Borgonovo and E. Plischke, "Sensitivity analysis: a review of recent advances," *European Journal of Operational Research*, vol. 248, no. 3, pp. 869–887, 2016.

[20] X.-Y. Zhang, M. N. Trame, L. J. Lesko, and S. Schmidt, "Sobol sensitivity analysis: A tool to guide the development and evaluation of systems pharmacology models," *CPT: Pharmacometrics & Systems Pharmacology*, vol. 4, no. 2, pp. 69–79, 2015.

[21] H. M. Wainwright, S. Finsterle, Y. Jung, Q. Zhou, and J. T. Birkholzer, "Making sense of global sensitivity analyses," *Computers & Geosciences*, vol. 65, pp. 84–94, 2014.

[22] X. Zhang and M. D. Pandey, "An effective approximation for variance-based global sensitivity analysis," *Reliability Engineering & System Safety*, vol. 121, pp. 164–174, 2014.

[23] A. Saltelli, "Making best use of model evaluations to compute sensitivity indices," *Computer Physics Communications*, vol. 145, no. 2, pp. 280–297, 2002.

[24] X. Zhang and M. D. Pandey, "Structural reliability analysis based on the concepts of entropy, fractional moment and dimensional reduction method," *Structural Safety*, vol. 43, pp. 28–40, 2013.

[25] J. M. P. Dias and M. S. Pereira, "Sensitivity Analysis of Rigid-Flexible Multibody Systems," *Multibody System Dynamics*, vol. 1, no. 3, pp. 303–322, 1997.

[26] K. D. Bhalerao, M. Poursina, and K. . Anderson, "An efficient direct differentiation approach for sensitivity analysis of flexible multibody systems," *Multibody System Dynamics*, vol. 23, no. 2, pp. 121–140, 2010.

[27] D. Bestle and J. Seybold, "Sensitivity analysis of constrained multibody systems," *Archive of Applied Mechanics*, vol. 62, no. 3, pp. 181–190, 1992.

[28] S. M. Mousavi Bideleh and V. Berbyuk, "Global sensitivity analysis of bogie dynamics with respect to suspension components," *Multibody System Dynamics*, vol. 37, no. 2, pp. 145–174, 2016.

[29] P. M. McKay, R. Carriveau, D. S.-K. Ting, and J. L. Johrendt, "Global sensitivity analysis of wind turbine power output," *Wind Energy*, vol. 17, no. 7, pp. 983–995, 2014.

[30] K. Dykes, A. Ning, R. King, P. Graf, G. Scott, and P. S. Veers, "Sensitivity Analysis of Wind Plant Performance to Key Turbine Design Parameters: A Systems Engineering Approach," in *Conference Paper, NREL/CP-5000-60920, 2014, Contract No. DE-AC36-08GO28308*, National Harbor, Maryland, USA, 2014.

[31] J. Helsen, P. Peeters, K. Vanslambrouck, F. Vanhollebeke, and W. Desmet, "The dynamic behavior induced by different wind turbine gearbox suspension methods assessed by means of the flexible multibody technique," *Journal of Renewable Energy*, vol. 69, pp. 336–346, 2014.

[32] B. Marrant, The validation of MBS multi-megawatt gearbox models on a 13.2 MW test rig, Simpack User Meeting, 2012, 3.

[33] S. Asadi, V. Berbyuk, and H. Johansson, "Vibration Dynamics of a Wind Turbine Drive Train High Speed Subsystem: Modelling

and Validatio," in *Proceedings of the ASME, International Design Engineering Technical Conferences and Computers and Information in Engineering Conference*, pp. C2015–46016, August, Boston, Massachusetts, USA, 2015.

[34] S. Asadi, V. Berbyuk, and H. Johansson, "Structural dynamics of a wind turbine drive train high speed subsystem: Mathematical modelling and validation," in *Proc. of the International Conference on Engineering Vibration, Ljubljana, 7 - 10 September; [editors Miha Boltezar, Janko Slavic, Marian Wiercigroch] - EBook - Ljubljana*, pp. 553–562, 2015.

[35] M. Todorov, I. Dobrev, and F. Massouh, "Analysis of torsional oscillation of the drive train in horizontal-axis wind turbine," in *Proceedings of the 2009 8th International Symposium on Advanced Electromechanical Motion Systems and Electric Drives Joint Symposium, ELECTROMOTION 2009*, France, July 2009.

[36] T. A. Silva and N. M. Maia, "Elastically restrained Bernoulli-Euler beams applied to rotary machinery modelling," *Acta Mechanica Sinica*, vol. 27, no. 1, pp. 56–62, 2011.

[37] T. Iwatsubo, Y. Yamamoto, and R. Kawai, "Start-up torsional vibration of rotating machine driven by synchronous motor," in *Proc. of the Int. Conference on Rotordynamics, IFToMM*.

[38] T. C. Kim, T. E. Rook, and R. Singh, "Effect of smoothening functions on the frequency response of an oscillator with clearance non-linearity," *Journal of Sound and Vibration*, vol. 263, no. 3, pp. 665–678, 2003.

[39] B. Ghalamchi, J. Sopanen, and A. Mikkola, "Simple and versatile dynamic model of spherical roller bearing," *International Journal of Rotating Machinery*, vol. 2013, Article ID 567542, 13 pages, 2013.

[40] S. P. Harsha, "Nonlinear dynamic analysis of an unbalanced rotor supported by roller bearing," *Chaos, Solitons & Fractals*, vol. 26, no. 1, pp. 47–66, 2005.

[41] L. Niu, H. Cao, Z. He, and Y. Li, "Dynamic modeling and vibration response simulation for high speed rolling ball bearings with localized surface defects in raceways," *Journal of Manufacturing Science and Engineering—Transactions of the ASME*, vol. 136, no. 4, Article ID 041015, 2014.

[42] Rolling bearings (general catalogue published by SKF group), 2012.

[43] A. Saltelli, P. Annoni, I. Azzini, F. Campolongo, M. Ratto, and S. Tarantola, "Variance based sensitivity analysis of model output. Design and estimator for the total sensitivity index," *Computer Physics Communications*, vol. 181, no. 2, pp. 259–270, 2010.

[44] A. Saltelli and P. Annoni, "How to avoid a perfunctory sensitivity analysis," *Environmental Modeling and Software*, vol. 25, no. 12, pp. 1508–1517, 2010.

[45] I. M. Sobol, "Theorems and examples on high dimensional model representation," *Reliability Engineering & System Safety*, vol. 79, no. 2, pp. 187–193, 2003.

[46] H. Rabitz and O. F. Alis, "General foundations of high-dimensional model representations," *Journal of Mathematical Chemistry*, vol. 25, no. 2-3, pp. 197–233, 1999.

[47] I. M. Sobol, "Global sensitivity indices for nonlinear mathematical models and their Monte Carlo estimates," *Mathematics and Computers in Simulation*, vol. 55, no. 1-3, pp. 271–280, 2001.

[48] H. Liu and W. Chen, "Probabilistic Sensitivity Analysis Methods for Design under Uncertainty, Tech B224," Tech. Rep., Integrated Design Automation Laboratory, Department of Mechanical Engineering, Northwestern University.

[49] A. Saltelli and I. M. Sobol', "About the use of rank transformation in sensitivity analysis of model output," *Reliability Engineering & System Safety*, vol. 50, no. 3, pp. 225–239, 1995.

[50] T. Homma and A. Saltelli, "Importance measures in global sensitivity analysis of nonlinear models," *Reliability Engineering & System Safety*, vol. 52, no. 1, pp. 1–17, 1996.

[51] B. Sudret, "Global sensitivity analysis using polynomial chaos expansions," *Reliability Engineering & System Safety*, vol. 93, no. 7, pp. 964–979, 2008.

[52] http://www.climatetechwiki.org/technology/offshore-wind.

17

Influence of End Structure on Electromagnetic Forces on End Winding of a 1550 MW Nuclear Generator

Chong Zeng,[1] **Song Huang,**[1] **Yongming Yang,**[1] **and Guanghou Zhou**[2]

[1]*Chongqing University, Chongqing 400044, China*
[2]*Research and Test Center, Dongfang Electrical Machinery Co., Ltd., Deyang, Sichuan Province 618000, China*

Correspondence should be addressed to Song Huang; s.huang@cqu.edu.cn

Academic Editor: Paolo Pennacchi

A 3D electromagnetic model of the end region of a 1550 MW nuclear generator is set up. The electromagnetic forces on the involute and nose parts of the end winding under a rated operation are obtained through the 3D time-step finite element method. The electromagnetic forces on different coils in the same phase are analyzed. By changing the rotor's relative length and stator coil's linear length in the 3D electromagnetic model, the electromagnetic force distributions on the end winding are obtained. The influence of each structure change on the electromagnetic force in different directions is studied in detail. Conclusions that can be helpful in decreasing the electromagnetic forces on the end winding through optimizing the end region design are presented.

1. Introduction

Vibrations caused by electromagnetic forces can severely damage end-winding insulations [1, 2]. Electromagnetic force increases significantly with the increase of generator capacity, thereby making large generators more vulnerable to vibration. Vibration should be decreased to ensure the security of end winding. Simply enhancing and increasing the support structures can decrease the damage of electromagnetic force but will lead to other problems, such as low utilization of materials and high manufacturing cost [3]. Optimizing the electromagnetic design for the end region can decrease electromagnetic force and vibration without extra cost. Research on the influence of end structure on the electromagnetic force on end winding can be a reference for optimization.

The accurate calculation of electromagnetic field in the end region and electromagnetic force on end winding is the basis of studies on electromagnetic force. Many studies focus on calculation.

In the early years, analytical methods were used to calculate the electromagnetic field in the end region. Methods based on the Biot-Savart law are the most popular among them [4–6]. However, the errors in the results obtained through analytical methods are large because the end region's structure is complex. To improve the accuracy, numerical methods are utilized in the calculation. Recently, 3D finite element (FE) method has become the most popular method. Wang et al. [7] proposed full 3D models and methods for coupled electromagnetic and temperature fields in the end region of a large hydrogenerator. They discussed the effect of mesh size on the result of eddy current loss and obtained several meshing principles. Waldhart et al. [8] analyzed the influence of several common simplifications on the accuracy of the 3D FE method's results and found that the geometric details on the inner diameter, the accurate modeling of the stator winding, and the correct BH-curve of the eddy current material are essential to the electromagnetic field. Liang et al. [9] built a fine 3D geometrical model and calculated the magnetic field and eddy current loss in the end region. Wang et al. [10–14] calculated the electromagnetic field and the eddy current losses in the end region of a 330 MW turbogenerator. The influence of underexcitation [10], copper shield [11–13], and clamping plate [14] on the electromagnetic field and eddy current losses was calculated and discussed. Using the 3D time-step FE method, Huang et al. calculated the magnetic field and the eddy current loss on the end components of a 1550 MW nuclear generator and analyzed the influences of different structural parameters on the magnetic field and the eddy current loss [15].

Relatively few studies focus on the electromagnetic force on end winding. Simplifying the end coil as straight line inductors, Richard et al. [16, 17] calculated the electromagnetic force on the end winding of a 600 MW turbogenerator using the 3D electromagnetic model of the stator end region. Senske et al. [18], Drubel et al. [19], and Grüning and Kulig [20] calculated the electromagnetic force on end winding based on the Biot-Savart law to line circuit segments. Albanese et al. [21] proposed a numerical approach based on an integral formulation to calculate the magnetic field and forces on the end winding of a large generator. Zhao et al. [22] built a 3D electromagnetic model of a 600 MW turbogenerator and calculated the electromagnetic force on the end winding using the 3D time-step FE method. Furthermore, the vibrations of the end winding were calculated and subjected to 3D FE structural analysis. None of the studies above analyzed the influence of the end structure on the electromagnetic force on end winding.

In this paper, a 3D FE model of the end region of a 1550 MW nuclear generator is set up. The electromagnetic force on the end winding at rated operation is obtained by the FE method. The rotor relative length and the stator coil linear length are set to different values in the FE model. The radial, tangential, and axial force density distributions in different parts of the end winding are presented. The influence of the two parameters on the electromagnetic force on end winding is analyzed. Conclusions that can help in decreasing the electromagnetic forces on end winding through optimal design are presented.

2. 3D Finite Element Model for Electromagnetic Forces Calculation

2.1. Physical Model. The end region of a half-speed nuclear generator is quite complex. The end part of cores and windings is 3D in structure. The end region contains three end components: copper shield, press plate, and press finger. The structure of the end region is shown in Figure 1.

2.2. Mathematical Model for Electromagnetic Forces Calculation. To improve computational efficiency, the following are assumed:

(i) The distribution of current in the stator and rotor windings is uniform.

(ii) The high-order harmonics of current and displacement current are ignored.

(iii) The hysteresis effect is ignored. The core material is isotropic [11].

(iv) The cross section of the solution domain of the magnetic field calculation is shown in Figure 2.

The solution domain is shown in Figure 2. S_1, S_2, and S_3 are the boundaries of the solution domain. By adopting A,

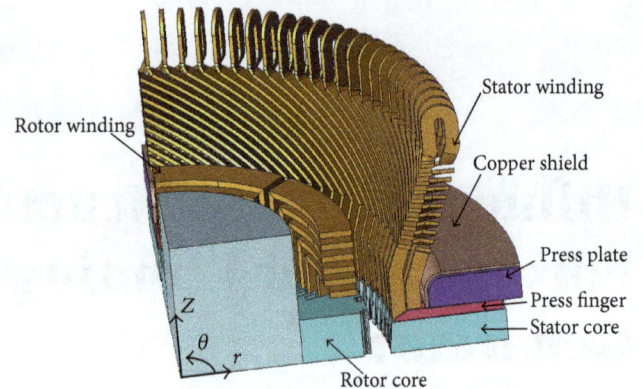

FIGURE 1: End region of a half-speed nuclear generator.

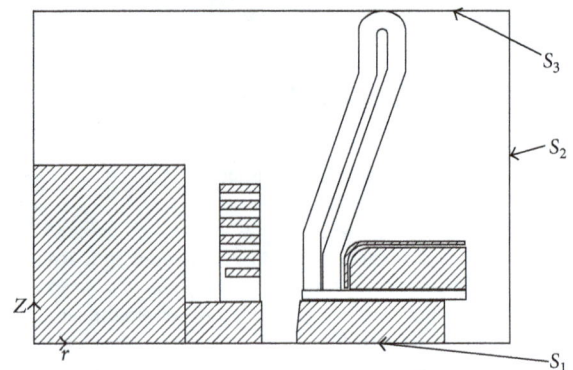

FIGURE 2: Cross section of the solution domain.

φ-A method, we can express the mathematical model of the end region as follows [23, 24]:

$$\nabla \times \left(\frac{1}{\mu} \nabla \times \mathbf{A} \right) + \sigma \frac{\partial \mathbf{A}}{\partial t} + \sigma \nabla \varphi = \mathbf{J}_s,$$

$$\nabla \cdot \left(-\sigma \frac{\partial \mathbf{A}}{\partial t} - \sigma \nabla \varphi \right) = 0, \qquad (1)$$

where $\mathbf{A}$ denotes the magnetic vector potential, φ is the electric scalar potential, σ is the electrical conductivity, ω is the angular frequency, and $\mathbf{J}_s$ is the source current density.

The axial flux is very small in S_1. Therefore, the symmetric boundary condition, which means no flux passes across the face, is set on S_1. S_2 and S_3 are far from the stator and rotor cores, and their magnetic fields are weak. Therefore, S_2 and S_3 satisfy the first boundary condition $\mathbf{A} = 0$.

The stator current cannot be calculated from the above model because the solution domain only contains the end region. Thus, an additional 2D model of the generator is required. The solution domain of the 2D model is shown in Figure 3, where A_1 is the inner area and E_1 is the outer edge of the solution domain.

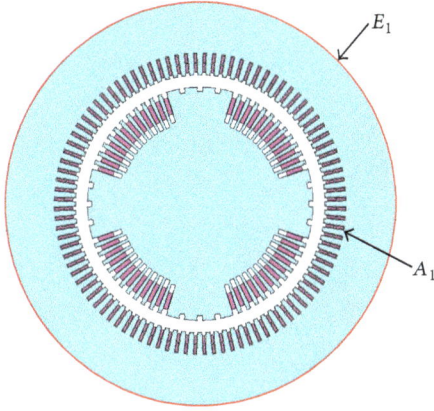

FIGURE 3: Solution domain of the 2D model.

TABLE 1: Parameters of the 1550 MW nuclear generator.

Parameters	Values
Power	1550 MW
Voltage	27 kV
Frequency	50 Hz
Number of poles	4
Current	36926.9 A
Power factor	0.9 (lagging)

TABLE 2: Iteration result of stator and rotor currents.

Parameters	Values
Rotor current	6668 A
θ	$-246°$

TABLE 3: Comparison between the calculated and measured value of magnetic density.

Probes	Calculated result	Measured result	Relative error
A	0.25 T	0.27 T	−7.41%
B	0.70 T	0.76 T	−7.89%
C	0.77 T	0.79 T	−2.53%

The mathematical model can be described as follows:

$$\frac{\partial}{\partial x}\left(v\frac{\partial \mathbf{A}}{\partial x}\right) + \frac{\partial}{\partial y}\left(v\frac{\partial \mathbf{A}}{\partial y}\right) = -\mathbf{J}_s$$

$$\sigma \nabla \varphi = \mathbf{J}_s \qquad (2)$$

$$(\text{in } A_1)$$

$$\mathbf{A} = 0 \quad (\text{on } E_1),$$

$\mathbf{J}_s$ is determined by the stator current and rotor current. When a generator operates at rated condition, the rotor current is constant and the 3-phase stator currents are given as follows:

$$i_A = I_m \sin(\omega t + \theta),$$

$$i_B = I_m \sin(\omega t + \theta - 120°), \qquad (3)$$

$$i_C = I_m \sin(\omega t + \theta - 240°),$$

where θ is the angle between i_A and d-axis, ω is the synchronous angular velocity, t is time, and I_m is the magnitude of stator phase current.

I_m can be calculated with the following formula:

$$I_m = \frac{P}{(3/\sqrt{2})\,U\cos\varphi}, \qquad (4)$$

where P is the rated power, U is the rated voltage, and φ is the power factor angle at rated condition.

Besides, θ and the rotor current can be calculated by iteration.

With the magnetic field in the end region, the electromagnetic force density can be calculated as

$$\mathbf{f} = \frac{I}{\mathbf{S}} \times \mathbf{B}, \qquad (5)$$

where $\mathbf{f}$ is the electromagnetic force density, I is the current in the coil, $\mathbf{S}$ is the cross-sectional area of the coil, and $\mathbf{B}$ is the magnetic flux density.

2.3. Modeling of the Proposed Generator. A 1550 MW nuclear generator is adopted because the electromagnetic force in a large generator is large. The basic parameters of the generator are shown in Table 1.

JMAG 12.1 is used to build the model and perform the calculation. A 3D FE model of the end region is established per the actual design of the proposed generator. The meshing of the FE model is shown in Figure 4. Using a sliding mesh, the rotation of the rotor is considered. The whole model contains 713386 elements and 129854 nodes.

A 2D FE model of the generator is set up with ANSYS Maxwell 16.1. The stator and rotor currents are obtained by iteration at a rated condition. The result is shown in Table 2.

2.4. Validation of the Model. The experiment is performed on an 1150 MW nuclear generator because the 1550 MW nuclear generator is not available for the experiment. The FE model of the 1150 MW generator is built on the same principles, and the structure of the 1150 MW nuclear generator is almost the same as but smaller than that of the 1550 MW generator. Therefore, the relative errors of the two FE models have a similar order. Three probes are set as shown in Figure 5. The comparison of the magnetic flux densities at no-load operation between the computational and experimental results is shown in Table 3. Table 3 shows that the calculation method is reasonable and the relative error of the calculated result is within 10%.

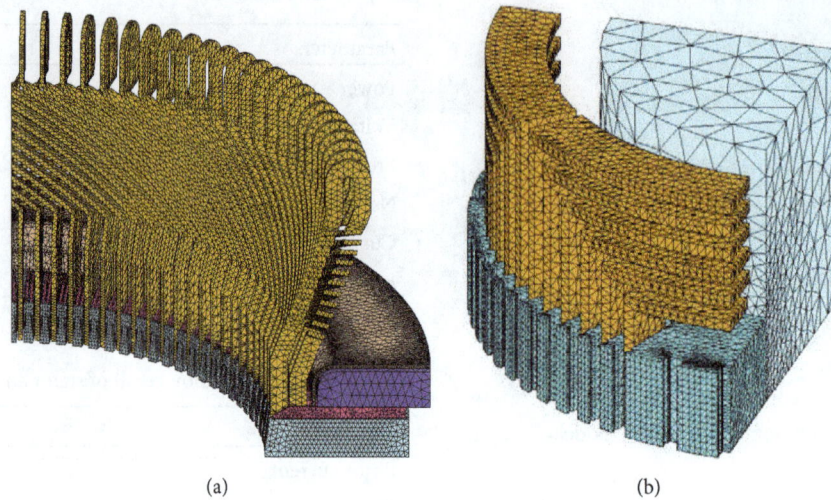

FIGURE 4: FE model mesh of the proposed generator end region. (a) Stator. (b) Rotor.

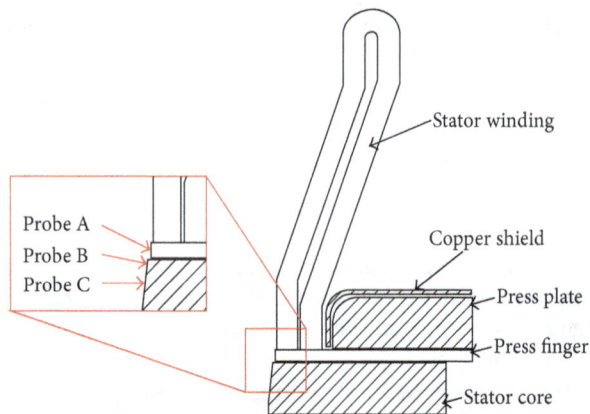

FIGURE 5: Distribution of the test points.

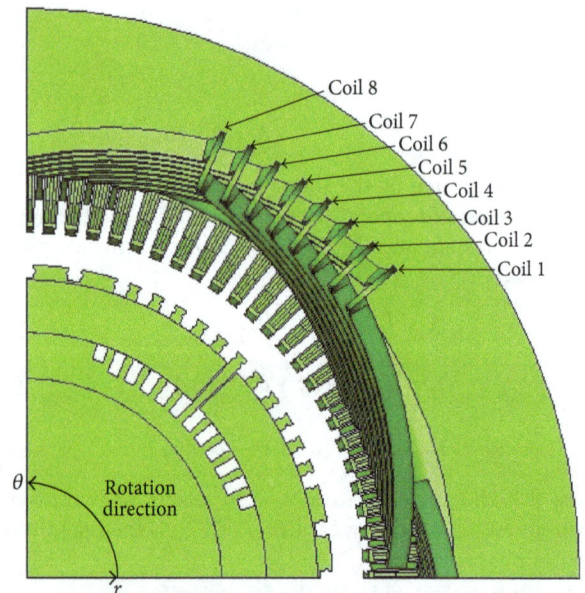

FIGURE 6: Numbering of phase A coils.

3. Electromagnetic Forces on End Winding

We should analyze only the electromagnetic force on one winding phase because the generator and the electromagnetic force on each winding phase are symmetric. The electromagnetic forces on phase A are analyzed in this paper. In the proposed generator, a winding phase has eight coils. The eight coils are numbered as shown in Figure 6. An end coil can be divided into several parts and the definition of these parts is shown in Figure 7.

The magnetic fields on different coils in the same phase are different. Therefore, the electromagnetic forces on different coils are also different, as shown in Figure 8. The forces on the involute parts change significantly. Generally, the forces on the upper involute parts increase with the coil number. Conversely, the forces on the lower involute parts decrease with the coil number. The upper involute part suffers the largest force among all the involute parts. The

electromagnetic forces on the nose parts are smaller than the forces on the involute parts, and they change slightly.

4. Parametric Study on Electromagnetic Forces on End Winding

Although the force on the nose part is smaller than that on the involute part, the vibration on the nose part is even larger than that on the involute part because of the weak constraint [22]. The influences of the end structure parameters on the electromagnetic force on the involute part and the force on

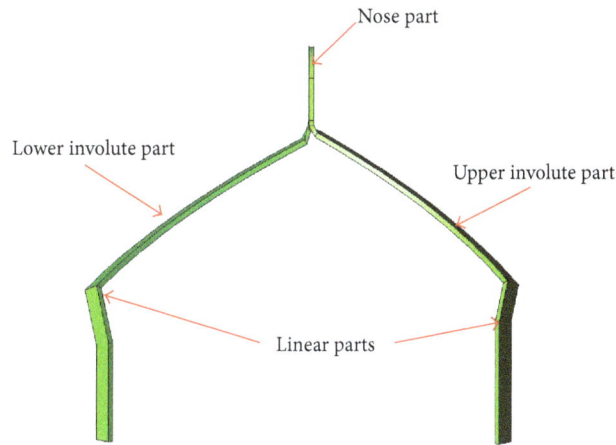

FIGURE 7: Definition of the parts of the end coil.

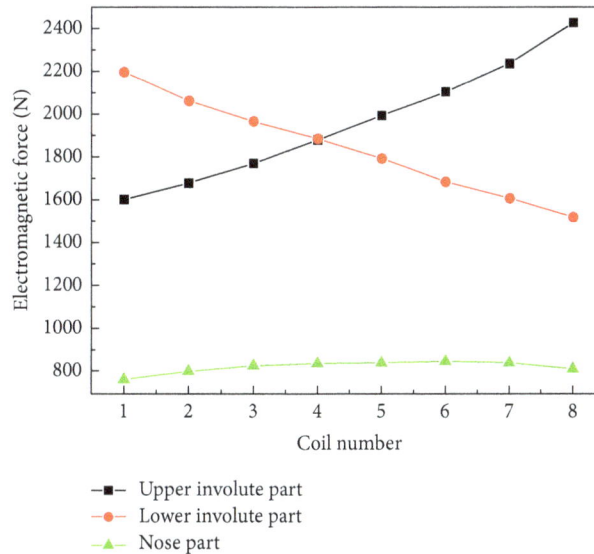

FIGURE 8: Electromagnetic forces on different coils of phase A.

the nose part are analyzed in this section. To simplify the analysis, only the forces on coil 8 are studied in detail. Only the upper involute part and the nose are analyzed because the constraints of the upper and lower involute parts are almost the same.

4.1. Influence of the Relative Length of Rotor.

Rotor relative length (l_r) is defined as shown in Figure 9. The lengths of the rotor core and winding change with l_r. When the rotor core is longer than the stator core, l_r is positive. Conversely, when the rotor core is shorter than stator core, l_r is negative. To study the influence of the rotor relative length on electromagnetic force, l_r is set to −75, 0, and 75 mm in the FE model.

The electromagnetic force distribution on the involute part at 0.039 s is shown in Figure 10. Most of the forces reach

their maximums at the same time because the current in the coil reaches the maximum at 0.039 s. Figure 10 shows that radial force is the largest and increases with l_r. The tangential and axial forces are smaller than the radial force and decrease with the increase of l_r. For the three components, the changes are more significant at the side close to the linear part (the right side of each involute part in Figure 10). When l_r varies from −75 mm to +75 mm, the maximum of the electromagnetic force density varies from 2.3×10^6 N/m^3 to 2.6×10^6 N/m^3, 1.2×10^6 N/m^3 to 1.0×10^6 N/m^3, and 1.8×10^6 N/m^3 to 1.4×10^6 N/m^3 in the radial, tangential, and axial directions, respectively.

Figure 11 shows the maximum electromagnetic forces on the involute part with different l_r. Figure 11 shows that the radial force is relatively larger than the tangential and axial

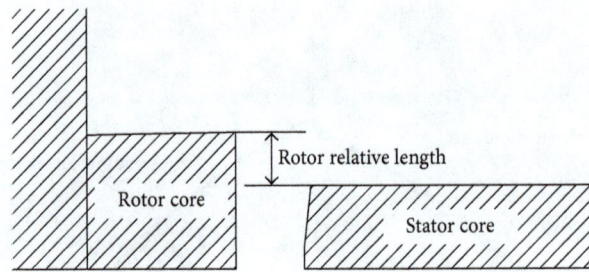

FIGURE 9: Definition of rotor relative length.

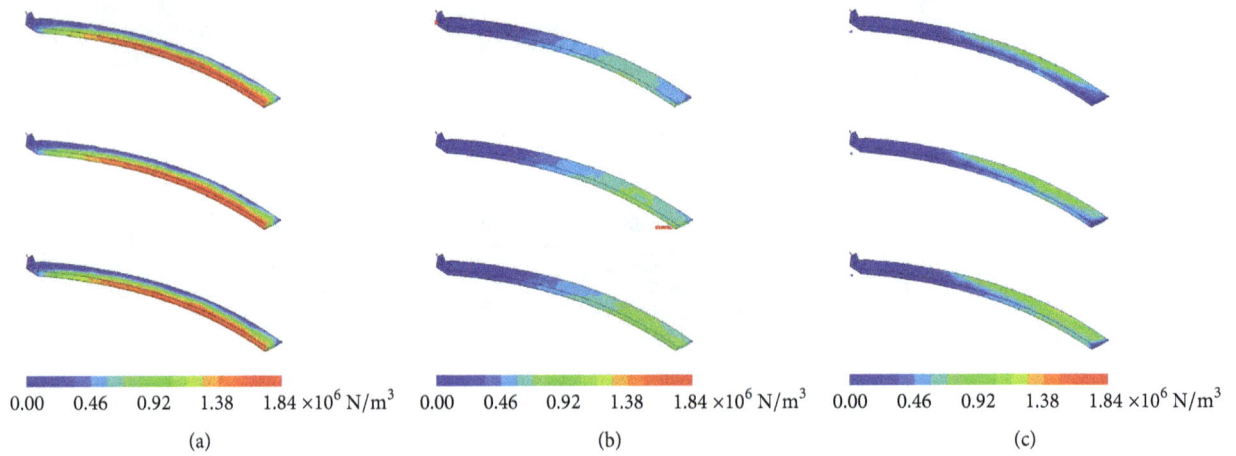

FIGURE 10: Electromagnetic force density distribution on the involute part at 0.039 s, when $l_r = -75$ mm (bottom), $l_r = 0$ mm (middle), and $l_r = 75$ mm (top). (a) Radial component. (b) Tangential component. (c) Axial component.

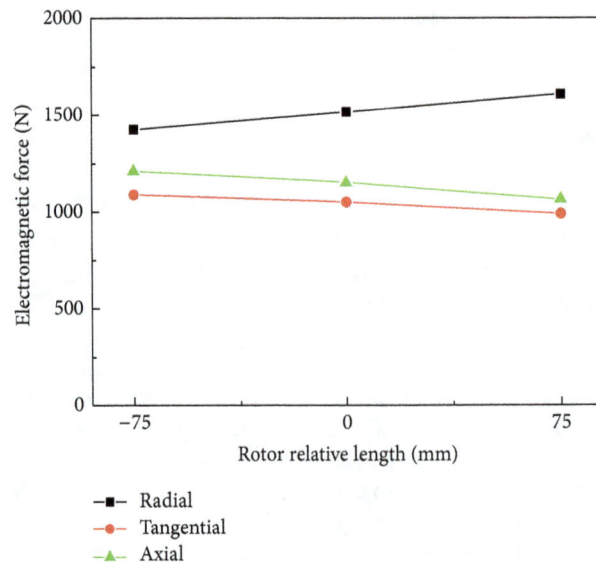

FIGURE 11: Electromagnetic forces on the involute part with different rotor relative lengths.

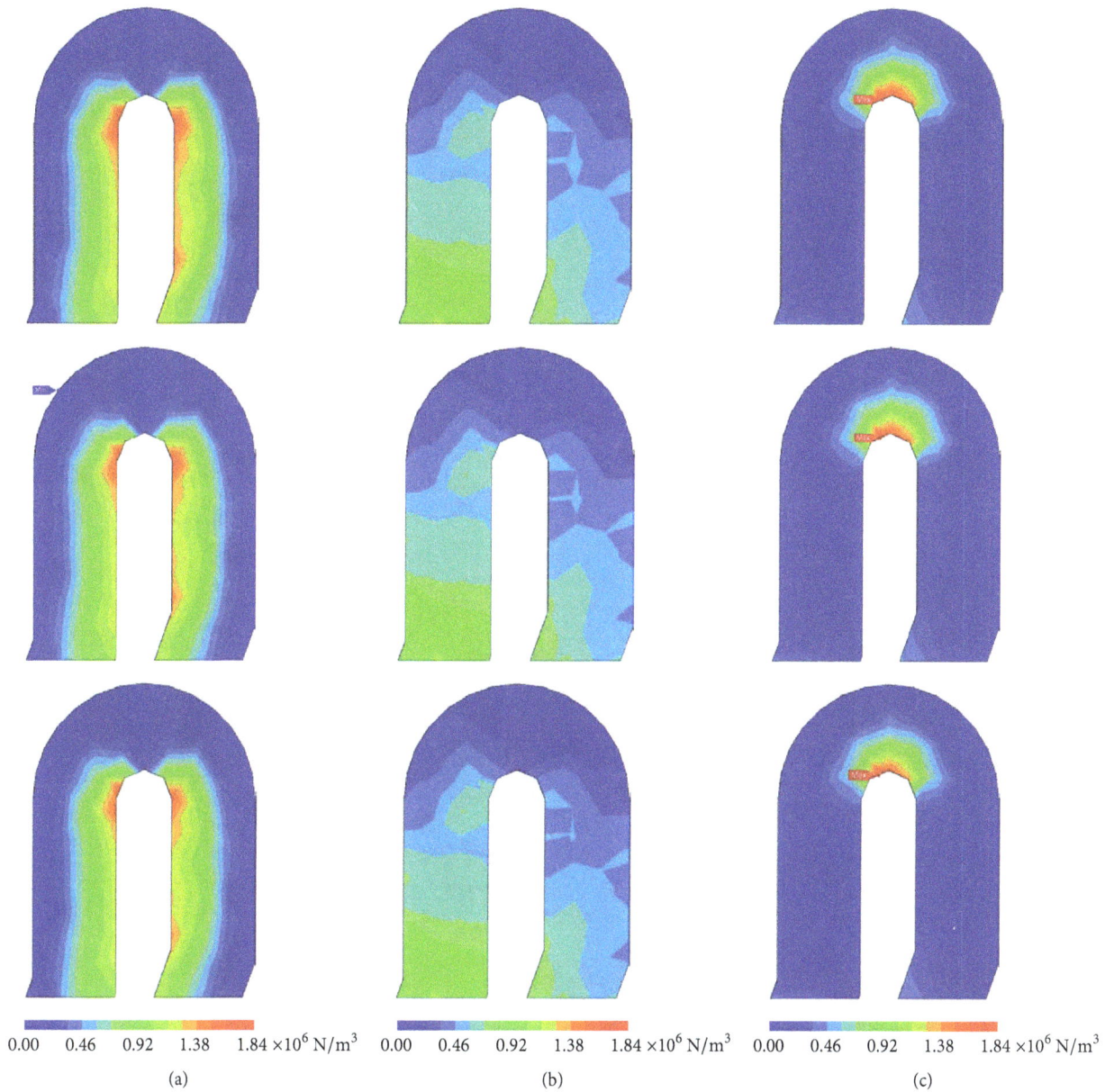

FIGURE 12: Electromagnetic force density magnitude distribution on the nose part at 0.039 s, when $l_r = -75$ mm (bottom), $l_r = 0$ mm (middle), and $l_r = 75$ mm (top). (a) Radial component. (b) Tangential component. (c) Axial component.

forces. When l_r increases from -75 mm to $+75$ mm, the radial force increases by 13% while the tangential and axial forces decrease by 9% and 12%, respectively.

The electromagnetic force distribution on the nose part is shown in Figure 12. The magnitude of the force density is used to show the change in force strength because the direction of the force density changes to the opposite direction of the axis on some areas of the nose part. Figure 12 shows that the radial and axial force densities increase with l_r. The radial force change mainly appears at the inner side of the nose part, and the axial force change mainly appears at the inner side of the top section of the nose part. When l_r varies from -75 mm to $+75$ mm, the maximum of the electromagnetic force density varies from 1.9×10^6 N/m^3 to 2.0×10^6 N/m^3, 1.0×10^6 N/m^3 to 1.0×10^6 N/m^3, and 2.1×10^6 N/m^3 to 2.2×10^6 N/m^3 in the radial, tangential, and axial directions, respectively.

The maximum electromagnetic forces on the nose part with different rotor relative lengths are shown in Figure 13. The radial force is smaller than the tangential and axial forces because the direction of the radial force on both sides of the nose part are opposite and almost offset each other. When l_r varies from -75 mm to $+75$ mm, the radial and axial forces increase by 8% and 23%, respectively, while the tangential force almost stays constant.

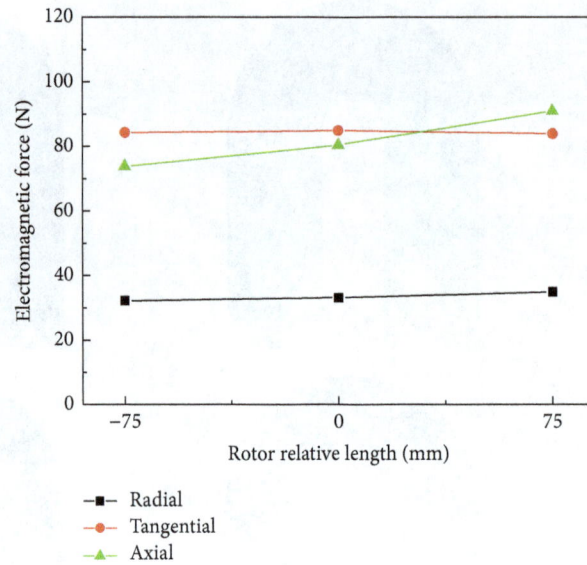

FIGURE 13: Electromagnetic forces on the nose part with different rotor relative lengths.

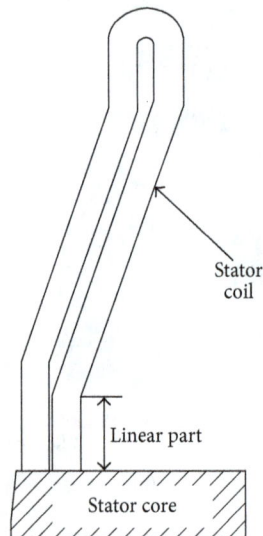

FIGURE 14: Definition of stator coil linear length.

For large generators, the radial vibration of end winding is the most serious [22]. A relatively short rotor can be adopted to reduce the radial vibration from the source. Both the electromagnetic force on the involute and nose parts are smaller with a relatively short rotor. The axial force on the nose part (especially on the top of the inner side) also decreases. However, the tangential and axial forces on the involute part (especially on the region close to the linear part of the end coil) increase at the same time.

4.2. Influence of Stator Coil Linear Length. The stator coil linear length (l_{sc}) is defined in Figure 14 as the length of the lower linear part beyond the stator core. To study the influence of l_{sc} on electromagnetic force, l_{sc} is set to 221, 271, 321, 371, and 421 mm in the FE model.

The electromagnetic force density distributions on the involute part at 0.039 s with different stator coil linear lengths are shown in Figure 15. Figure 15 shows that the radial force decreases slightly with the increasing l_{sc}, and the maximum of the radial electromagnetic force is reduced to 2.3×10^6 N/m^3 when l_{sc} is 421 mm from 2.7×10^6 N/m^3 when l_{sc} is 221 mm. The tangential and axial forces increase significantly with l_{sc}. The maximums of the tangential and axial forces increase to 1.1×10^6 N/m^3 and 1.8×10^6 N/m^3, respectively, when l_{sc} is

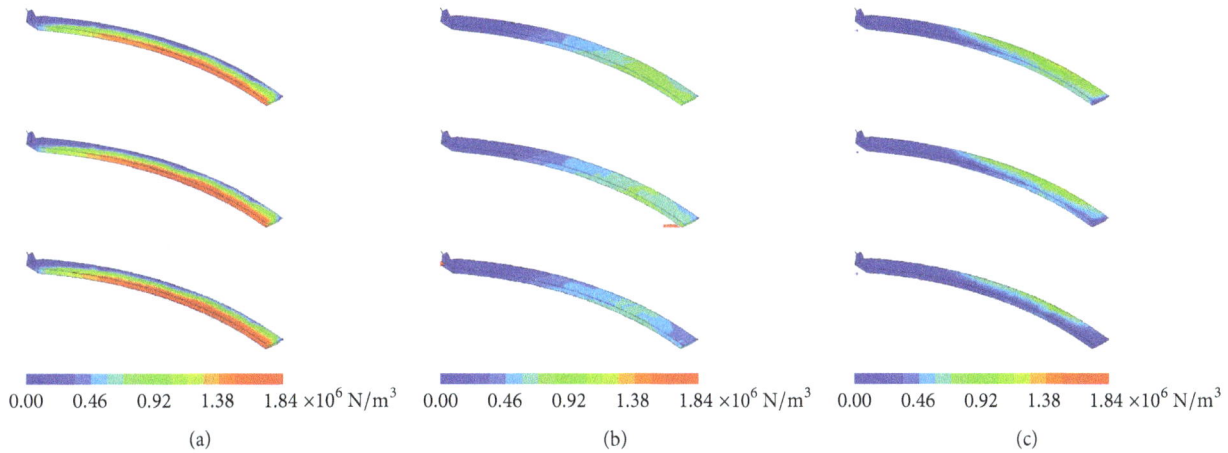

FIGURE 15: Electromagnetic force density distribution on the involute part at 0.039 s, when l_{sc} = 221 mm (bottom), l_{sc} = 321 mm (middle), and l_{sc} = 421 mm (top). (a) Radial component. (b) Tangential component. (c) Axial component.

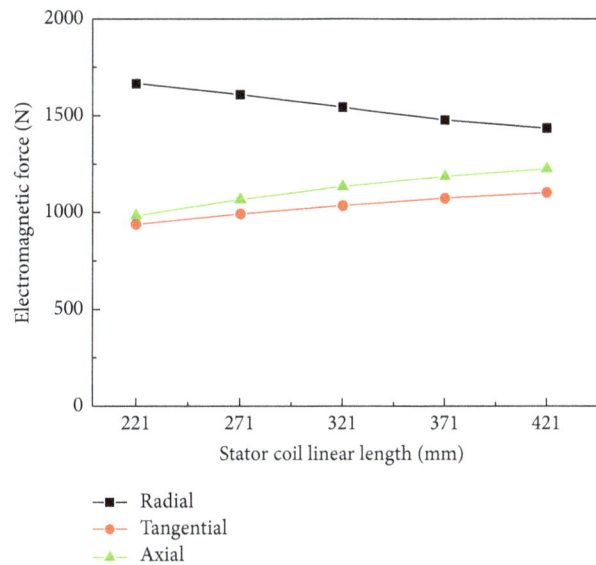

FIGURE 16: Electromagnetic forces on the involute part with different stator coil linear length.

421 mm from 0.9×10^6 and 1.3×10^6 N/m^3, respectively, when l_{sc} is 221 mm.

Figure 16 shows the maximum electromagnetic forces on the involute part with different l_{sc}. Figure 16 shows that the radial force is relatively larger than the tangential and axial forces. When l_{sc} increases from 221 mm to 421 mm, the radial force decreases by 14% while the tangential and axial forces increase by 17% and 24%, respectively.

The distribution of the electromagnetic force density magnitude on the nose part with different l_{sc} is shown in Figure 17. The influence of the stator coil linear length on electromagnetic force on the nose part is small. Generally, the radial force decreases with the increase of the stator coil's linear length. Conversely, the tangential force increases with the stator coil's linear length. The axial force is constant against the changes in stator coil linear length.

Figure 18 shows the maximum electromagnetic forces on the nose part with different l_{sc}. When l_{sc} increases from 221 mm to 421 mm, the radial and axial forces decrease by 9% and 17%, respectively, while the tangential force remains almost constant.

To reduce the radial vibration on end winding, a relatively large stator coil linear length can be adopted. Consequently, the radial force on the involute and nose parts and the axial force on the nose part will decrease. In contrast, the tangential and axial forces on the involute part increase.

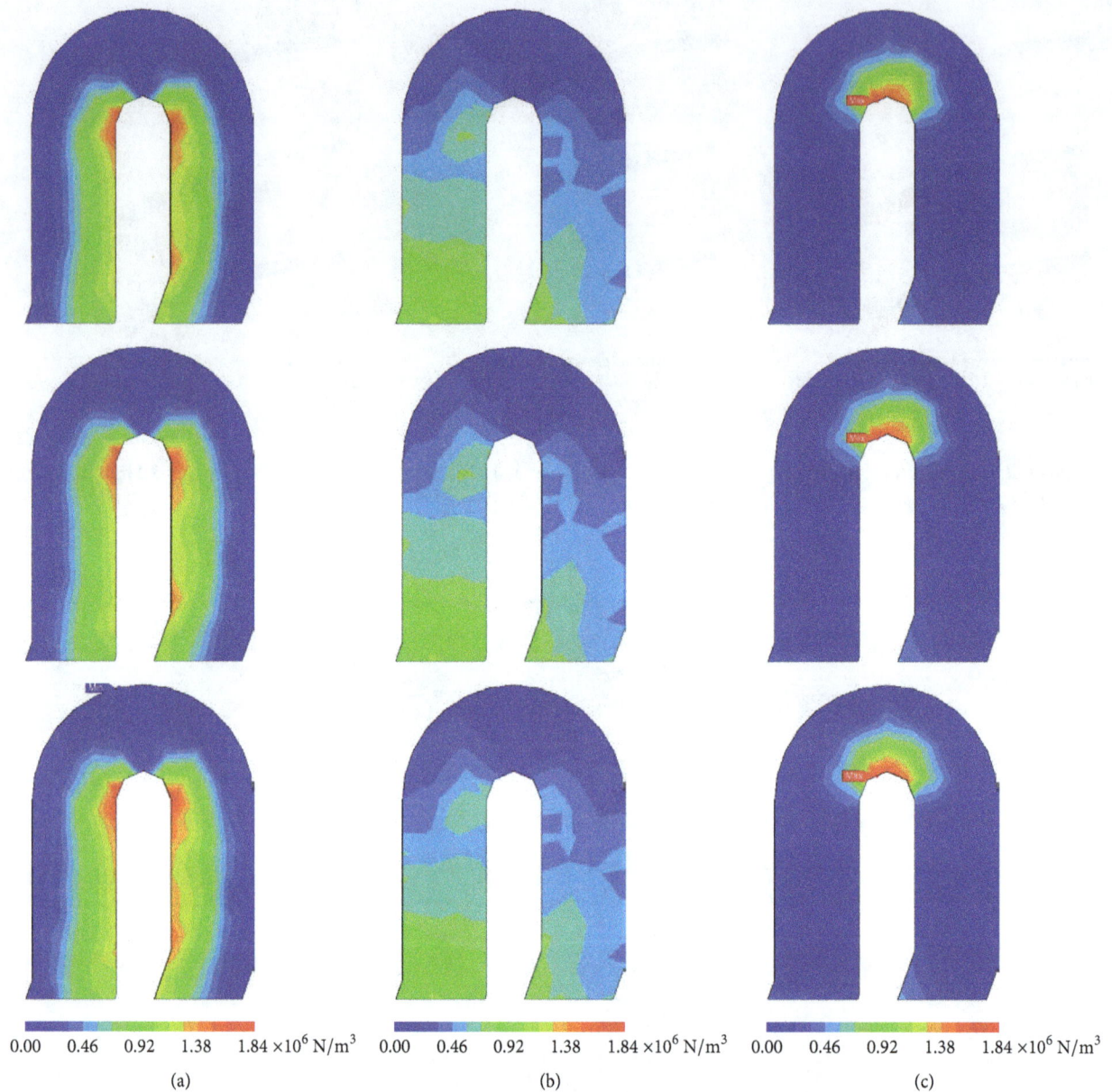

FIGURE 17: Electromagnetic force density magnitude distribution on the nose part at 0.039 s, when l_{sc} = 221 mm (bottom), l_{sc} = 321 mm (middle), and l_{sc} = 421 mm (top). (a) Radial component. (b) Tangential component. (c) Axial component.

5. Conclusions

In this paper, a 3D FE model of the end region of a 1550 MW nuclear generator is set up. The electromagnetic force on the stator end winding at a rated operation is calculated. The electromagnetic forces on different coils in the same phase are presented. By changing the rotor relative length and stator coil linear length in the FE model, the influence of these parameters on electromagnetic force is analyzed in detail. The following conclusions are drawn:

(1) The electromagnetic force on the involute part varies in different coils at the same phase and the upper involute part of the last coil in the rotation direction suffers the largest force. The electromagnetic force on the nose part is smaller than that on the involute part and changes slightly in different coils.

(2) Decreasing the rotor relative length decreases the radial forces on the involute and nose parts. However, decreasing the rotor relative length increases the tangential and axial forces on the involute parts.

(3) The stator coil linear length can be increased to decrease the radial forces on the end winding. The radial force on the involute and nose parts decreases significantly but the tangential and axial forces on the involute parts simultaneously increase.

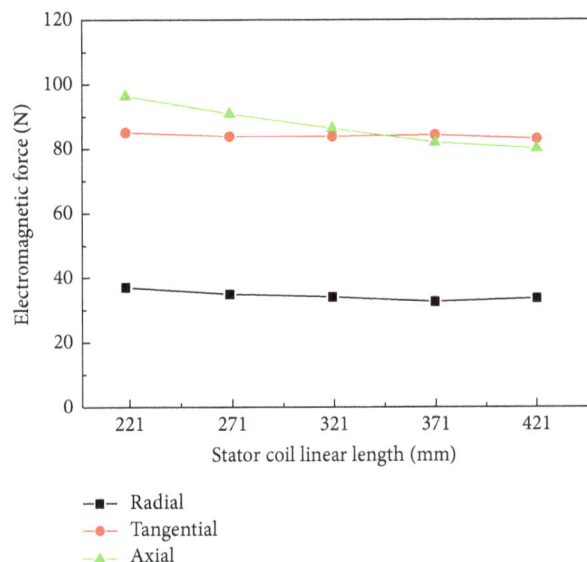

FIGURE 18: Electromagnetic forces on the nose part with different stator coil linear lengths.

Conflicts of Interest

The authors declare that they have no conflicts of interest.

Acknowledgments

The work is financially supported by the project supported by the National Natural Science Foundation of China (Grant no. 51477015).

References

[1] D. Shally, M. Farrell, and K. Sullivan, "Generator end winding vibration monitoring," in *Proceedings of the 43rd International Universities Power Engineering Conference (UPEC '08)*, pp. 1–5, Padova, Italy, September 2008.

[2] S. Wan, C. Zhan, X. Yao, Y. Deng, and G. Zhou, "Investigation on electromagnetic forces and experimental vibration characteristics of stator end windings in generator," *Transactions of the Canadian Society for Mechanical Engineering*, vol. 38, no. 3, pp. 331–345, 2014.

[3] H. Yuda, Q. Jiajun, and Q. Guanghui, "Calculation of electromagnetic force and vibration on end winding of turbo generators," in *Proceedings of the 10th National Academic Conference of Structure Engineering*, Nanjing, China, 2001 (Chinese).

[4] J. F. Calvert, "Forces in turbine generator stator windings," *Transactions of the American Institute of Electrical Engineers*, vol. 50, no. 1, pp. 178–194, 1931.

[5] D. Harrington, "Forces in machine end windings," *Electrical Engineering*, vol. 72, no. 2, pp. 153-153, 1953.

[6] M. Ohtaguro, K. Yagiuchi, and H. Yamaguchi, "Mechanical behavior of stator endwindings," *IEEE Transactions on Power Apparatus and Systems*, vol. 99, no. 3, pp. 1181–1185, 1980.

[7] H. Wang, R. Zong, L. Liu, and S. Yang, "Full 3D eddy current and temperature field analysis of large hydro-generators in different operating conditions," in *Proceedings of the 17th International Conference on Electrical Machines and Systems (ICEMS '14)*, pp. 3599–3603, Hangzhou, China, October 2014.

[8] F. J. Waldhart, J. P. Bacher, and G. Maier, "Modeling eddy current losses in the clamping plate of large synchronous generators using the finite element method," in *Proceedings of the 21st International Symposium on Power Electronics, Electrical Drives, Automation and Motion (SPEEDAM '12)*, pp. 1468–1473, Sorrento, Italy, June 2012.

[9] Y.-P. Liang, H.-H. Wang, J.-T. Zhang, and J. Liu, "Research of end fields and eddy current losses for air-cooling steam-turbogenerator," *Journal of Electric Machines and Control*, vol. 14, no. 1, pp. 29–34, 2010 (Chinese).

[10] L. Wang, W. Li, F. Huo, S. Zhang, and C. Guan, "Influence of underexcitation operation on electromagnetic loss in the end metal parts and stator step packets of a turbogenerator," *IEEE Transactions on Energy Conversion*, vol. 29, no. 3, pp. 748–757, 2014.

[11] H. Feiyang, L. Weili, W. Likun, G. Chunwei, Z. Yihuang, and L. Yong, "Influence of copper screen thickness on three-dimensional electromagnetic field and eddy current losses of metal parts in end region of large water-hydrogen-hydrogencooled turbogenerator," *IEEE Transactions on Industrial Electronics*, vol. 60, no. 7, pp. 2595–2601, 2013.

[12] H. Feiyang, H. Jichao, L. Weili et al., "Influence of copper shield structure on 3-D electromagnetic field, fluid and temperature fields in end region of large turbogenerator," *IEEE Transactions on Energy Conversion*, vol. 28, no. 4, pp. 832–840, 2013.

[13] L. Wang, F. Huo, W. Li et al., "Influence of metal screen materials on 3-D electromagnetic field and eddy current loss in the end region of turbogenerator," *IEEE Transactions on Magnetics*, vol. 49, no. 2, pp. 939–945, 2013.

[14] W. Likun, L. Weili, X. Yi, and G. Chunwei, "Influence of clamping plate permeability and metal screen structures on three-dimensional magnetic field and eddy current loss in end region of a turbo-generator by numerical analysis," *Journal of Applied Physics*, vol. 114, no. 19, Article ID 193904, 2013.

[15] S. Huang, C. Zeng, Y. Yang, and G. Zhou, "Influence of the end structure on the leakage magnetic field and eddy current loss in a 1,550 MW nuclear generator," *Proceedings of the Chinese Society of Electrical Engineering*, vol. 36, pp. 200–205, 2016.

[16] N. Richard, F. Duffeau, A. C. Leger, and N. Szylowicz, "Computation of forces and stresses on generator end windings using a 3d finite element method," *IEEE Transactions on Magnetics*, vol. 32, no. 3, pp. 1689–1692, 1996.

[17] N. Richard, "Calculation of electromagnetic forces on large generator end-windings under fault conditions using a three-dimensional finite element method," *Mathematics and Computers in Simulation*, vol. 46, no. 3-4, pp. 257–263, 1998.

[18] K. Senske, S. Kulig, J. Kauhoff, and D. Wünsch, "Vibrational behaviour of the turbogenerator stator end winding in case of electrical failures," in *Proceedings of the Conférence Internationale des Grands Rèseaux Electriques*, pp. 1–12, Yokohama, Japan, October 1997.

[19] O. Drubel, S. Kulig, and K. Senske, "End winding deformations in different turbo generators during 3-phase short circuit and full load operation," *Electrical Engineering*, vol. 82, no. 3, pp. 145–152, 2000.

[20] A. Grüning and S. Kulig, "Electromagnetic forces and mechanical oscillations of the stator end winding of turbogenerators," in *Recent Developments of Electrical Drives*, pp. 115–126, Springer, The Netherlands, 2006.

[21] R. Albanese, F. Calvano, G. Dal Mut et al., "Coupled three dimensional numerical calculation of forces and stresses on the end windings of large turbo generators via integral formulation," *IEEE Transactions on Magnetics*, vol. 48, no. 2, pp. 875–878, 2012.

[22] Y. Zhao, B. Yan, C. Zeng, S. Huang, C. Chen, and J. Deng, "Dynamic response analysis of large turbogenerator stator end structure under electromagnetic forces," *Transactions of China Electrotechnical Society*, vol. 31, no. 5, pp. 199–206, 2016.

[23] J. Cheaytani, A. Benabou, A. Tounzi, and M. Dessoude, "Stray load losses analysis of cage induction motor using 3-D finite element method with external circuit coupling," in *Proceedings of the 17th Biennial IEEE Conference on Electromagnetic Field Computation (CEFC '16)*, Miami, Fla, USA, November 2016.

[24] K. Yamazaki, S. Kuramochi, N. Fukushima, S. Yamada, and S. Tada, "Characteristics analysis of large high speed induction motors using 3-D finite element method," *IEEE Transactions on Magnetics*, vol. 48, no. 2, pp. 995–998, 2012.

Permissions

List of Contributors

Yasuyuki Nishi
Department of Mechanical Engineering, Ibaraki University, 4-12-1 Nakanarusawa-cho, Hitachi-shi, Ibaraki 316-8511, Japan

Hikaru Fushimi
Graduate School of Science and Engineering, Ibaraki University, 4-12-1 Nakanarusawa-cho, Hitachi-shi, Ibaraki 316-8511, Japan

Kazuo Shimomura
Yoshida Seiko, Ltd., 660 Yakimaki, Namegata-shi, Ibaraki 311-3506, Japan

Takeshi Hasegawa
The Yoshida Dental MFG Co., Ltd., 1-3-6 Kotobashi, Sumida-ku, Tokyo 130-8516, Japan

L. X. Hou and A. K. Hu
School of Naval Architecture and Ocean Engineering, Dalian Maritime University, No. 1 Linghai Street, Ganjingzi District, Dalian, Liaoning 116026, China

Fang Feng
College of Science, Northeast Agricultural University, Harbin, China
Heilongjiang Provincial Key Laboratory of Technology and Equipment for Utilization of Agricultural Renewable Resources in Cold Region, Harbin 150030, China

Chunming Qu, Shouyang Zhao, Yuedi Bai and Wenfeng Guo
College of Engineering, Northeast Agricultural University, Harbin, China

Yan Li
Heilongjiang Provincial Key Laboratory of Technology and Equipment for Utilization of Agricultural Renewable Resources in Cold Region, Harbin 150030, China
College of Engineering, Northeast Agricultural University, Harbin, China

Kaprawi Sahim, Dyos Santoso and Dewi Puspitasari
Mechanical Engineering Department, Sriwijaya University, Sumatera Selatan, Indonesia

Tilahun Nigussie and Edessa Dribssa
School of Mechanical and Industrial Engineering, Addis Ababa Institute of Technology, Addis Ababa, Ethiopia

Abraham Engeda
Department of Mechanical Engineering, Michigan State University, East Lansing, USA

Hui Li and Dian-Gui Huang
School of Energy and Power Engineering, University of Shanghai for Science and Technology Shanghai 200093, China
Shanghai Key Laboratory of Power Energy in Multiphase Flow and Heat Transfer, Shanghai 200093, China

Xiao Zhou, Yuejin Tang and Jing Shi
State Key Laboratory of Advanced Electromagnetic Engineering and Technology, R&D Center of Applied Superconductivity, Huazhong University of Science and Technology, Wuhan 430074, China

Yu Zhang, Chris Bingham and Miguel Martínez-García
School of Engineering, University of Lincoln, Lincoln LN6 7TS, UK

Darren Cox
Siemens Industrial Turbomachinery Ltd., Lincoln LN5 7FD, UK

Huiwen He, Lei Wang, Peihong Zhou and Fei Yan
State Key Laboratory of Power Grid Environmental Protection, China Electric Power Research Institute, Wuhan 430074, China

Igor Shevchenko, Nikolay Rogalev, Andrey Rogalev, Andrey Vegera and Nikolay Bychkov
National Research University Moscow Power Engineering Institute, 14 Krasnokazarmennaya Street, Moscow 111250, Russia

Ilia K. Marchevsky and Valeria V. Puzikova
Applied Mathematics Department, Bauman Moscow State Technical University (BMSTU), 5 2nd Baumanskaya, Moscow 105005, Russia

Yuliang Zhang
College of Science, Northeast Agricultural University, Harbin 150030, China

Shouyang Zhao, Chunming Qu and Yuedi Bai
College of Engineering, Northeast Agricultural University, Harbin 150030, China

Yan Li
Heilongjiang Provincial Key Laboratory of Technology and Equipment for Utilization of Agricultural Renewable Resources in Cold Region, Harbin 150030, China
College of Engineering, Northeast Agricultural University, Harbin 150030, China

Fangyuan Lou, John Charles Fabian and Nicole Leanne Key
Purdue University, West Lafayette, IN 47907, USA

Younes Ait El Maati, Lhoussain El Bahir and Khalid Faitah
National School of Applied Sciences, LGECOS Laboratory, Cadi AyyadUniversity, Marrakech, Morocco

Feng Lin and Jingyi Chen
Institute of Engineering Thermophysics, Chinese Academy of Sciences, Beijing, China

Saeed Asadi, Viktor Berbyuk and Håkan Johansson
Department of Mechanics and Maritime Sciences, Chalmers University of Technology, 412 96 Göteborg, Sweden

Chong Zeng, Song Huang and Yongming Yang
Chongqing University, Chongqing 400044, China

Guanghou Zhou
Research and Test Center, Dongfang Electrical Machinery Co., Ltd., Deyang, Sichuan Province 618000, China

Index

www.ingramcontent.com/pod-product-compliance
Lightning Source LLC
Chambersburg PA
CBHW050454200326
41458CB00014B/5173